4·13·78

GROUND-WATER HYDROLOGY AND HYDRAULICS

by

David B. McWhorter

and

Daniel K. Sunada

Water Resources Publications
P. O. Box 303
Fort Collins, Colorado 80522 USA

For information and correspondence:

WATER RESOURCES PUBLICATIONS
P. O. Box 303
Fort Collins, Colorado 80522

GROUND-WATER HYDROLOGY AND HYDRAULICS

ISBN-0-918334-18-7

U. S. Library of Congress Catalog Card Number 77-074259

This publication is printed and bound by LithoCrafters, Ann Arbor, Michigan, U.S.A.

To

Arthur T. Corey

An Exemplary Scholar and Teacher

PREFACE

This book is an outgrowth of a course in ground-water hydrology taught for seniors and first-year graduate students in Agricultural Engineering, Civil Engineering, Geology, and Watershed Science. The selection of subject material and presentation has been favorably received by both students and professionals in these disciplines. The book is designed, primarily, for use as a text, but the emphasis on basic physical and mathematical concepts should be found useful to professional hydrologists and geohydrologists as well. It has been our experience, both in teaching and in professional practice, that a solid background in fundamentals and a thorough understanding of ground-water phenomena in idealized cases contributes immeasurably to imaginative and successful analysis, synthesis, and solution of field problems; even when the results from idealized situations are not directly applied.

Presentation of the subject matter begins with an overview of ground-water hydrology with a definition of some basic terms and a limited discussion of the role of geology as a fundamental supporting science. The role of ground-water in the hydrologic cycle is implicit in the material of Chapter II where ground-water storage and water budgets are discussed. Fluid flow through porous media is presented in Chapter III with emphasis on Darcy's Law and its significance. The discussions of steady and unsteady flow in Chapters IV and V emphasize local ground-water problems such as well flow, drainage, mine-inflow, etc.. A chapter on finite difference techniques is included to provide the student with an introduction to methods and tools available for basin-wide or regional analysis. One feature of the book that we hope will be especially useful to the reader is the numerous worked examples and the problem sets at the end of each chapter. Many of the examples and problems are designed to help elucidate principles and several are straight-forward applications of the text material, but in contexts that deviate to some degree from the context of the original presentation. We hope that this will reduce the tendency of the reader to associate a particular development only with the context in which it was derived.

Our selection and presentation of subject material has been influenced by authors of similar books and by our former teachers, and we are indebted to these many people. We have included many other books, both texts and reference books, in our list of references in anticipation that the reader will often desire to study other summaries and digests of the literature as well as the original papers.

We are grateful for comments and suggestions received from colleagues and students who have reviewed and studied the material from which this book evolved. In particular, we wish to thank Dr. H. R. Duke, Agricultural Research Service, USDA; and Dr. A. T. Corey, Professor of Agricultural Engineering for their review and comments. Thanks are also due Ms. Diane English for typing the manuscripts and Mr. J. A. Brookman for the drafting. Also, we would like to express our deep appreciation to our families who sacrificed evening and weekend activities on numerous occasions because of our involvement in the preparation of this text.

Finally, we wish to express our appreciation to Colorado State University and particularly to the Agricultural and Chemical Engineering and the Civil Engineering Departments. It was through the direct and indirect support of the University that this undertaking became possible.

April, 1977
Fort Collins, Colorado
U.S.A.

David B. McWhorter
Associate Professor
Agricultural and Chemical Engineering

Daniel K. Sunada
Associate Professor
Civil Engineering
Colorado State University

TABLE OF CONTENTS

Chapter		Page
	Preface	iii
I	INTRODUCTION	1
	1.1 Scope	1
	1.2 Historical Perspective	2
	Extraction of Ground Water	2
	Ground-Water Hydrology	5
	1.3 Ground-Water Geology -- An Overview	6
	Geologic Strata -- Ground-Water Vessels	6
	Confined and Unconfined Ground Water	9
	Observation of Ground Water	11
	REFERENCES	12
	PROBLEMS AND STUDY QUESTIONS	14
II	GROUND-WATER STORAGE AND SUPPLY	15
	2.1 Storage In Water-Table Aquifers	15
	Porosity and Representative Volume Elements	15
	Example 2-1	19
	Example 2-2	20
	Desaturation and Specific Retention of Aquifer Materials	20
	Example 2-3	24
	Distribution of Water Content and Pressure Above a Water Table	25
	Specific Yield and Apparent Specific Yield	28
	Example 2-4	30
	Example 2-5	31
	2.2 Storage In Confined Aquifers	34
	Stress in Aquifers	34
	Specific Storage and Storage Coefficient	35
	Example 2-6	39
	Example 2-7	39
	2.3 Water Level Fluctuations	40
	Piezometric Surface and Barometric Pressure	40
	The Water Table and Barometric Pressure	43
	Water Levels, Tides and Other External Loads.	45
	Water Levels and Evapotranspiration	47
	2.4 Hydrologic Budgets	48
	Components of a Surface-Water Budget	49
	Components of a Soil-Water Budget	51
	Example 2-8	56
	The Ground-Water Budget	56
	Example 2-9	57
	REFERENCES	58
	PROBLEMS AND STUDY QUESTIONS	61
III	DARCY'S LAW AND BASIC DIFFERENTIAL EQUATIONS	65
	3.1 Darcy's Empirical Equation	65
	Forces on Fluids in Porous Solids	67

TABLE OF CONTENTS - Continued

Chapter Page

Example 3-1 71
Example 3-2 71
Example 3-3 73
Force Potential, Velocity Potential, and Darcy's Law 73
Example 3-4 75
Example 3-5 76
Example 3-6 76
Example 3-7 78
Laboratory Determination of Hydraulic Conductivity 79
3.2 Non-Homogeniety And Anisotropy 82
Non-Homogeneious Aquifers 82
Example 3-8 86
Example 3-9 86
Anisotropic Aquifers 87
3.3 Flow In Confined Aquifers 90
A Differential Mass Balance 91
Differential Equations For Confined Flow . . 92
3.4 Flow In Water-Table Aquifers 95
The Dupuit-Forchheimer Approximation 96
The Boussinesq Equation 97
3.5 Equations For Flow With Vertical Accretion . 99
REFERENCES 101
PROBLEMS AND STUDY QUESTIONS 103

IV STEADY GROUND-WATER HYDRAULICS 110
4.1 Equipotential Contours And Streamlines . . . 110
Equipotential and Stream Surfaces in Relation to Physical Boundaries 113
Example 4-1 113
4.2 Elementary Solutions For Confined Flow . . . 115
One-Dimensional Flow 115
Radial Flow 116
Example 4-2 120
4.3 Superposition Of Elementary Solutions 121
Drawdown in a Confined Aquifer Due to a Well Field 121
Example 4-3 122
Pumping Near Hydro-Geologic Boundaries . . . 124
Example 4-4 128
Example 4-5 128
Example 4-6 130
A Pumped Well in Uniform Flow 131
Example 4-7 133
Partially Penetrating Wells 134
Example 4-8 135
Concluding Remarks on Superposition of Solutions 136

TABLE OF CONTENTS - Continued

Chapter | Page

4.4 Flow Nets 136
Flow Nets in Homogeneous Aquifers 136
Example 4-9 139
Example 4-10 140
Flow Nets in Non-Homogeneous Aquifers 141
4.5 Flow In Confined Aquifers With Vertical Accretion 142
One-Dimensional Flow in a Leaky Aquifer . . . 142
Radial Flow in a Leaky Aquifer 143
Example 4-11 145
4.6 Steady Flow In Unconfined Aquifers 146
One-Dimensional Unconfined Flow 146
One-Dimensional Unconfined Flow With Vertical Accretion 148
Example 4-12 150
Radial Unconfined Flow 151
Example 4-13 154
4.7 Hydraulics Of Two Fluids In Aquifers 156
Two-Fluid Idealization of a Nonhomogeneous Fluid 156
The Ghyben-Herzberg Equation 159
Fresh-Water Flow in a Coastal Aquifer 160
Example 4-14 161
Interface Upconing Beneath Pumping Wells . . 162
Example 4-15 164
The Island Fresh-Water Lens 165
Example 4-16 166
REFERENCES 168
PROBLEMS AND STUDY QUESTIONS 170

V UNSTEADY GROUND-WATER HYDRAULICS 177
5.1 Radial Flow 177
Flow Toward a Fully Penetrating Well 177
Example 5-1 179
The Effective Radius of Influence 182
Example 5-2 183
The Pseudo-Steady State 184
Example 5-3 185
Response of an Unconfined Aquifer Near a Pumped Well 185
Radial Flow in a Leaky Aquifer 188
Drawdown With Variable Pumping Rates 190
5.2 Aquifer Test Applications 194
Preliminary Considerations in Aquifer Testing 195
Analysis of Aquifer Test Data Using the Theis Solution 198
Example 5-4 200
Example 5-5 202
Example 5-6 203

TABLE OF CONTENTS - Continued

Chapter | Page

Analysis of Data Influenced by Delayed Water-Table Response 204
Example 5-7 205
Analysis of Recovery and Slug Test Data . . . 206
Example 5-8 207
Example 5-9 208
5.3 Pumping Near Hydro-Geologic Boundaries . . . 209
Pumping Near a Stream or Impermeable Boundary 210
Example 5-10 211
Aquifer Tests in the Vicinity of Hydro-Geologic Boundaries 213
5.4 One-Dimensional Flow 214
Flow Toward a Plane on Which the Piezometric Head is Prescribed 214
Example 5-11 217
Flow Toward a Plane at Which the Discharge is Prescribed 218
Example 5-12 220
5.5 One-Dimensional Flow With Distributed Recharge 221
Flow Between Parallel Drains 221
Example 5-13 224
Example 5-14 225
Flow in Response to Continuously Varying Recharge 226
Example 5-15 227
REFERENCES 228
PROBLEMS AND STUDY QUESTIONS 231

VI FINITE DIFFERENCE METHODS 240
6.1 One-Dimensional Flow - Confined Aquifer . . . 240
Forward Difference Equation - Explicit Solution 243
Example 6-1 244
Backward Difference Equation - Implicit Solution 246
Example 6-2 246
Crank-Nicholson Approximation 247
6.2 Two-Dimensional Flow - Confined And Unconfined Aquifer 249
Alternating Direction Implicit Procedure - Two-Dimensional Case 249
Backward Difference Equation - Two-Dimensional Case 250
Example 6-3 255
6.3 Sensitivity To Solution Parameters 257
REFERENCES 258
PROBLEMS AND STUDY QUESTIONS 260

TABLE OF CONTENTS - Continued

Chapter | Page

APPENDIX A - *Systems of Units and Conversion Tables* 261
APPENDIX B - *Gauss Elimination Scheme for Example 6-2 in Text* 263
APPENDIX C - *Finite Difference Ground-Water Model* 265
Procedure for Analysis 265
Data Input 268
Description of Subprograms 270
Computer Program Listing 272
APPENDIX D - *Additional Solutions* 282
Index 287

Symbols

Major definitions of symbols are presented here. Secondary definitions or constants are defined where used.

<u>Symbol</u>

Symbol	Definition
a	Perpendicular distance from source or sink to hydrogeologic boundaries
AW	Available water for plants
AW_m	Maximum available water for plants
b	Thickness of confined aquifer or saturated thickness of unconfined aquifer
BE	Barometric efficiency
C_p	Pan coefficient for evaporation
C_T	Temperature coefficient for evaporation
d	Grain diameter
$\bar{d}$	Characteristic dimension of flow space in porous media
D_r	Rooting depth of plant
E	Evaporation rate
ET	Evapotranspiration rate
E_{pan}	Pan Evaporation rate
ET_p	Potential evapotranspiration rate
F	Force
g	Gravitational constant
h	Vertical height of water table or piezometric head
i	Hydraulic gradient
I	Infiltration rate
k	Intrinsic permeability
K	Hydraulic conductivity
$K_o(r/\mathcal{B})$	Modified Bessel function of the second kind, order zero
L_γ	Latent heat of vaporization of water
ℓ	Length of perforated well bore

Symbol

M	Mass of fluid
n_s	Number of streamtubes
n_ℓ	Number of equipotential drops
NC	Number of columns in grid system
NR	Number of rows in grid system
p	Fluid pressure
p_a	Atmospheric pressure
p_c	Capillary pressure
q	Darcy velocity
$\vec{q}$	Velocity vector
Q	Flow rate, discharge
r	Radial distance
r_e	Radius of influence
R	Gas constant
R_s	Solar radiation
s	Drawdown
S	Storage coefficient
S_r	Specific retention
S_s	Specific storage
S_y	Specific yield (effective porosity)
S_{ya}	Apparent specific yield
t	Time
T	Transmissivity
TE	Tidal efficiency
u	Boltzman variable = $r^2/(4\alpha t)$
v	Seepage velocity
V	Control volume
W	Vertical seepage rate
$W(u)$	Well function
x,y,z	Rectangular coordinate system
α	T/S or T/S_{ya}
α_p	Pore volume compressibility

Symbol

β	Compressibility of water
$\mathcal{B}$	Leakage factor for leaky aquifer
δ	Angle - defined where used
η	Distance to fresh water - salt water interface
θ	Volumetric water content
θ_{wp}	Volumetric water content at permanent wilting
μ	Dynamic viscosity
ρ	Density of fluid
σ	Surface tension
σ_z	Intergranular stress
τ	Dummy time variable
ϕ	Porosity
Φ	Velocity potential
ψ	Stream function

Chapter I

INTRODUCTION

1.1 SCOPE

Water contained in the voids of the geologic materials that comprise the crust of the earth have been variously classified according to origin and mode of occurrence. In conformance with common practice, the term ground water is used herein to denote subsurface water that exists at pressures greater than or equal to atmospheric pressure. Pressures of subsurface water in the capillary fringe and above are less than atmospheric and typically capillary water is not included as ground water. This distinction between subsurface waters that exist at pressures greater or less than atmospheric pressure is artificial in the sense that no physical boundary separates them, their properties are identical, they do not behave independently and the same principles of physics apply to both. On the other hand, the experimental and mathematical methods required for analyses are distinctly different as are the methods by which the two waters are exploited and used in the course of human affairs.

The subject matter of this book deals almost exclusively with ground water as defined above. Some 97 percent of the world's potable water supply available at any moment in time occurs as ground water, and the importance of ground water in the total water resource is unquestioned (Ground Water and Wells, 1972). Proper evaluation of ground-water resources requires thorough hydrologic, geologic and hydraulic investigation over areas that may range in size from a few hundred hectares to entire basins or even countries. Use of simulation and management models is widespread in such studies. On the other end of the spectrum are ground-water studies that may be quite local in extent and only indirectly motivated by ground water as a resource. Examples include aquifer testing, drainage of agricultural land, dewatering for construction, inflow to a mine shaft or pit, seepage beneath or through a dam, subsurface return flow from irrigation, seepage from canals and reservoirs, intrusion of saline water, and subsurface disposal of liquid wastes.

It is beyond the scope of this basic text to cover all aspects of hydrology, geology, and hydraulics that are pertinent to the analysis and solution of the above problems. Especially notable omissions are treatments of ground-water exploration, ground-water quality and an adequate discussion of the role of geology. The hydraulics of ground water receive the most emphasis, but hydrological aspects are introduced through discussion of ground-water storage, supply, and water budgets.

1.2 HISTORICAL PERSPECTIVE

Extraction of Ground Water

Ground water has been a source of water supply since the dawn of recorded history. The first use of ground water was from springs, but dug wells were widely used in the earliest of Biblical times. Dug wells, with diameters from 2 to 10 meters, lined with rock or brick are still in use today in many parts of the world. The time and place at which the first drilled well (distinct from dug well) was constructed is lost in unrecorded history, but the ancient Chinese are generally regarded as the inventors of drilled and cased wells. Tolman (1937) reports that the Chinese sunk bamboo-cased wells to depths of more than 1500 meters, requiring three generations to complete.

Extraction of ground water from wells required a water-lift method. The earliest water-lift methods consisted of little more than a means of raising and lowering a container on a rope. Apparently the Egyptians used a lever, fulcrum and counterweight in conjunction with the rope and container as early as 2000 B.C. (Fig. 1-1). Animal power was, and still is, used to lift water from dug wells by means of Persian wheels (Fig. 1-2). Containers, formed from gourds, clay, and more

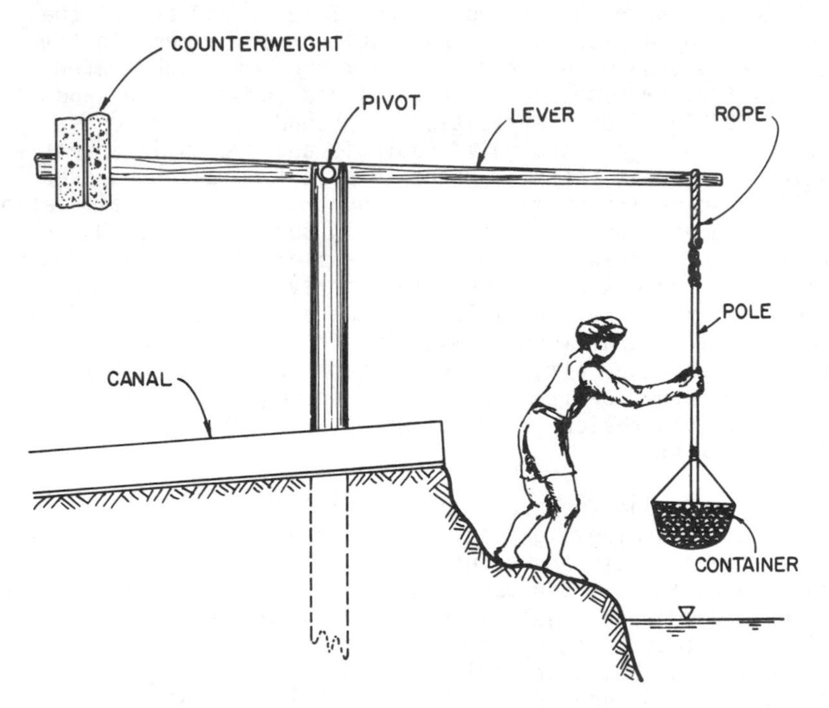

Figure 1-1. Counterpoise lift.

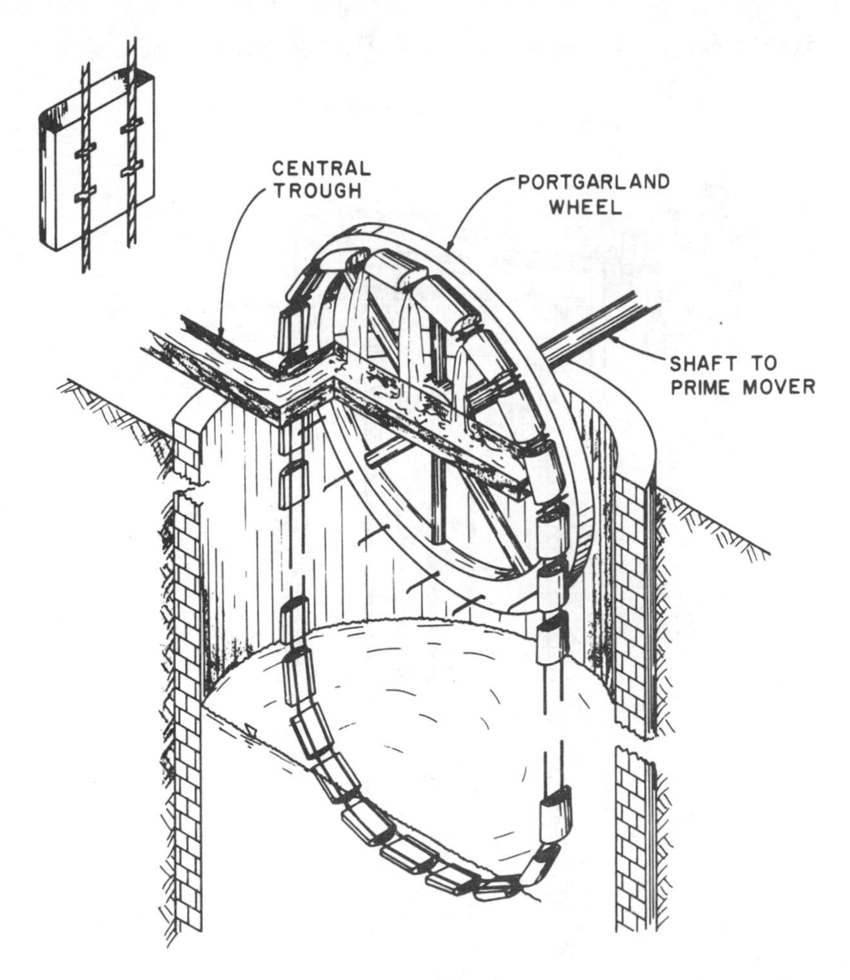

Figure 1-2. Persian wheel with portgarland drive wheel and horizontal drive shaft (After Wood, 1976).

recently from metal, were attached to a continuous loop of rope or belting that was suspended into the well over a drive wheel at the surface. The arrangement was not dissimilar to a modern chain and sprocket mechanism. The lower end of the loop extended to below water level in the well. The chain of containers was rotated by the sprocket wheel so that the containers acted as dippers at the bottom of the well and emptied into a trough as they became inverted at the top of the loop. The dates of first use of Persian wheel seem to be in dispute, but they were in use at least by 200 B.C. (Wood, 1976).

The Chinese water ladder, a device used primarily to raise surface water to increased elevations, was modified during the last centuries B.C. to provide a vertical lift mechanism suitable for use in wells (Fig. 1-3). Circular discs were

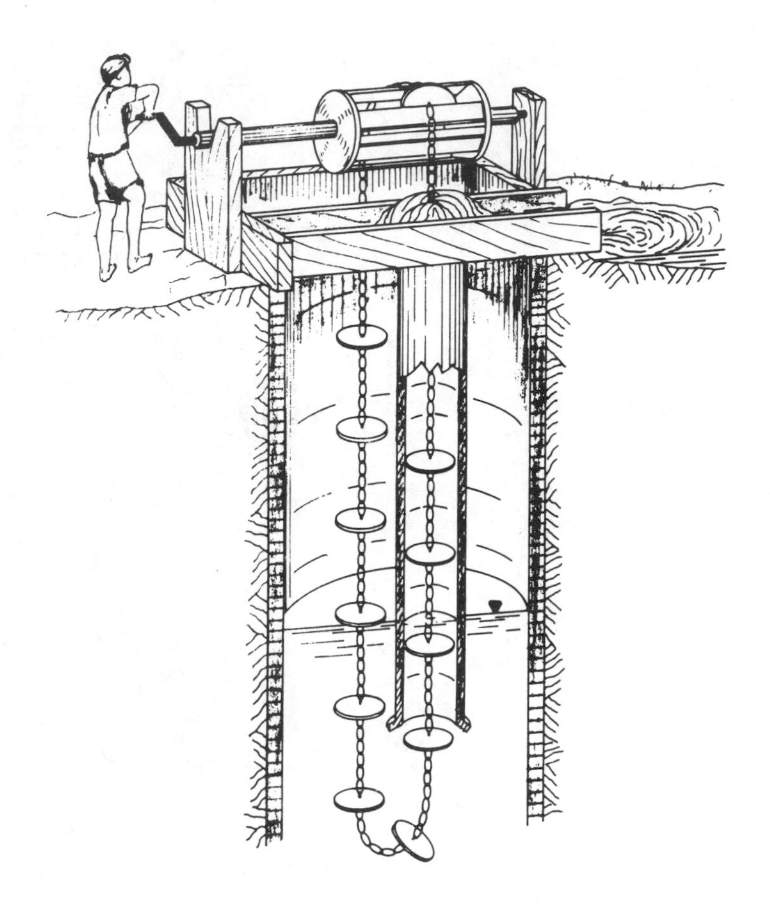

Figure 1-3. Chain pump with manual windlass (After Wood, 1976).

attached to a continuous loop of rope suspended in the well over a windlass. The discs on the rising side of the loop passed upward through a vertical cylinder of slightly larger diameter than that of the discs. The rising discs displaced the water in the cylinder and discharged it at the surface. This device is known as a chain pump.

The first centuries A.D. saw the development of the piston pump. The forerunners of modern centrifugal and other rotary-type pumps were developed in the 16th, 17th and 18th centuries. A unique lift method was used at the Saint Patrick's well at Orvieto, Italy. This 16th century well was constructed with two spiral staircases that permitted donkeys to carry water on their backs from a depth of some 60 meters.

The most remarkable of the ancient methods of ground water extraction did not require lifting devices. The Persians developed ground water as early as 800 B.C. from alluvial outwash adjacent to the mountains by means of huge gravity drains called kanats. Kanats consist of tunnels, dug on an approximate grade, that connect vertical construction shafts through which spoil from the tunnels was lifted and light was reflected on the digging operations. Kanats intercepted ground water and delivered it by gravity to lower elevations. Ground water development by kanats spread to Egypt, where it is believed that tunnels may have exceeded 100 km in length (Tolman, 1937). Kanats exist today in many areas including Pakistan, Afganistan and Iraq, in addition to Iran and Egypt.

Ground-Water Hydrology

Ground-water exploitation preceded the establishment of the rudiments of ground-water science by many centuries. The source of ground water remained unproven, if not undisclosed, until the latter part of the 17th century when a French scientist, Pierre Perrault, concluded on the basis of measurements that annual runoff from the Seine River catchment above Paris was less than one-sixth of the annual volume of precipitation. A few years later, Edme Mariotte verified that the annual precipitation on the Seine River catchment was approximately six times the annual discharge of the river. Further, he established that the discharge of seeps and springs was supplied by precipitation (DeWiest, 1965). Prior to these studies, it was widely believed that precipitation was entirely inadequate to account for the flow of streams and springs. The earth was thought to be practically impenetrable by rain water; a belief that, in itself, precluded a correct explanation for the source of ground water. The prevailing theory, articulated by many of the most prominent thinkers of the time, held that ocean water migrated inland through subsurface caverns and was eventually returned to the surface through springs. Elaborate corollary theories were advanced to explain how the springs could exist at elevations above sea level and how the sea water was purified.

The 19th century saw the development of the basis for the quantitative description of ground-water motion. Hagen (1839) and Poiseuille (1840) derived, independently, the equations for viscous flow in capillary tubes and Henry Darcy (1856), published his now famous empirical equation for flow of water through sand. Darcy's law, in a generalized form, remains today the fundamental rate equation in the analysis of ground-water motion.

The study of ground-water hydrology expanded rapidly in the last decade of the 19th century and the first part of the 20th century. In particular, O. E. Meinzer of the United States Geological Survey was instrumental in the integration of elements of surface hydrology, geology and hydraulics into the scientific study of ground water. The literature is now composed of hundreds of articles and books with important contributions from geologists, hydrologists, engineers, chemists, mathematicians and petroleum and agricultural scientists.

The cumulative knowledge of ground-water phenomena now encompasses an understanding of the role of ground water in the hydrologic cycle, a unifying theory for the motion of fluids in porous media, and procedures for field investigation that reflect the central role of geology in the occurrence and distribution of ground water. Should the study of ground-water hydrology be confined to the desks and laboratories of researchers and teachers, where knowledge of the hydro-geologic characteristics can be presupposed, one might argue that the science is relatively well developed. Ground water, however, occurs in geologic formations whose hydrologic, hydraulic, chemical and geometric properties exhibit all of the vagaries of the natural processes from which they result. While in principle the theoretical and computational tools available are applicable to such complicated cases, the practical difficulty of adequately characterizing the ground-water reservoirs remains a significant constraint. The practioner in ground-water hydrology may be obliged, in some cases, to use less than the full potential of his supporting science simply because the available data or financial resources do not justify additional refinements.

1.3 GROUND-WATER GEOLOGY -- AN OVERVIEW

Geologic Strata -- Ground-Water Vessels

The sequence of sedimentary rocks extending upward from metamorphic or igneous basement rock is the *stratagraphic column*. The column contains a widely variable sequence of sandstone, conglomerates, shales, siltstones, limestones, dolostones, and evaporites, among others. An assemblage of loose material, known as *regolith*, mantles the upper-most stratum in the stratagraphic column in most areas, but bare rock exposed at the surface is not an uncommon occurrence. The stratagraphic section can be thousands of meters thick in large depositional basins or completely absent where erosion has exposed the basement rock.

The rocks in the stratagraphic column and the overlying regolith are the hosts of ground water. The water is contained in the voids formed by the interstices between individual grains or aggregates of grains, in channels and vugs resulting from dissolution of carbonates or evaporites, and in fractures or other openings. A measure of the relative void volume is *porosity* defined as the ratio of the void volume to the bulk volume of the material. Porosity is an important hydrogeologic parameter that will be discussed in more detail later.

Another important hydrogeologic parameter is hydraulic conductivity. A precise technical definition of hydraulic conductivity will be provided in a subsequent chapter. For the purposes of the present discussion, the *hydraulic conductivity* is loosely defined as a measure of the ease with which water can be transmitted through a porous material. A material which permits water to easily flow through it has a higher hydraulic conductivity (is more permeable) than a material that more severely impedes water movement, all other conditions being equal.

A geologic stratum that exhibits porosity and hydraulic conductivity sufficient to store and transmit water in significant quantities is called an *aquifer*. This definition, while imprecise, is nevertheless useful. Regolith in the form of unconsolidated deposits of sand and gravel are probably the most important aquifers from the standpoint of water supply. More than 90 percent of the ground water used in the United States is pumped from such deposits (Walton, 1970). Sand and gravel aquifers occur as alluvium adjacent to and beneath streams and rivers, in buried valleys and abandoned river courses, and in plains and inter-montane valleys. The river deposited material adjacent to the South Platte River in Colorado is an example of an alluvial aquifer. Clastic deposits in the alluvial fans of the Sierra Nevada Range are a second example. Aquifers in the Basin and Range region of Nevada are examples of the inter-montane aquifers. The Ogallala formation which immediately underlies the High Plains of Nebraska, Colorado and Texas is a partially cemented alluvial aquifer. The tremendous alluvial deposits formed by the Indus River in Pakistan and India form a huge aquifer. Water courses, long since abandoned by the Indus, are to be found buried in the flood plain. Often, it is within these buried water courses that the most prolific wells and the best quality water can be found.

Sandstones and conglomerates are the cemented equivalents of unconsolidated sands and sand-gravel-boulder mixtures. These geologic materials also constitute important aquifers. The Dakota sandstone is one example and the Foxhills formation in the Dakotas and in eastern Montana another.

Carbonate rocks are sometimes important aquifers. Limestones and dolomites may exhibit an interstitial porosity and hydraulic conductivity. Often, however, the porosity and permeability of carbonate rocks are secondary in the sense they result from channels formed by dissolution of the rock by percolating waters. Huge caverns may result in some instances.

Igneous and metamorphic rocks form aquifers at some locations. The porosity and permeability are provided by fractures, faults and other openings and not by interstitial void space.

Rarely, if ever, are geologic materials homogeneous over large areas or to great depths. Alluvial deposits are, normally, characterized by relatively coarse material near the sediment source which grades into increasingly finer material at larger distances from the source. This is the result of the progressive decrease in carrying capacity of the streams as their gradient becomes smaller with increasing distance from the mountains. The same process often causes a stratum to exhibit a vertical gradation of particle size. The finest sediments, deposited farthest from the mountains, may be so fine that they no longer qualify as aquifers. This is true because the dimensions of the void space are too small to permit significant flow, even though the porosity may be as large or larger than that of the coarser sediments. Materials which have sufficient porosity to store water but a very small capacity to transmit water are called *aquicludes*. Clays and shales are examples of aquicludes. The occurrence and distribution of aquicludes are important in ground-water analyses because they form the boundaries of aquifers in many cases. Thus, aquicludes often dictate the type and geometry of boundary conditions that must be considered by the hydrologist.

The term *aquitard* refers to a geologic material, the hydraulic conductivity of which is too small to permit the development of wells or springs but may be sufficiently large to significantly influence the hydraulics of aquifers adjacent to it. Another term used by practioners in ground-water hydrology is *aquifuge* which refers to a geologic material with no significant porosity or permeability. Unweathered and unfractured granite and some carbonates afford examples of aquifuges. Aquifuges are important as boundaries of aquifers and sources of clastic material for aquifer formation.

Clearly, the stratagraphic column has the important hydrologic significance of a sequence of potential aquifers, aquicludes and aquitards in addition to its more traditional significance in geology. The actual order in which strata occur in a particular column influences the vertical distribution of ground water. Aquicludes may isolate potential aquifers from sources of recharge, for example. On the other hand, an aquifer that may appear isolated in a particular section can be exposed at other locations via the processes of geologic uplift, folding, faulting or erosion. Geologic maps and cross-sections, which depict the lateral distribution of strata, are also indispensable aids to the ground water hydrologist, therefore.

While the occurrence of ground water in regolith and the upper portions of the stratagraphic column is problematical at any particular location, ground water can nearly always be found at depth in a thick sequence of sedimentary rocks. Water saturated rocks more than 8 km beneath the earth's surface have been penetrated by wells in the petroleum industry. Water at such great depths is usually completely isolated from surface water and occurs as a result of entrapment in the interstices when the sediments were originally deposited beneath the sea. Such water is known as *connate* water and is highly mineralized and saline. The temperature of water generally increases with depth and in some areas water temperatures are high enough to provide useful energy. Such geothermal activity occurs in many areas of the world and sometimes is apparent at the ground surface as geysers such as is evident at Yellowstone National Park, USA.

Confined and Unconfined Ground Water

A *water table*, also known as the phreatic surface, is the surface upon which the water pressure is equal to atmospheric pressure. An *unconfined* or water-table aquifer is one in which a water table exists, and a *confined* or artesian aquifer contains water at pressures greater than atmospheric pressure. While it is traditional to use the words confined and unconfined as modifiers of the term aquifer, the meaning should not be extended to the geologic stratum as a whole. As will be seen presently, it is quite possible for a particular stratum to be properly classified as a confined aquifer at one location and as unconfined in another. The distinction between confined and unconfined aquifers is made because there are substantial differences in their hydraulic behavior, particularly, their capacity to take up or release water from storage.

In order that water exist in a confined aquifer at pressures greater than atmospheric, the aquifer must be bounded by relatively impervious materials on the top and bottom. A confined aquifer is shown in Fig. 1-4. The shale formations bounding the limestone aquifer act as walls of a container permitting the water in the aquifer to exist at elevated pressures. Well A is a flowing artesian well, the pressure in the confined aquifer being sufficient to cause the water to rise above the land surface. The source of the elevated pressure observed at the location of well A is the hydrostatic pressure developed between the elevation of the recharge area and the elevation of the aquifer at point A. It is only necessary for the water to rise above the top of the aquifer to be classified as a confined or artesian aquifer, however. The aquifer is artesian, for example, at the location of well C where the water level is below the ground surface.

An imaginary surface passing through all points to which water will rise in wells penetrating a confined aquifer is called the *piezometric surface*. The water levels in wells A

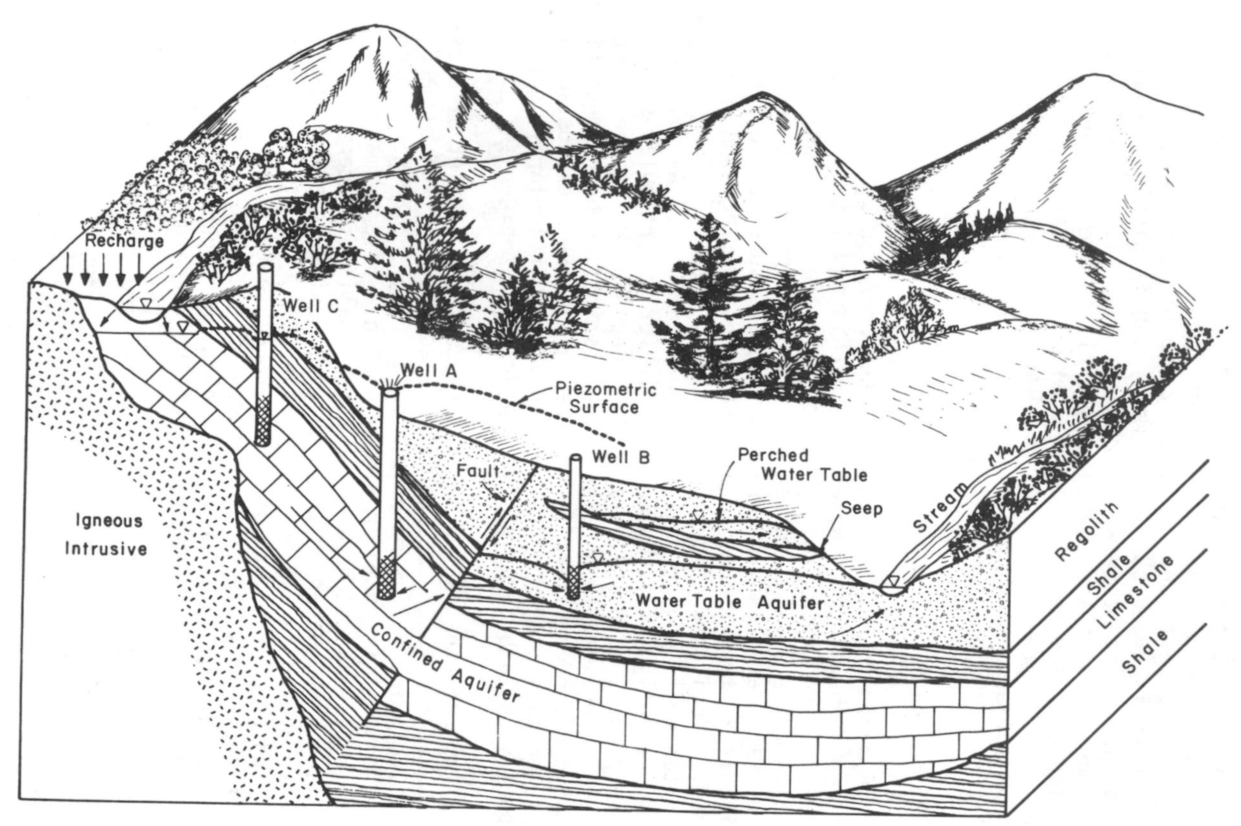

Figure 1-4. Example of ground-water occurrence.

and C are points on the piezometric surface. This surface plays a central role in the identification, description and analysis of flow in confined aquifers. Notice that, should the piezometric surface fall below the top of the aquifer, the aquifer becomes unconfined at that point and the piezometric surface and the water table coincide.

An unconfined or water-table aquifer is also shown in Fig. 1-4. The water level in well B indicates the position of the water table. Water levels in wells A and B are not equal because the aquiclude tends to hydraulically isolate the aquifers, one from the other. The fault zone, however, permits water from the confined aquifer to seep into the water-table aquifer, and the aquifers are not entirely independent. The perched water table in Fig. 1-4 is the result of the collection, on a clay lens, of water which has percolated from the ground surface. Shallow ground water of local occurrence that returns to the land surface along lithic contacts above the main aquifers is called *interflow*. Interflow is often responsible for small springs and seeps that flow only during wet periods.

Observation of Ground Water

The occurrence, distribution, and movement of ground water must be detected and analyzed by means other than direct visualization. Detailed geologic study is one method by which the required data can be obtained, but important synergistic effects accrue to the analysis that combines both geologic and hydraulic data. By far the most important sources of hydraulic data are the water levels in observation wells and piezometers.

A *piezometer* is a device which indicates the water-pressure head h_p at a "point" in the aquifer (Fig. 1-5). The piezometer consists of a casing, perforated near the terminal point only, that is installed in such a way that the casing fits tightly against the geologic formation. A common method of installation is to place the casing in a drilled hole, backfilling the annulus between the casing and the wall of the bore hole with clay. The height to which water rises in the piezometer is the water-pressure head at the terminal point of the piezometer.

The *observation well* of Fig. 1-5 is a perforated casing simply placed in a bore hole with no attempt to provide a seal between the casing and the aquifer. In the case shown, the observation well properly indicates the water table position. Should the observation well penetrate into the underlying confined aquifer, however, the observed water level would be meaningless. This is true because the water level would reflect neither the water table nor the piezometric surface elevation, but a combination of the two since the well bore provides a flow conduit between the two aquifers.

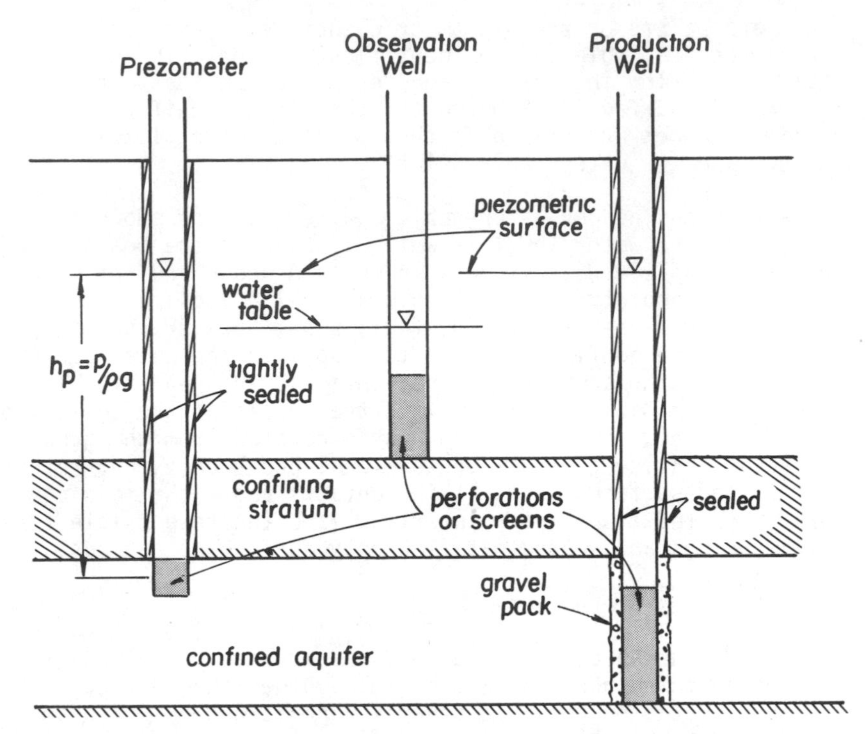

Figure 1-5. Common facilities for observing water levels in aquifers.

The production well on the right side of Figure 1-5 can function as an observation well in the confined aquifer, because the annulus is sealed in the confining layer. The production well is not a piezometer, strictly speaking, because the well screen extends over an appreciable fraction of the aquifer. Thus, the water level indicates an average water-pressure head over the screened interval. The water levels in the production well and in the piezometer will be identical only under special conditions, one of which is no flow of water in the aquifer.

REFERENCES

Darcy, H. (1856), Les fontaines publiques de la ville de Dijon, V. Dalmont, Paris, 674 p.

DeWiest, R. J. M. (1965), Geohydrology, John Wiley and Sons, Inc., New York, 366 p.

Hagen, G. (1839), Ann. Physical Chemistry, Vol. 46, pp. 423-442.

Johnson Div., Universal Oil Products (1972), Ground Water and Wells, Second Printing, Saint Paul, Minnesota, 440 p.

Poiseuille, J. L. (1840), Compte Rendus, Vol. 11, pp. 961 and 1041.

Tolman, C. F. (1937), Ground Water, McGraw-Hill Book Co., Inc., New York, 593 p.

Walton, W. C. (1970), Groundwater Resource Evaluation, McGraw-Hill Book Co., Inc., New York, 664 p.

Wood, A. D. (1976), Water lifters and pumps for the developing world, Unpublished MS Thesis, Colorado State University, Fort Collins, Colorado, 303 p.

PROBLEMS AND STUDY QUESTIONS

1. Identify locally a groundwater basin and provide references for available basic data on the area.

2. Why is it important to identify and distinguish between an aquifer, aquiclude, aquitard and aquifuge?

3. Identify groundwater problems which are of primary concern to:

   a. civil engineers,
   b. geologists,
   c. petroleum engineers,
   d. agricultural engineers,
   e. agronomists,
   f. lawyers,
   g. physicists,
   h. chemists,
   i. mathematicians.

Chapter II

GROUND-WATER STORAGE AND SUPPLY

Ground-water storage is no less important to the ground-water hydrologist than above-ground storage is to hydrologists and engineers concerned with surface-water supplies. The aquifer is an underground storage reservoir and also serves as the conduit through which water moves to streams, wells, drains or other extraction points. Just as the change of stage is an indicator of the change of storage in a surface reservoir, fluctuations of the water table or piezometric surface elevation are indicators of the changes in the relative volume of ground-water storage. The relative volume of water taken into or released from storage is related to the corresponding water-level change through two important hydrologic parameters; the apparent specific yield for water-table aquifers, and the specific storage for confined aquifers. The relationships between aquifer storage and water-level changes are discussed in this chapter without regard for the capacity of the aquifer to also act as a conduit. The latter subject and its relationship to ground-water storage and supply are dealt with in detail in subsequent chapters.

2.1 STORAGE IN WATER-TABLE AQUIFERS

Porosity and Representative Volume Elements

Porosity, being a measure of the relative void volume of a porous medium, is a fundamental hydrologic parameter of aquifers. Porous materials occasionally contain voids that are isolated from adjacent voids and, therefore, do not participate in the storage and interchange of fluids. An example is some volcanic rocks that contain isolated void space formed by the evolution of gases. These voids are of little importance in hydrology and are excluded from the following discussion.

Porosity is the fraction of a representative volume element of porous medium that is interconnected void space. Viewed on a scale of the same order as the grain size, granular aquifer materials exhibit a profoundly irregular variation of solids and voids that is virtually impossible to describe in a detailed manner. This difficulty is circumvented by considering average values of hydrologic parameters and variables for volume elements which are large relative to the grain size but are small relative to the dimensions of the aquifer (or sample thereof) as a whole. Such a volume is called a representative volume element. The representative volume element is the smallest element for which gradual or smooth changes in average values of hydrologic parameters and variables are observed. Further reduction in the element of volume results in large, abrupt changes in the average values owing to the irregularity of porous materials on a small scale. Hydrologic parameters and variables defined for representative volume elements are *macroscopic* quantities.

A graphic illustration of the variation of the ratio of void volume to element volume, as the volume of the element increases incrementally, is presented in Fig. 2-1. The extreme

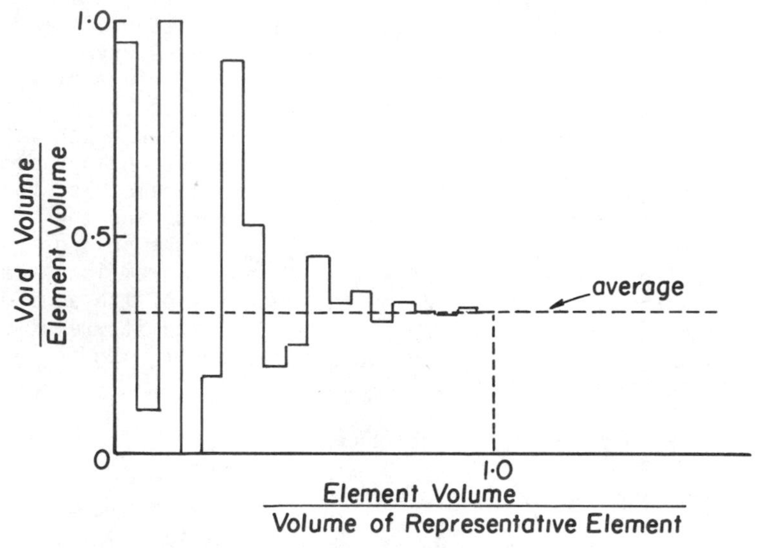

Figure 2-1. Variation of void volume with element size in a homogeneous porous medium.

variation shown for small volume elements results because the volume element is so small that it may be entirely filled with solids, entirely void, or consist of both solids and voids in any proportion depending upon the location of the element. As the size of the volume element is increased beyond the volume of individual grains, it will always encompass some solids and some void space, but the proportion that is void space will still vary abruptly as the location of the element is changed. The variation becomes smaller as the volume of the element is further increased. In a homogeneous porous medium, the fraction of the volume element that is void space approaches a constant as shown. The representative volume element will exhibit the same porosity, regardless of location in a homogeneous aquifer. In an unstratified, nonhomogeneous aquifer, smooth variations of porosity are observed as the location of the element is changed. Abrupt changes in porosity, owing to textural discontinuities, may occur at the interfaces between different strata. At a particular location in a porous material, the porosity is the same regardless of the orientation of the volume element. Porosity does not exhibit directional properties and is a scalar quantiy.

The porosity can also be defined as the void area per unit of bulk area. Figure 2-2 is a photograph of a cut-section of

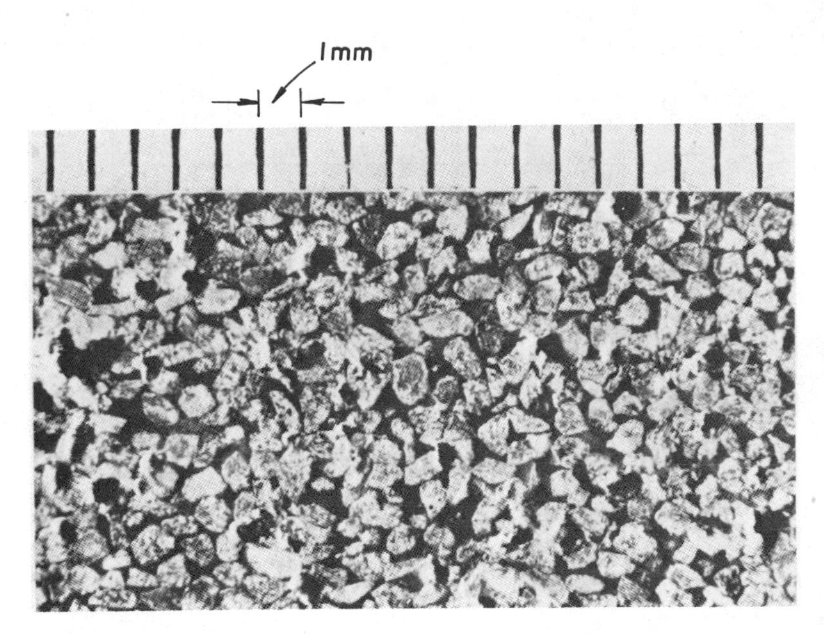

Figure 2-2. Typical photograph of a cut-section of porous medium.

a porous medium, showing the solids and the voids. Statistical techniques can be used to determine porosity from such a section. For example, one may drop pins at random on a photograph of a cut section and count the number of times n the ends of the pin fall within a void area and the total number of drops N. The porosity ϕ for large N is

$$\phi = n/2N \quad . \tag{2-1}$$

It is, of course, very difficult to establish numerical values for the dimensions of representative volume elements. Fortunately, the volume of the representative element is much smaller than the minimum volume within which changes of properties are important in most cases of practical interest. Only in the laboratory or when the hydrologic properties are dominated by sparse fractures, solution channels, or by very coarse material is the size of the representative volume element of practical concern. The utility of the concept of the representative volume element is that it provides a rational basis for the use of continuum mathematics for the description and interpretation of the storage and flow of fluids in porous media.

The porosity of aquifer materials ranges from near zero in sparsely fractured rocks and in some limestone formations to 50 percent or more in well aggregated clays. Typical values of and osity for various geologic materials are presented in Table 2-1.

Table 2-1. Porosity of Aquifer Materials (Adapted from Morris & Johnson, 1967)

Aquifer Material	No. of Analyses	Range	Arithmetic Mean
Igneous Rocks			
Weathered granite	8	0.34-0.57	0.45
Weathered gabbro	4	0.42-0.45	0.43
Basalt	94	0.03-0.35	0.17
Sedimentary Materials			
Sandstone	65	0.14-0.49	0.34
Siltstone	7	0.21-0.41	0.35
Sand (fine)	243	0.26-0.53	0.43
Sand (coarse)	26	0.31-0.46	0.39
Gravel (fine)	38	0.25-0.38	0.34
Gravel (coarse)	15	0.24-0.36	0.28
Silt	281	0.34-0.61	0.46
Clay	74	0.34-0.57	0.42
Limestone	74	0.07-0.56	0.30
Metamorphic Rocks			
Schist	18	0.04-0.49	0.38

The porosity of clastic materials is affected by:
1) particle shape (angularity and roundness),
2) degree of compaction and cementation,
3) particle size distribution.

Particle shape influences porosity because particle shape affects the manner in which particles arrange themselves (Fig. 2-2). Highly angular and irregularly shaped, noncemented particles tend to result in a larger porosity than smooth, regularly shaped particles, although the difference is not always a substantial one. The degree of compaction or cementation changes the porosity significantly, however. Cement, formed by iron and calcium compounds (among others), at interparticle contacts, occupies a portion of the interstitial volume that would otherwise be void. Comparison of the porosities in Table 2-1 for sandstone and siltstone with their uncemented equivalents, sand and silt, illustrates the effect of cementation.

Aquifers composed of granular materials of nearly uniform particle size exhibit a larger porosity than materials with a wide distribution of particles because small particles occupy a portion of the volume between the larger particles in graded materials (Fig. 2-2). Fine gravel and sands often have larger porosities than their coarse counterparts for this reason. The porosity of clastic materials with a nearly uniform particle size is nonsensitive to actual particle size. In the extreme case of spheres with equal diameters, the porosity is completely independent of the size of the spheres.

The data in Table 2-1 indicate that the porosity of limestone ranges widely. Porosities in the upper portion of the range result from interstitial void space bounded by the solid crystals of the rock. Vugs and solution channels usually constitute a small fraction of the aquifer volume, notwithstanding the fact that their dimensions are quite large relative to interstitial-pore space in crystalline limestone.

Volume methods for measurement of porosity requires that any two of the following three volumes be determined: the bulk volume of the sample, the particle volume, and the pore volume. A variety of methods are available by which these volumes can be determined. Inasmuch as the more common methods require nothing more than gravimetric and/or volumetric measurements and simple manipulations of the definition of porosity, the details of porosity determination are best illustrated by calculations. Several methods are illustrated in the problem set at the end of this chapter.

EXAMPLE 2-1

Uniform, solid spheres of radius R are packed into a cube of dimensions $20R$. Packing is accomplished in tiers so that the spheres in each tier are directly centered between the spheres of adjacent tiers (cubic packing). Calculate the porosity.

Solution:

Number of spheres per tier = (10)(10) = 100

Number of tiers = 10

Number of spheres = 1000

$$\text{Volume of spheres} = (1000)(\tfrac{4}{3}\pi R^3)$$

$$\text{Volume of cube} = (20R)^3$$

$$\text{Volume of voids} = (20R)^3 - (1000)(\tfrac{4}{3}\pi R^3)$$

$$\phi = \frac{\text{Volume of voids}}{\text{Volume of cube}} = 0.476$$

Notice that the porosity for this simple case is independent of the size of the spheres. The representative volume element in this case is 2R by 2R by 2R or $8R^3$ and is much smaller than the volume used in this example. Clearly, the representative volume element is highly dependent upon the sphere size.

EXAMPLE 2-2

A container with a volume of 44.0 cm^3 is filled with a loose sand. When the sand is poured into a graduated cylinder partially filled with water, it is observed that 25.7 cm^3 of water are displaced. Calculate the porosity of the sand in the container.

Solution:

The volume of water displaced is the volume of solids V_s, and the volume of the container is the bulk volume V_b.

$$\phi = \frac{V_b - V_s}{V_b} = 1 - \frac{25.7}{44.0} = 0.42 \quad .$$

Desaturation and Specific Retention of Aquifer Materials

Removal of water from the pore space of aquifers and replacement by air is a fundamental process that accompanies the reduction of the water-table level in unconfined aquifers. Displacement of water by air (desaturation) occurs because the pressure of the water in the pores becomes less than the local air pressure. Thus, air and water exist simultaneously in the voids where the air pressure exceeds the water pressure sufficiently to displace a portion of the water. Water, however, adheres to aquifer solids more strongly than does air which causes the air-water interfaces to become curved, resulting in interfacial forces that oppose the force due to pressure difference. Interfacial tension σ is an interfacial force per unit length that acts along the perimeter of the interface in directions tangent to the curved surface. At equilibrium, the force due to interfacial tension balances the displacement force due to pressure difference.

Figure 2-3 shows an idealized, hemi-spherical interface across the throat or connecting channel between two pores. On the left, the pore is filled with water at pressure $-p$, and the right-hand pore is filled with air at zero gage (atmospheric) pressure. The difference between the air pressure and the water pressure is called *capillary pressure*, p_c. In this case, $p_c=p$. The forces due to interfacial tension are preventing the displacement of the water in the left pore. A force balance on the hemi-spherical interface leads to

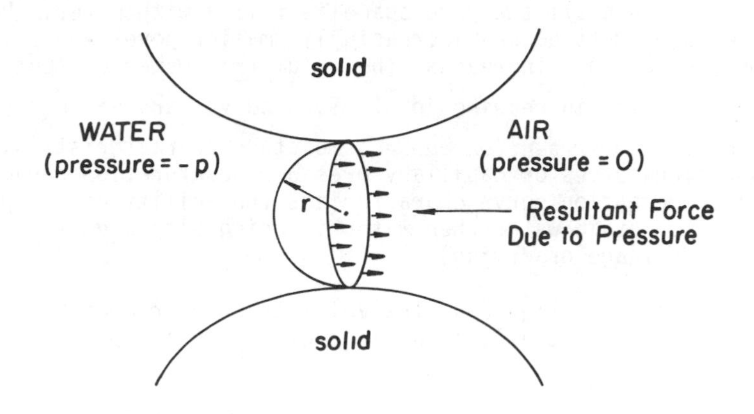

Figure 2-3. Idealized air-water interface.

$$r = \frac{2\sigma}{p_c} \tag{2-2}$$

obtained by equating the force due to interfacial tension, $2\pi r\sigma$, and the force due to pressure difference, $\pi r^2 p_c$.

Equation 2-2 was obtained for a highly idealized interface geometry, but nevertheless, the conclusion that the equilibrium radii of curvature of air-water interfaces must decrease as the capillary pressure increases remains valid. The fact that air-water interfaces must become more sharply curved with increasing p_c means that water must occupy increasingly smaller pores and void subspaces as p_c is increased. For example, should the water pressure in the left pore of Figure 2-3 be reduced slightly (i.e., p_c increased), the force tending to displace the water will increase. The radius of curvature of the interface at the throat is already minimum, however, and cannot decrease further as required to balance the increased pressure force. The result is a detachment of the interface from its position across the throat and displacement of water in the left pore occurs. The displacement ceases when the air-water interface recedes to other pores or void subspaces that are sufficiently small to support an interface with the radius required to balance the force of displacement.

The fact that both air and water may exist in the pore space requires a parameter other than porosity to characterize the relative volume of water in the aquifer. The fraction of a representative volume element that contains water is the *volumetric-water content*, θ . Volumetric-water content ranges from zero in a completely dry porous medium to a maximum, equal to porosity, when all the pore space is filled with water. Because water must occupy increasingly smaller pores and void subspaces as p_c increases, the volumetric-water content decreases with increasing p_c. Such curves are variously known as *water-retention curves*, moisture characteristics, desorption curves or capillary pressure-desaturation curves. A water-retention curve characterizes the ability of the porous medium to retain water when water is being displaced (i.e., during drainage or drying).

As shown in Fig. 2-4, the volumetric-water content changes very little, if at all, in a range of p_c near $p_c=0$. Even

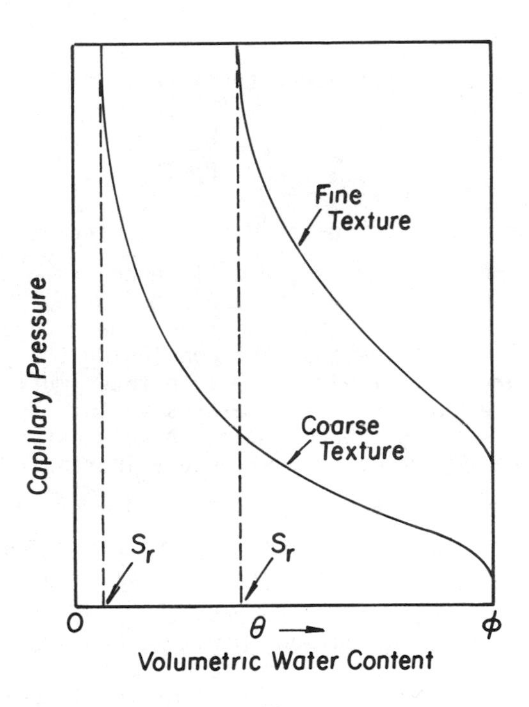

Figure 2-4. Typical water retention curves.

the largest pores are sufficiently small to resist desaturation in this range of p_c. Significant desaturation occurs only after the difference between the air and water pressures is sufficient to displace water from the largest pores. At a particular value of p_c, the radii of curvature of air-water interfaces are equal in both materials, but the volumetric-water content differs significantly because the fine textured material contains a larger volume of pores and subspaces that are sufficiently small to prevent desaturation.

The volumetric-water content tends toward a constant value at large p_c for both materials in Fig. 2-4. The value of θ for which $d\theta/dp_c$ tends to zero is the *specific retention*, S_r. The specific retention is also known as field capacity or water-holding capacity in other fields of study. Specific retention is a reasonably characteristic parameter of porous solids, especially in sands and gravels in which $d\theta/dp_c$ tends toward zero rather abruptly at low values of p_c. A very gradual change of θ with increasing p_c is often observed in very fine-grained materials, making it difficult to establish a value of θ at which $d\theta/dp_c$ approaches zero. As a practical matter, the volumetric-water content at a capillary pressure of 1/3 bar is often taken as being the specific retention.

The relationship between p_c and θ observed when the porous material is absorbing water (wetting) is substantially different from that shown in Fig. 2-4. The dependence of θ upon p_c for a material that was originally saturated, then desaturated (curve 1), allowed to absorb water again (curve 2), and, finally, desaturated a second time (curve 3) is shown in Fig. 2-5. It is apparent that $p_c(\theta)$ is not a single-valued

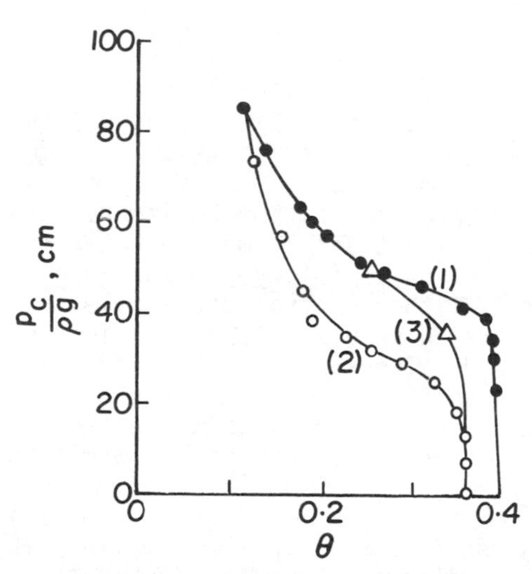

Figure 2-5. Capillary pressure - water content curves with different pressure history (after McWhorter, 1971).

function but depends upon the pressure (or water content) history. One reason for the hysteresis of the $p_c(\theta)$ function is entrapment of air during the wetting cycle. The presence of entrapped air accounts for the observation that $\theta=\theta_m<\phi$ at $p_c=0$ for curve 2. In the field, air is entrapped during infiltration and during periods of rising water tables. For this reason, the pore space is rarely, if ever, completely filled with water even when $p_c=0$.

EXAMPLE 2-3

Figure 2-6 is a schematic diagram of a device that is used to measure the functional relationship between volumetric-water content and capillary-pressure head.

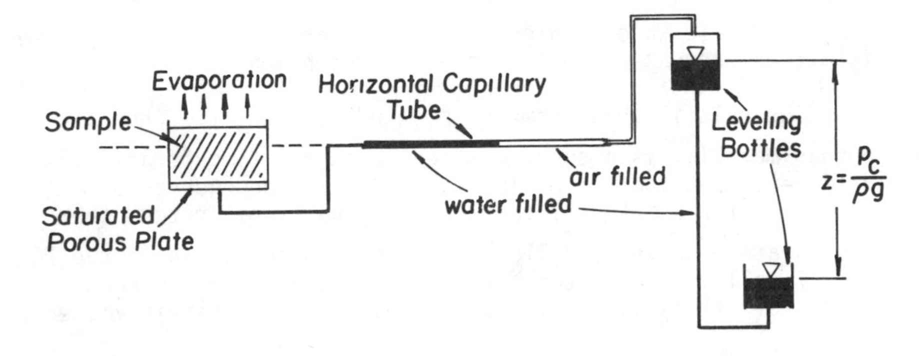

Figure 2-6. Measurement of relationship between θ and $p_c/\rho g$.

The experiment is begun with a completely saturated sample (i.e., $\theta=\phi$). A suction (capillary pressure) is applied to the sample by lowering the leveling bottle in which the water level is exposed to the atmosphere. Except for a small correction at the air-water interface in the horizontal capillary tube, the difference in elevation z of the water in the leveling bottles is the average capillary-pressure head applied to the sample. Water is withdrawn from the sample through the saturated porous plate in response to an applied increment of capillary-pressure head. The size of the pores in the porous plate are too small to desaturate over the range of capillary pressures encountered in the experiment, and therefore, the porous plate remains impermeable to air while permitting water to flow from the sample.

The air-water interface in the capillary tube moves outward as water is withdrawn from the sample following an incremental increase of z . Evaporation from the surface of the sample is also increasing the capillary pressure, however, and eventually the air-water interface ceases to move outward and begins to recede toward the sample. The average capillary-pressure head in the sample, at the moment the interface begins to recede, is very nearly equal to z. When the interface begins to recede, the sample holder is detached and weighed. The weight of the sample and holder, the tare weight of the holder, and the dry weight of sample permit computation of the volumetric-water content. The procedure is repeated through successively higher values of z.

A procedure similar to that described above was used to obtain the following data for a fragmented, partially weathered shale.

$p_c/\rho g$ - cm	0	2.2	4.2	6.2	7.7	9.7
θ	0.430	0.425	0.421	0.414	0.408	0.401
	12.7	15.8	20	25	32	42
	0.392	0.384	0.374	0.362	0.348	0.331
	56	76	106	147	198	282
	0.313	0.300	0.281	0.264	0.251	0.242

Distribution of Water Content and Pressure Above a Water Table

The pressure of pore water, as measured in representative volume elements, is distributed in static subsurface water systems in exactly the same manner as in any other static hydraulic system. The pressure in static ground water increases with depth precisely as the pressure increases with depth in a tank or other container. The differential equation of fluid statics

$$-\frac{dp}{dz} = \rho g \quad , \tag{2-3}$$

derived from a balance of forces on static fluid elements in elementary hydraulics, also applies to subsurface water, therefore. In the above equation, p is the average pore-water pressure in a representative volume element, dz is a differential displacement of the center of the volume element, ρ is the water density, and g is the gravitational constant. Equation 2-3 states that the downward force per unit volume of water due to weight is balanced by an upward force due to the gradient of water pressure.

Integration of Eq. 2-3, and division by ρg, gives

$$\frac{p}{\rho g} = -z + \text{constant} \tag{2-4}$$

wherein z is the vertical coordinate measured positive upward from an arbitrary datum. The quantity $p/\rho g$ is termed *pressure head*. For the purposes of this discussion, it is convenient to measure z from the water table where $p=0$ relative to the local atmospheric pressure. The immediate conclusion is that the gage pressure of water at points above the water table is negative (Fig. 2-7). For this reason, water above the water

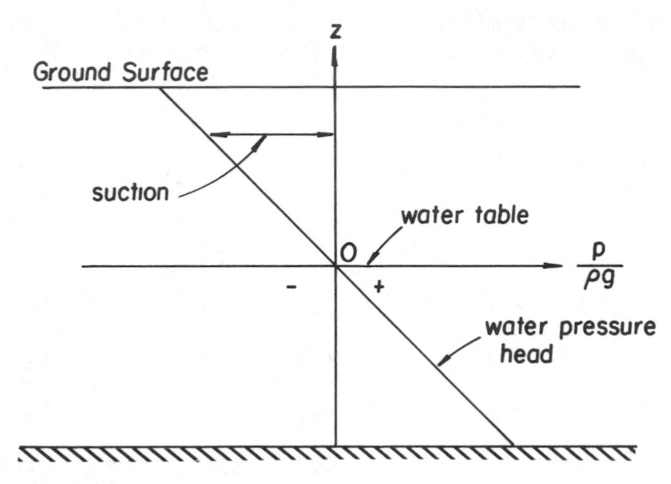

Figure 2-7. Distribution of water-pressure head in static subsurface water.

table has been called suspended water, and the negative pressure head called tension or suction, although it is clear from Eq. 2-3 that the water is not actually suspended by tension forces. Because the water pressure above the water table is less than atmospheric, water from this zone will not enter wells, drains, or open holes. Water pressures less than atmospheric pressure must be sensed by special devices known as tensiometers which function in a manner similar to the porous plate in Example 2-3.

The negative water pressure at points above the water table is simply the capillary pressure. Thus,

$$\frac{p_c}{\rho g} = z \qquad (2\text{-}5)$$

is valid for static subsurface water, independent of the porous solid. Equation 2-5, used in conjunction with the appropriate $p_c(\theta)$ function, is used to deduce the distribution of water content above the water table. Example distributions of θ in a homogeneous and in a stratified aquifer are depicted in Fig. 2-8. The capillary fringe, shown in Fig. 2-8a, is the increment

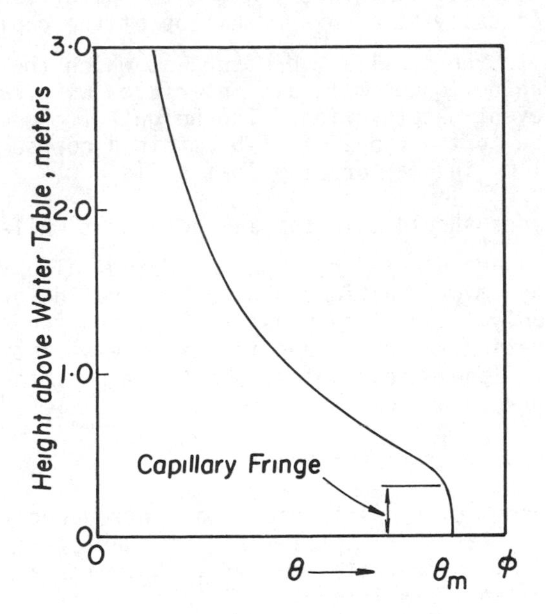

Figure 2-8a. Typical equilibrium distribution of water above a water table in homogeneous materials.

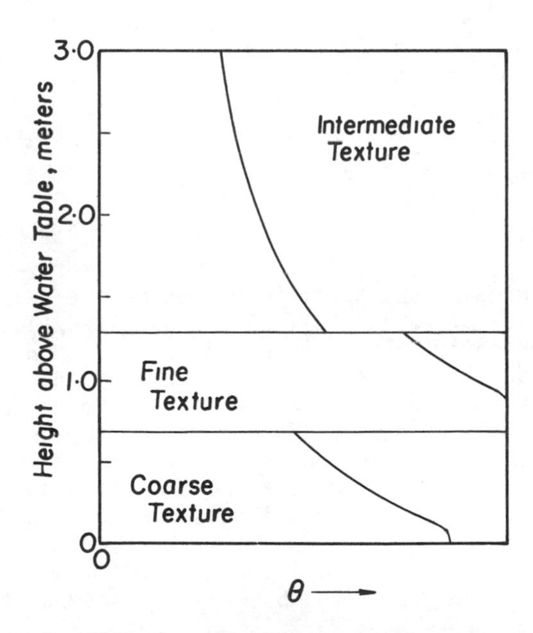

Figure 2-8b. Equilibrium distribution of water above a water table in stratified aquifer.

of aquifer material immediately above the water table that remains practically at $\theta=\theta_m$. The top of the capillary fringe corresponds to the capillary pressure at which the largest of the pores can no longer maintain interfaces with radii sufficient to prevent desaturation. The height of the capillary fringe may be in the range of 1-5 cm in a coarse sand and in excess of 1 m in clay or clay loam soils.

The reader should take special note that while $p_c(\theta)$ relationships are usually regarded as being valid for both static and dynamic situations, Eq. 2-5 is valid for equilibrium conditions only. Capillary-pressure head bears no simple relation to elevation z under conditions of evaporation, recharge, or rapidly falling water tables, and therefore, the distribution of θ is correspondingly more complex.

Specific Yield and Apparent Specific Yield

The concept of specific yield was introduced in groundwater hydrology as a practical means of characterizing the storage capacity of water-table aquifers. Notwithstanding its practical utility, specific yield is a somewhat ambiguous concept (Duke, 1972; Youngs, 1969). To some degree, the ambiguity results from the use of the term -- specific yield -- as both an objective property of porous solids and as a characterization of the storage behavior of aquifer systems as a whole. In the latter case, many factors, other than properties of the aquifer material itself, may dominate the changes in storage associated with water-table fluctuations.

Specific yield, S_y , as used herein, is defined as the difference between porosity and specific retention and is

$$S_y = \phi - S_r \quad . \tag{2-6}$$

Insofar as porosity and specific retention are relatively objective properties of porous solids, the specific yield is a characteristic property. Specific retention is dependent upon the fluids used during its laboratory determination, however, so neither specific retention nor specific yield is entirely characteristic of the porous solids for which they are measured. Specific yield is a dimensionless parameter, often interpreted as being the ratio of the drainable volume to the bulk volume of the medium. Specific yield is also known as *effective porosity*.

Apparent specific yield, S_{ya} , is defined as the ratio of the volume of water added or removed directly from the saturated aquifer to the resulting change in the volume of aquifer below the water table. The apparent specific yield, when determined directly from its definition and field measurement, is a bulk parameter that incorporates the influences of such factors as air entrapped near the water table, stratification of materials above the water table, water table position and the rate of change of water table elevation. Although in principle the apparent specific yield can be treated as a variable in both time and space, it is nearly always assumed that S_{ya} is a constant in time. This assumption implies, for example, an instantaneous release of water from storage when the water-table level falls. Instantaneous release of water from storage does not occur, of course, but is an acceptable approximation when the changes in water-table elevation occur slowly.

Because apparent specific yield depends heavily on factors other than the physical properties of porous solids, values of S_{ya} may bear little or no relation to the specific yield, S_y. The upper limit of S_{ya} is S_y, however. One field condition can be visualized in which $S_y = S_{ya}$; that being the situation in a homogeneous aquifer in which a slowly falling water table is everywhere at depths much greater than the height of the capillary fringe. Figure 2-9 shows the distribution of volumetric-water content between the ground surface and two successive depths to the water table in a homogeneous material. Provided that the water table falls from z_1 to z_2 very slowly, the water-content distributions are approximately those that would exist at equilibrium. The volume of water released per unit area, as the water table falls from z_1 to z_2, is $(\phi - S_r)(z_2 - z_1)$ and the apparent specific yield is $\phi - S_r$, equal to the specific yield. Note that the apparent specific yield is equal to the air content at the surface. This is also true in any homogeneous material, regardless of water table depth, provided the water content distribution is the equilibrium distribution.

Had the material above the water table been stratified as shown in Fig. 2-8b, the apparent specific yield would have been less than the specific yield of the material in which the water table is located. Also, drainage from the pores cannot keep pace with a rapidly declining water table, causing the water content to be greater at points above the water table than predicted for the equilibrium condition. Therefore, the apparent specific yield for rapidly declining water tables is substantially less than S_y. Air entrapped in the zone below the water table results in a further reduction of the apparent specific yield. Hansen (1977) reports measured quantities of entrapped air in the field averaging 20 percent of the pore volume. Aquifer stratification, slow drainage of materials

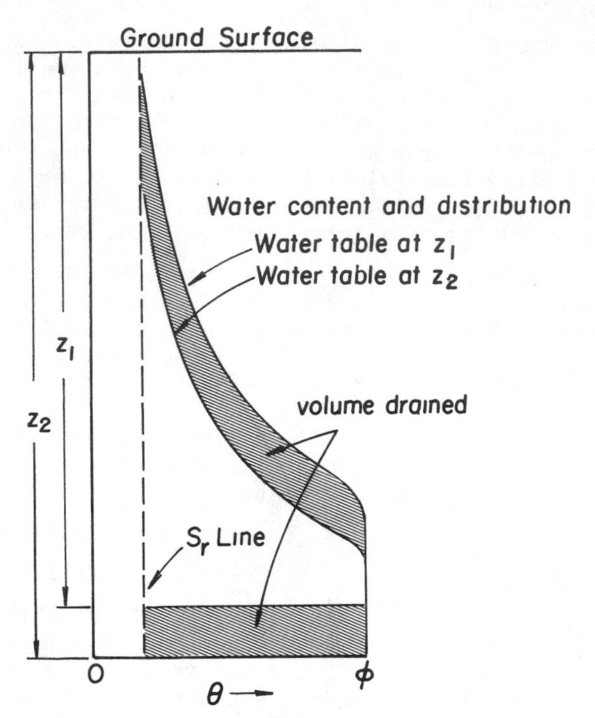

Figure 2-9. Volume drained when a deep water table is lowered in a homogeneous aquifer.

above the water table, and air entrapment often result in values of S_{ya} on the order of one-third S_y .

The hydrologist may sometimes be obliged to use laboratory values of S_y as an approximation for S_{ya} in the absence of field determinations. Values of specific yield, as determined in accordance with Eq. 2-6, are tabulated for several aquifer materials in Table 2-2. Specific yield is most likely to be a fair approximation for apparent specific yield in homogeneous aquifers, composed of relatively coarse materials with a small capillary fringe, in which the water table is at a depth equal to several times the height of the capillary fringe.

EXAMPLE 2-4

Figure 2-10 shows a plot of the capillary pressure - desaturation data for the fragmented shale of Example 2-3. Determine ϕ, S_r and S_y.

Solution:

From Fig. 2-10

$$\phi = 0.43 \quad , \quad S_r = 0.24 \quad , \quad S_y = 0.19 \quad .$$

Table 2-2. Specific Yield of Aquifer Materials
(Adapted from Morris & Johnson, 1967)

Aquifer Material	No. of Analyses	Range	Arithmetic Mean
Sedimentary Materials			
Sandstone (fine)	47	0.02-0.40	0.21
Sandstone (medium)	10	0.12-0.41	0.27
Siltstone	13	0.01-0.33	0.12
Sand (fine)	287	0.01-0.46	0.33
Sand (medium)	297	0.16-0.46	0.32
Sand (coarse)	143	0.18-0.43	0.30
Gravel (fine)	33	0.13-0.40	0.28
Gravel (medium)	13	0.17-0.44	0.24
Gravel (coarse)	9	0.13-0.25	0.21
Silt	299	0.01-0.39	0.20
Clay	27	0.01-0.18	0.06
Limestone	32	$\sim$0 -0.36	0.14
Wind-Laid Materials			
Loess	5	0.14-0.22	0.18
Eolian Sand	14	0.32-0.47	0.38
Tuff	90	0.02-0.47	0.21
Metamorphic Rock			
Schist	11	0.22-0.33	0.26

EXAMPLE 2-5

The initial water levels (with respect to an arbitrary datum) and the levels after 8 hrs and 22 min of pumping are tabulated below for six observation wells in the vicinity of the pumped well. The average pumping rate for the test period is 0.0312 m^3/s . The aquifer is a mixture of clayey sand and gravel. The effective radius of the pumped well is 0.3 m . Calculate the apparent specific yield using the data provided.

Observation Well Number	HF1	HF2	HF3	HF4	HF5	HF6
Distance from Pumped Well - m	4.6	10.4	19.8	34.5	65.8	91.8
Initial Water Level - m	96.49	96.52	96.53	96.55	96.56	96.57
Water Level After 8 hr-22 min	95.14	95.58	95.88	96.14	96.36	96.43

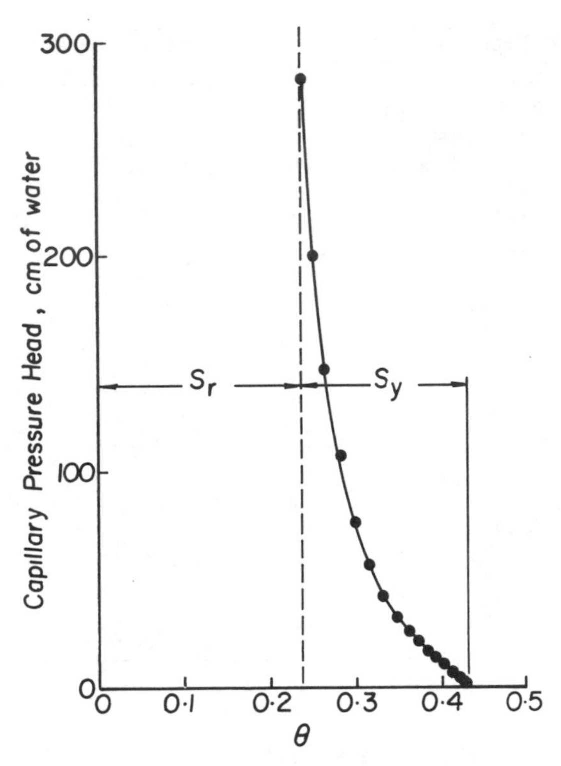

Figure 2-10. Water-retention curve for fragmented shale.

Solution:

The total volume of water removed from the aquifer during the test period is (0.0312)(30120) = 939.7 m^3 . The change in aquifer volume below the water table is equal to the volume of aquifer through which the water table has passed during the test. A plot of the water levels, initially and at the end of the test, is shown in Fig. 2-11. The data indicate a slightly lower initial water level near the pumped well than at some distance away. This is probably residual drawdown from a previous pumping period. The average initial water level is taken as 96.55 m .

It is assumed that the water-table profile in all vertical planes passing through the well is identical to the measured profile. This is the assumption of radial symmetry. Therefore, the volume of aquifer material between the initial and final water table positions is

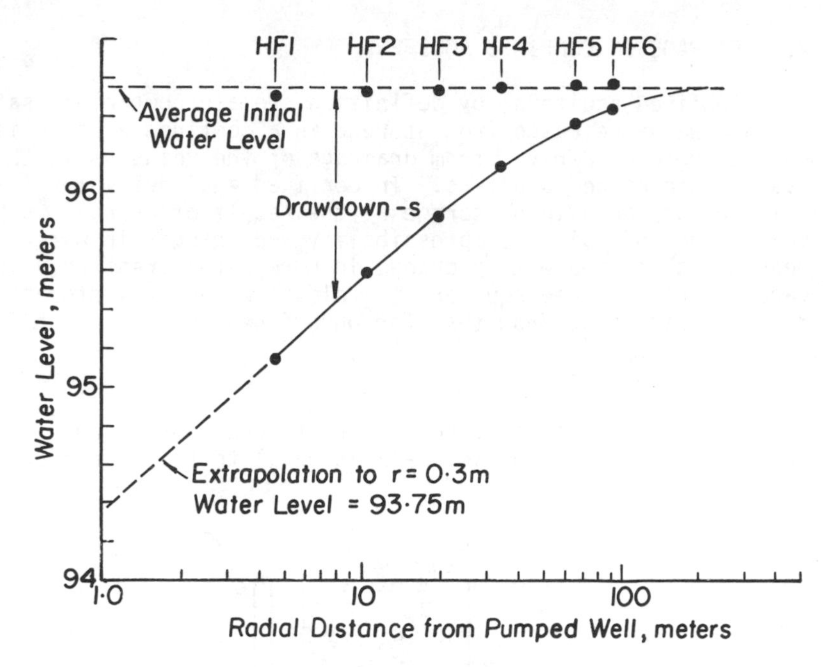

Figure 2-11. Distribution of water level in vicinity of a pumped well.

$$V = 2\pi \int_{0.3}^{\infty} (sr)dr \quad ,$$

where s is the difference between the initial and final water table positions (i.e., drawdown), and r is the radial distance measured from the axis of the pumped well. The dependence of s on r is determined from the measured water table profile extrapolated to $r=0.3$ m and $r=200$ m. The indicated integration may be accomplished by any one of several standard rules for numerical integration. The trapezoidal rule

$$\int_a^b (rs)dr = \Delta r\{ \frac{(rs)_0}{2} + (rs)_1 + \ldots + (rs)_{n+1} + \frac{(rs)_n}{2} \}$$

was used to determine $V=10260\ m^3$. The trapezoidal rule was applied in segments, thus permitting the use of large Δr at greater distances from the pumped well.

From the definition of apparent specific yield

$$S_{ya} = \frac{932.7}{10260} = 0.09 \quad .$$

2.2 STORAGE IN CONFINED AQUIFERS

Confined aquifers, by definition, remain completely saturated. Water released from storage in a confined aquifer is not, therefore, derived from drainage of the voids as is the case in unconfined aquifers. In confined aquifers water is released or taken into storage as the result of changes in pore volume due to aquifer compressibility and changes in water density associated with a change in pore-water pressure. The capacity of confined aquifers to release water from storage is markedly different from that for unconfined aquifers, therefore.

Stress in Aquifers

The total vertical force per unit area exerted upward on the confining layer in Fig. 2-12 is equal to the combined

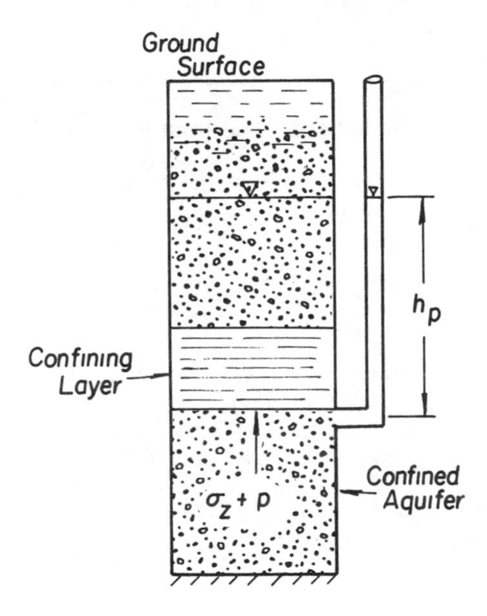

Figure 2-12. Stress in a confined aquifer.

weight per unit area of the geologic materials and water in the column extending from the top of the aquifer to ground surface. A portion of the weight of overburden and its contained water is supported by intergranular stress σ_z and a part by the pore water pressure p . The intergranular stress is a macroscopic quantity defined as the difference between the total stress and the pore-water pressure.

$$\sigma_z = \text{total stress} - p \quad . \tag{2-7}$$

Suppose that water is withdrawn from the piezometer in Fig. 2-12, thereby lowering the pore-water pressure. The total stress in the confined aquifer remains constant because the total weight of the overburden does not change. Thus, the reduction in pore-water pressure in the confined aquifer is accompanied by a corresponding increase in intergranular stress. From Eq. 2-7,

$$d\sigma_z = -dp \quad . \tag{2-8}$$

The increased stress in the granular skeleton is, in turn, accompanied by compaction of the skeleton (referred to as consolidation by soils engineers). The thickness and pore volume of the confined aquifer are thus actually reduced. There is a tendency for the aquifer to deform laterally, as well as vertically, when the pressure of water is reduced. However, lateral deformation is usually negligible with respect to the vertical deformation and will not be considered further.

Specific Storage and Storage Coefficient

Of interest here is the volume of water released from a unit volume of confined aquifer when the water-pressure head is reduced by one unit. Both aquifer compaction and decreased water density contribute to the volume released. Therefore, the approach is one of mass balance rather than volume balance. The latter is permissible only when the water can be considered incompressible as in the case of unconfined aquifers.

The mass of water in the saturated element of aquifer in Fig. 2-13 is

$$M = \rho\phi\Delta x\Delta y\Delta z \quad . \tag{2-9}$$

The change of mass

$$dM = \{\rho d(\phi\Delta z) + \phi\Delta z d\rho\}\ \Delta x\Delta y \tag{2-10}$$

follows from the chain rule for differentiation of a product and the assumption that the aquifer deforms only in the vertical direction. The quantity $dM_1 = \rho d(\phi\Delta z)$ is the contribution per unit area due to a change in pore volume at constant density, and $dM_2 = \phi\Delta z d\rho$ is the contribution due to a change in water density at constant pore volume.

The change in pore volume, due to vertical compression, per unit pore volume per unit change in intergranular stress is the pore-volume compressibility α_p and is defined as

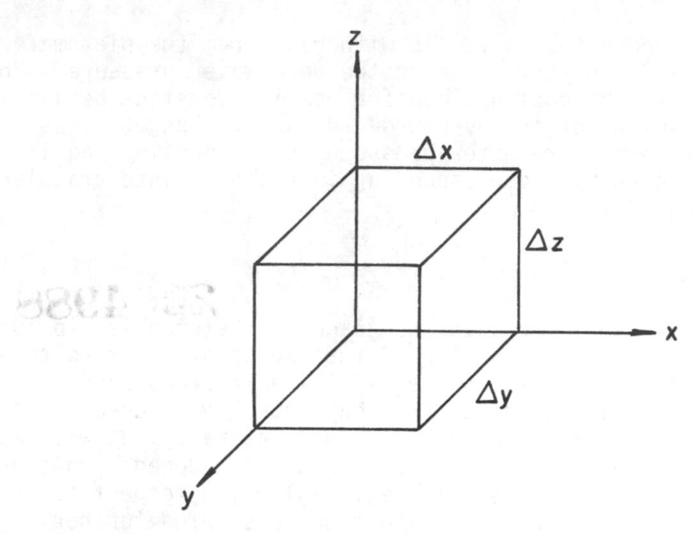

Figure 2-13. Control volume for material balance in a confined aquifer.

$$\alpha_p \equiv - \frac{1}{\phi \Delta z} \frac{d(\phi \Delta z)}{d\sigma_z} = \frac{1}{\phi \Delta z} \frac{d(\phi \Delta z)}{dp} \tag{2-11}$$

from which

$$dM_1 = \rho \alpha_p \ \phi \Delta z \ dp \quad . \tag{2-12}$$

The second equality in Eq. 2-11 holds only because the total stress is constant. It can be shown that $\alpha_p \phi$ is equal to the bulk vertical compressibility α_b (defined as the change in pore volume per unit bulk volume per unit of change in σ_z) provided that the compressibility of individual grains is zero (Geertsma, 1957; Bear, 1972). The pore-volume compressibility can be regarded as a constant over the range of pressure changes normally encountered in ground-water hydrology. The pore-volume compressibility for loose sand is on the order of 4.4×10^{-9} cm/dyne (3×10^{-5} vol/vol/psi) at low overburden pressures and decreases sharply with increasing overburden stress (Fatt, 1958). Sandstone pore-volume compressibility is greater than that for loose sand at equal overburden pressures.

Definitions for compressibility of the form in Eq. 2-11 imply an elastic behavior of the aquifer. Actually, aquifer materials are not completely elastic, the compaction process being a combination of both elastic and viscous phenomena (Corapcioglu, 1975). Furthermore, the relationship between aquifer volume and pore pressure during compaction is not identical to that during expansion (increasing pore pressure). These facts become important in the study of land subsidence.

The compressibility of water β can be expressed as

$$\beta = -\frac{1}{V_w}\frac{dV_w}{dp} \tag{2-13}$$

in which V_w is the volume of water. Equation 2-13 expresses the relationship between a change of volume and a change of pressure on a constant mass of water. Since the mass is constant, the definition of density yields

$$d\rho = -\rho\ \frac{dV_w}{V_w} = \rho\beta dp \tag{2-14}$$

in which Eq. 2-13 was used to obtain the right-hand equality. The change of mass per unit area due to water compressibility becomes

$$dM_2 = \phi\Delta z\beta\rho dp \quad . \tag{2-15}$$

The total change in mass per unit volume of aquifer is

$$\frac{dM}{\Delta x\Delta y\Delta z} = \phi\rho(\alpha_p + \beta)dp \tag{2-16}$$

from Eqs. 2-10, 2-12, and 2-15.

In the practice of hydrology, changes in ground water storage are observed in units of volume rather than mass. A more convenient form of Eq. 2-16 is obtained by dividing both sides by ρ to obtain

$$\frac{d\bar{V}_w}{\Delta x\Delta y\Delta z} = \phi(\alpha_p + \beta)dp \quad , \tag{2-17}$$

in which $d\bar{V}_w$ is the change in water volume in the element due to both water and aquifer compressibility. It is also true that hydrologists observe changes of pressure head $h_p = p/\rho g$ at a point in the aquifer rather than changes of pressure directly. It is left to the reader to show that

$$dp = \rho g\ dh_p \quad , \tag{2-18}$$

to a close approximation, for water with a compressibility of 4.8×10^{-10} cm^2/dyne (3.3×10^{-6} per psi).

Replacement of dp in Eq. 2-17 by its equivalent from Eq. 2-18 and division by dh_p gives

$$S_s = \frac{1}{\Delta x \Delta y \Delta z} \frac{d\bar{V}_w}{dh_p} = \rho g \phi (\alpha_p + \beta) \quad . \qquad (2\text{-}19)$$

The parameter S_s is the *specific storage*. It is to be noted that S_s is the volume of water released from storage per unit volume of aquifer per unit decline in pressure head and has the dimension L^{-1}.

The specific storage is regarded as a constant; a property of the aquifer material, its contained water, and the overburden stress. The phenomena of water expansion and aquifer compression occur in water-table aquifers as well as in confined aquifers. Their contribution to the volume of water released is negligible with respect to that derived from drainage of the voids, however, and the concept of specific storage is used, almost exclusively, in confined aquifer analysis.

A second parameter for confined aquifers, the *storage coefficient* S,

$$S = S_s b \qquad (2\text{-}20)$$

also finds widespread use for the case where the confined aquifer has constant thickness b. The storage coefficient is a dimensionless parameter that can be interpreted as the volume of water released from a column of unit area and height b per unit decline of pressure head. It should be understood, however, that specific storage is the more fundamental parameter, storage coefficient being a parameter that depends upon both the specific storage and the aquifer geometry.

The perceptive reader may question Eq. 2-20 in view of the fact that the aquifer undergoes compaction, thus rendering the thickness b a variable. More rigorous discussions of storage in confined aquifers are presented by DeWiest (1966) and Cooper (1966). The above discussion follows closely the derivation of Jacob (1950). Even though the aquifer deformation is very important from the viewpoint of storage changes, it is rarely necessary to otherwise consider the change of thickness as the change of b is usually small relative to b. An exception is the case of land-surface subsidence caused by ground-water withdrawals where vertical displacements can seriously disrupt sewers, waterlines, building foundations, and other facilities. Poland et al. (1973) report surface subsidence on the order of 9 m in an area in California, for example.

It has not been the purpose of the discussions in this section to derive equations from which the specific storage and storage coefficient can be calculated from other aquifer properties. Just as in the case of water-table aquifers, the storage parameters for confined aquifers can best be obtained from indirect field measurements, a subject that must be postponed until unsteady flow in aquifers is studied. The student should have gained an appreciation for the fundamental differences in the storage parameters for unconfined and confined aquifers, however, and an understanding of the reasons why the specific storage can be orders of magnitude smaller than the apparent specific yield.

EXAMPLE 2-6

Estimate the specific storage and storage coefficient for a confined aquifer for which $b=40$ m and $\phi=0.32$. Assume

$$\beta = 4.8\times10^{-10}\ \mathrm{cm^2/dyne}$$
$$\alpha_p = 4.4\times10^{-9}\ \mathrm{cm^2/dyne}\quad .$$

Solution:

From Eq. 2-19,

$$S_s = \left(\frac{980\ \mathrm{dynes}}{\mathrm{cm}^3}\right)(0.32)(4.4\times10^{-9} + 4.8\times10^{-10})\ \frac{\mathrm{cm}^2}{\mathrm{dyne}}$$
$$= 1.53\times10^{-6}\ \text{per cm}$$
$$S_s = 1.53\times10^{-4}\ \text{per m}$$
$$S = bS_s = (40)(1.53\times10^{-4}) = 6.12\times10^{-3}\quad .$$

EXAMPLE 2-7

The average volume of a confined aquifer per km^2 is $3\times10^7\ m^3$. The storage coefficient, determined from a pumping test at a location where $b=50$ m is 3.4×10^{-3}. Estimate the volume of water recovered per km^2 by reducing the pressure head 25 m in the aquifer. Assume no recharge.

Solution:

$$S_s = \frac{S}{b} = \frac{3.4\times10^{-3}}{50} = 6.8\times10^{-5}\ \text{per m}$$

$$\text{Volume recovered per km}^2 = 6.8\text{x}10^{-5} \text{ x } 3\text{x}10^7 \text{ x } 25$$
$$= 5.1\text{x}10^4 \text{ m}^3 \quad .$$

Note that if it were assumed that $S=3.4\text{x}10^{-3}$ applied to the entire aquifer, the estimated recovered volume is $3.4\text{x}10^{-3}$ x 1 km^2 x 25 m = $8.5\text{x}10^4$ m^3 . This is an incorrect estimate because it does not account for the influence of aquifer thickness on the storage coefficient (i.e., average thickness of the aquifer is 30 m). Also note that the recovered water is only 0.57 percent of the volume of water contained in the voids, assuming $\phi=0.30$.

2.3 WATER LEVEL FLUCTUATIONS

Changes in ground-water storage due to pumping and recharge are reflected by corresponding changes in the water table and the piezometric surface elevations. Factors other than pumping such as barometric pressure changes, ocean tides, and use of ground water by plants also influence water levels. An appreciation for the water-level fluctuations induced by these factors is required lest observed changes in water level be erroneously interpreted.

Piezometric Surface and Barometric Pressure

The elevation of the piezometric surface in confined aquifers is indicated by the water level in piezometers. Of interest here is the change in the piezometric level associated with a change in barometric pressure. Consider the situation in Fig. 2-14 and suppose that the barometric pressure p_a^o increases by dp_a^o . The increase of atmospheric pressure is transmitted directly to the water surface in the piezometer, tending to displace water from the piezometer into the aquifer. On the other hand, the increased atmospheric pressure also increases the load on the confined aquifer which tends to displace water from the aquifer into the piezometer. Part of the increased load is born by the aquifer skeleton, however, and the net result of the increase in barometric pressure is to decrease h_p . The absolute value of the ratio of dh_p to $dp_a^o/\rho g$ is the *barometric efficiency*, BE , given by

$$BE = \left| \frac{dh_p}{dp_a^o/\rho g} \right| \quad . \tag{2-21}$$

The barometric efficiency of a confined aquifer depends upon the compressibility of the aquifer and its contained water and upon the degree to which the increased load dp_a^o on the

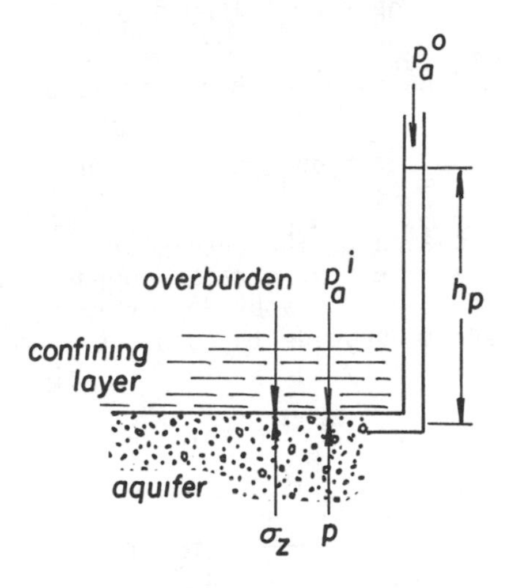

Figure 2-14. Stress balance on the interface between the aquifer and confining layer.

ground surface is transmitted to the interface between the aquifer and the upper confining layer. Expressing all fluid pressures as absolute pressures,

$$\sigma_z + p = \text{overburden pressure} + p_a^i \tag{2-22}$$

and

$$p = \rho g\, h_p + p_a^o \tag{2-23}$$

where p is the pore-water pressure at the top of the confined aquifer, p_a^o is the atmospheric pressure at ground surface, and p_a^i is that part of p_a^o which is transmitted to the interface between the aquifer and the confining layer (Fig. 2-14). The pressures p_a^i and p_a^o are not necessarily equal because the strata above the confined aquifer may be capable of supporting a part of the atmospheric load by a bridging effect.

The total derivatives of Eqs. 2-22 and 2-23 are

$$dp = dp_a^i - d\sigma_z$$

and (2-24)

$$dp = dp_a^{\,o} + \rho g \, dh_p \quad ,$$

respectively. For the purposes here, it is assumed that

$$dp_a^{\,i} = f \, dp_a^{\,o} \qquad 0 \leq f \leq 1 \tag{2-25}$$

wherein f is a constant. The parameter f is a measure of the rigidity of the overlying strata, approaching zero for rigid formations that act as a bridge over the aquifer. Thin, soft, confining strata provide the other extreme condition in which f approaches unity.

Equations 2-24 and 2-25 are combined to yield

$$\frac{\rho g \, dh_p}{dp_a^{\,o}} = \frac{dp_a^{\,i} - dp_a^{\,o} - d\sigma_z}{dp_a^{\,o}} = f - 1 - f \frac{d\sigma_z/dp}{1 + d\sigma_z/dp} \quad . \tag{2-26}$$

Replacing the quantity $\phi \Delta z$ in Eq. 2-11 with the pore volume V_p , the change in pore volume is

$$dV_p = - \alpha_p V_p d\sigma_z \quad . \tag{2-27}$$

Also, the change in water volume is

$$dV_w = -\beta V_w dp \tag{2-28}$$

from Eq. 2-13. But $dV_p = dV_w$ and $V_p = V_w$ so that

$$\frac{d\sigma_z}{dp} = \frac{\beta}{\alpha_p} \quad . \tag{2-29}$$

Substitution of Eq. 2-29 into Eq. 2-26 results in

$$\frac{\rho g \, dh_p}{dp_a^{\,o}} = f - 1 - f\left(\frac{\beta}{\alpha_p + \beta} \right) \quad . \tag{2-30}$$

Recalling that f is always less than or equal to unity and that β and α_p are positive, it is apparent that $dh_p/dp_a^{\,o}$ is negative, thus verifying the previous statement that water levels h_p decrease when the barometric pressure increases. Note that the absolute value of Eq. 2-30 is the barometric efficiency. Should the pore-volume compressibility be zero, the barometric efficiency is unity regardless of the value of f. The barometric efficiency is also unity when $f=0$. This case is approached when the overlying strata are thick, rigid formations that transmit only a small part of the increased load dp_a^o to the top of the confined aquifer. The other extreme occurs where the entire additional load is transmitted to the confined aquifer, in which case $f=1$,

$$BE = \frac{\beta}{\alpha_p + \beta} \quad , \qquad (2\text{-}31)$$

and the barometric efficiency is characteristic of the aquifer itself. Provided that $f=1$, the barometric efficiency observed in the field is inversely proportional to the specific storage or the storage coefficient. It is left to the student to derive the relationship between BE, S_s and S.

The Water Table and Barometric Pressure

In contrast to the case of the confined aquifer, a change in barometric pressure is transmitted equally to the water in an unconfined aquifer and to the water surface in a piezometer or observation well penetrating the aquifer. One might suppose, therefore, that no change in water level in the piezometer or well should be observed. Indeed, such would be the case but for the presence of entrapped air (isolated air bubbles) in the water above and below the water table. Air entrapment results during periods of infiltration and rising water tables. The entrapped air exists in discontinuous assemblages that are isolated from the continuous external air phase.

An increase in barometric pressure increases the pressure in the entrapped air and a decrease in air volume results. On the other hand, the water table elevation decreases because of the reduction of the volume of entrapped air. The decrease in water level reduces the hydrostatic pressure on the air and tends to offset the increase in barometric pressure. Peck (1960) has studied the fluctuation of the water table in response to changes in atmospheric pressure, and his analysis illustrates the essential features of the phenomena.

Let V_a be the volume of entrapped air per unit volume of aquifer. In a vertical column of unit cross-sectional area that extends from the floor of the aquifer to the ground surface (Fig. 2-15), the total volume of entrapped air is the integral of V_a over z from 0 to H. Any change in the total volume of entrapped air is reflected by a change in the water table elevation h according to

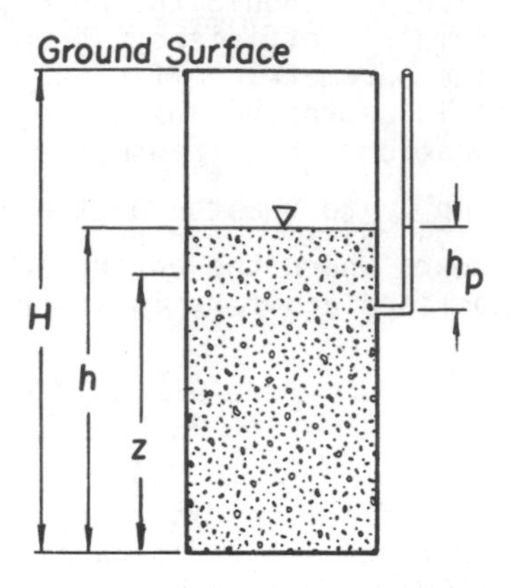

Figure 2-15. Definition of symbols used in the calculation of water table fluctuations in response to barometric pressure changes.

$$S_{ya} dh = d\left(\int_0^H V_a dz\right) \quad . \qquad (2\text{-}32)$$

The volume V_a of entrapped air per unit volume of aquifer undoubtedly varies with elevation z in a very complex way in the field due to diffusion, gas solubility, and biological activity (Hansen, 1977). The problem of evaluating the right side of Eq. 2-32 becomes tractable if it is assumed that the mass of entrapped air per unit volume is a constant independent of z and that the water is static. Hence the ideal gas law can be used to deduce the function $V_a(z)$. The absolute pressure of the entrapped air at a particular elevation is the sum of the atmospheric pressure p_a , the hydrostatic gage pressure of the adjacent water $\rho g(h-z)$, and the capillary pressure p_c across the air-water interfaces of the discontinuous air bubbles. The latter can be neglected relative to the sum of the hydrostatic and atmospheric pressures, provided that the analysis is restricted to water table depths less than about 9 m (Peck, 1960). From the ideal gas equation,

$$V_a(p_a/\rho g + h-z) = \frac{MRT}{\rho g} = c \qquad (2\text{-}33)$$

wherein M is the mass of entrapped air per unit of aquifer volume, R is the gas constant, and T is temperature.

With $V_a(z)$ known, Eq. 2-32 yields

$$S_{ya} dh = d\left\{\int_0^H \frac{c\, dz}{\frac{p_a}{\rho g} + h - z}\right\} = d\left\{c\, \ell n\left[\frac{(p_a/\rho g + h)}{(\frac{p_a}{\rho g} + h - H)}\right]\right\}, \tag{2-34}$$

from which

$$\frac{dh}{dp_a/\rho g} = -\frac{c\, H}{S_{ya}(\frac{p_a}{\rho g} + h)(\frac{p_a}{\rho g} + h - H) + c\, H} . \tag{2-35}$$

Equation 2-35 is an expression analogous to the barometric efficiency defined for confined aquifers. Notice that if there is no entrapped gas (i.e., $c=0$), the water-table elevation (the water level in a piezometer or well) does not change in response to a change in barometric pressure. However, when $c \neq 0$ the water level falls when the atmospheric pressure increases. The special case in which the top of the capillary fringe is coincident with the ground surface yields

$$\frac{dh}{dp_a/\rho g} = -1 \tag{2-36}$$

because the apparent specific yield is zero under this circumstance. It is further recognized that thick aquifers should exhibit large water-level changes relative to that in thin aquifers.

A closely related case is that of an unconfined aquifer for which the interchange of air above the water table with the atmosphere is restricted by overlying strata. Restricted flow of air between the atmosphere and the aquifer causes the equilibration of air pressure in the aquifer with the atmosphere to lag a change in barometric pressure and the response of the water levels in piezometers may indicate substantial barometric efficiencies. In some cases, the most direct connection between air in the aquifer and the atmosphere is provided by well bores, in which case, air passes through the well bore at very substantial velocities following a change in barometric pressure.

Water Levels, Tides and Other External Loads

The water level in wells and piezometers responds to external loads other than atmospheric pressure. Because the change in load due to fluctuating tides or the passage of a train, for example, are applied only to the aquifer and not to the water surface in the piezometer, the water level response is opposite that observed for changes in barometric pressure. In other words, the increased load produces a rise in water levels.

The response of water levels to slowly changing external loads can be analyzed in a manner similar to that for barometric loading. For example, consider a confined aquifer extending beneath the ocean floor or beneath a river or estuary in which the water stage, H , changes with the tide. Provided that the entire change in pressure $\rho g\, dH$ is transmitted to the confined aquifer,

$$\frac{dh_p}{dH} = \frac{1}{1 + d\sigma_z/dp} \quad . \qquad (2\text{-}37)$$

From Eqs. 2-29 and 2-37, the tidal efficiency, TE , is

$$TE = \frac{dh_p}{dH} = \frac{\alpha_p}{\alpha_p + \beta} \quad . \qquad (2\text{-}38)$$

Note that the sum of the tidal and barometric efficiencies is unity when $f=1$.

Water level changes, in response to tidal fluctuations, are observed in both confined and unconfined aquifers that outcrop in the sea. In this case, the change in pressure head due to tides is transmitted directly to the water in the aquifer at the outcrop, however, and the tidal efficiency near the outcrop is unity. A pressure wave is transmitted through the aquifer, the amplitude of which decreases with increasing distance from the outcrop. A change in stage of a river in which an aquifer is exposed produces a similar effect. A thorough discussion of these phenomena requires, as a prerequisite, a knowledge of ground-water hydraulics and will therefore be treated in a subsequent chapter.

A local, external load, applied rapidly on a small portion of the aquifer and held constant thereafter (Fig. 2-16) causes a sharp increase in water levels followed by a gradual decline. A part of the load is born by the water in the aquifer in the vicinity of the application, thus increasing the pressure head there. The pressure head is greater than at locations more remote from the load, therefore. The result is a period of water flow away from the location of the load, during which the granular skeleton of the aquifer bears an increasingly larger portion of the load and the pressure head decreases. The pressure head eventually approaches its original value in all but very small aquifers.

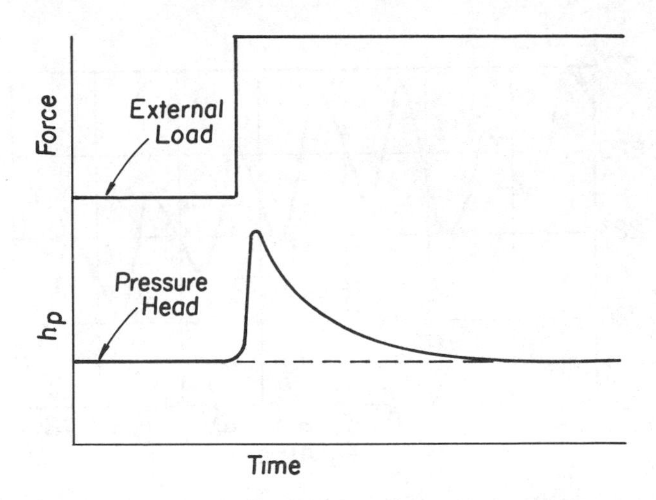

Figure 2-16. Variation of pressure head in an aquifer subjected to an external load.

A similar phenomenon occurs in highly compressible unconfined aquifers. The load causes the pore volume to be reduced, expelling water. If the load is applied rapidly, the water table near the site of loading rises and eventually dissipates by flow away from the loaded area. For a given load the rise of the water level in the unconfined aquifer is not as pronounced as in the confined aquifer, because the ground water, being free to move upward, does not support as great a portion of the applied load as in the confined case. Notwithstanding the fact that the aquifer is unconfined, the water does, indeed, bear a portion of the applied load because the water experiences a resistance to flow and thereby resists compression of the aquifer.

Water Levels and Evapotranspiration

Evapotranspiration is the combined processes of evaporation from the soil and the evaporation from the stoma in plant leaves. The latter is known as transpiration. Evapotranspiration is an important component of the hydrologic water balance that is discussed in the following section. Plants that take up water directly from the capillary fringe and the water table are *phreatophytes*. Among the most common phreatophytes are saltcedar, willow, saltgrass, cottonwood, and alfalfa. Because they extract water from the saturated zone, phreatophytes produce daily water-table fluctuations in response to their transpiration (Fig. 2-17). This is in contrast to plants that derive their water from partially saturated soil high above the water which have no direct, immediate effect on the water-table elevation.

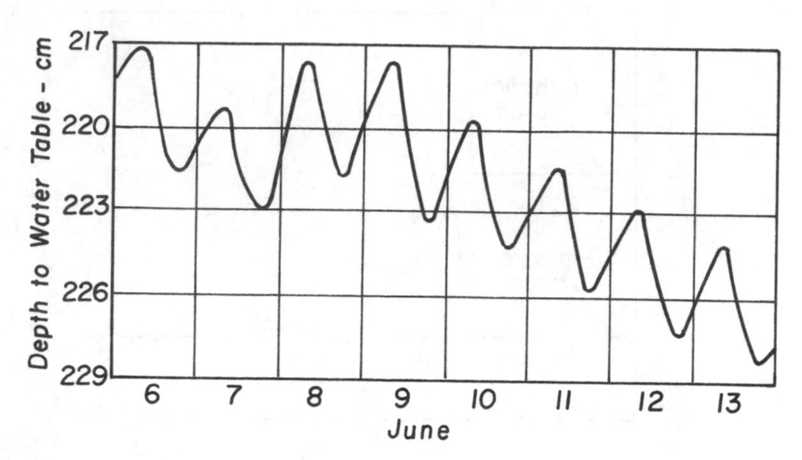

Figure 2-17. Water level fluctuation due to phreatophyte transpiration (After Robinson, 1950).

Water use by phreatophytes is especially important in the flood plains adjacent to streams and rivers. In some instances, the effect of phreatophyte transpiration is so pronounced that daily fluctuations in river stage are directly attributable to ground-water extraction by these plants. Annual water use is sometimes as great as $2\ m^3/m^2$, but depends upon the type of plant, climatic and meteorological conditions, depth to ground water and availability of soil water.

In Fig. 2-17, it is apparent that the water level falls during the daylight hours, reaching a minimum in the evening, after which a rise is observed that persists until early morning. The water-table decline is caused by transpiration at a rate that exceeds the rate of lateral ground-water inflow to the phreatophyte area. During the night, transpiration is negligible and lateral ground-water inflow occurs in response to the depressed water table. Thus, the water table rises during the night. Notice also, that there is a general decline of water level, upon which the diurnal fluctuations are superimposed. The decline of daily average water level is not necessarily a result of evapotranspiration only, but may be influenced by natural drainage, pumping, or a number of other factors not related to phreatophyte water use.

2.4 HYDROLOGIC BUDGETS

O. E. Meinzer wrote in 1932: "the most urgent problems in ground-water hydrology at the present time are those relating to the rate at which rock formations will supply water to wells in specified areas -- not during a day, a month, or a year, but perennially." The problem became known as the determination of "safe yield", a concept that has been the object of considerable controversy. It has not been possible to divorce from safe yield the practical aspects of extraction methods, distribution, legal considerations, and costs. Thus, safe yield as a hydrologic characteristic of aquifers has fallen into disuse. Even so, Meinzers' problem of ground-water supply remains a central issue.

An indispensable tool in the study of ground-water supply is the hydrologic budget. The hydrologic budget is nothing more than a material balance that accounts for all inputs, outputs, and changes in storage within a system defined by prescribed boundaries. A judicious selection of the system boundaries to correspond with physical locations where sources and discharges are known helps to maximize the uesfulness of the budget. For example, lateral surface inputs and outputs are zero across surface water divides and likewise in the ground water. The ground water catchment does not always correspond in geometry to the overlying surface catchment, however, and it is often convenient to define subsystems that are related to one another by the exchange of water across their common boundaries. The time period over which the water balance is made can be selected so as to simplify the analysis in many cases. For example, it is sometimes possible to select a time period over which the net change of water storage is practically zero, in which case the inputs equal the outputs.

The following treatment of the hydrologic budget is presented by considering four subsystems: the atmosphere, the surface water subsystem, the soil-water subsystem, and the ground-water subsystem. These subsystems are not independent but are linked by inflows and outflows across their boundaries as shown in Fig. 2-18. It is anticipated that the reader can combine or further divide these subsystems in any manner appropriate for the problem under his consideration. It is further expected that the inputs and outputs described herein will be lumped together or other ones added so that they become commensurate with the degree of detail required by the problem at hand. The scope of this book does not permit a thorough consideration of surface water components in the surface water subdivision. Such important factors as precipitation and surface runoff are treated in this text as known quantities with no discussion of factors which influence them or methods by which they can be determined.

Components of a Surface-Water Budget

The lower boundary of the surface-water subsystem is the ground surface. Waters in this subsystem include overland flow, depression storage, interception storage, and waters in

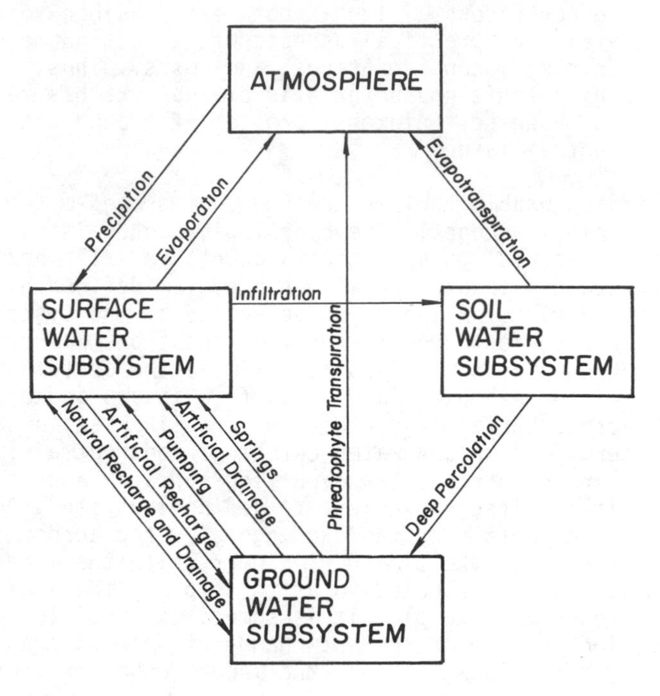

Figure 2-18. Linkages between subsystems in the hydrologic cycle.

streams, canals, reservoirs, lakes and ice. Inputs include precipitation and water transported into the subsystem by streams, canals, and pipelines. Typical depletions of water in the surface subsystem result from evaporation on water surfaces, infiltration of water into the soil, seepage, surface runoff, transport by canals and pipelines and consumptive use by industry and municipalities. Any natural or man-made feature that retains water, either temporarily or permanently, may require inclusion in the budget as a storage term.

A grouping of all inflows, outflows, and storage terms permits the material balance to be written as

$$(\text{Inflow} - \text{Outflow})\Delta t = \text{change of storage} \quad (2\text{-}39)$$

in which the inflow and outflow are average volume flow rates over the time period, Δt. The time period can range from seconds or minutes to a year or more, depending upon the purpose for which the equation is used.

Components of a Soil-Water Budget

The soil-water subsystem is related to the surface subsystem by the interchange of water across their common boundary, the ground surface. For example, infiltration, an outflow from the surface water, is inflow to the soil-water subsystem. Outflows from the soil-water zone include evaporation, transpiration and flow into the ground water below. Storage changes in the soil-water zone are recognized by changes in the volumetric-water content. A soil-water budget can be expressed by

$$(I - E - T - W)\Delta t = \Delta\left(\int_0^D \theta dz\right) \qquad (2\text{-}40)$$

where I is the infiltration rate, E is evaporation rate of soil water, T is transpiration rate, W is the flow rate across the lower boundary of the soil-water zone, and D is the depth of the soil-water zone. Flow rates in Eq. 2-40 are expressed as volumes per unit area per unit time. Evaporation and transpiration are normally treated as one variable, evapotranspiration, ET . The quantity W can be either positive or negative, being positive as an outflow quantity. The right-hand side of Eq. 2-40 is the change of water stored in the soil-water zone.

The change in soil-moisture storage must be evaluated by determining the depth distribution of volumetric moisture content. A common method by which $\theta(z)$ is measured is the neutron attenuation technique. In this method an access tube, usually 3.8 - 5 cm in diameter, is placed in the soil into which is lowered a radioactive source of fast neutrons and a detector of slow (thermal) neutrons. The neutrons, being of nearly the same mass as hydrogen, are slowed and reflected by impact with hydrogen atoms. The so-called thermal neutrons resulting from numerous collisons are detected and counted. The number of neutrons counted in a time interval correlates with the volumetric water content, thereby providing an indirect (but non-destructive) method of determining the soil moisture as a function of depth. Another widely used method is to extract soil samples at various depths and determine the volumetric water content by gravimetric procedures. All that is required is to measure the bulk volume of the sample, the initial weight, and the oven-dried weight. The difference between the initial weight and the oven-dried weight is converted to a volume of water, then divided by the bulk volume to obtain θ .

The average infiltration rate or the cumulative infiltration $I\Delta t$ is often determined from the surface-water budget by measuring, or otherwise determining, all other variables in Eq. 2-39. When a surface-water budget is not available, the hydrologist may be obliged to use one of several infiltration formulas. While it is true that the physics of infiltration are well understood and most infiltration formulas are based on sound theory, the variability of factors affecting infiltration make accurate estimation of infiltration via infiltration formulas a difficult task for all but small homogeneous areas.

Evapotranspiration ET represents the combined processes of evaporation from the surface of the soil and transpiration by the plants. Normally, the single largest outflow from the soil-water subsystem occurs by the evapotranspiration process. Evapotranspiration depends upon wind, temperature, humidity, albedo, solar radiation, type of plant, plant cover, and the availability of soil moisture. Because of its major importance in agriculture, evapotranspiration has been intensively studied, particularly for irrigated crops. A relatively small body of knowledge exists that pertains exclusively to watersheds covered with natural vegetation.

One method of estimating ET is to compute it as the residual required to obtain a material balance in the soil-water subsystem. The ground-water hydrologist may make an independent estimate of ET so that the soil water budget can be used to calculate the deep percolation W. Estimation of W by calculating it as the residual in the soil-water budget is not without difficulties, however, because this method lumps all errors in the other components into W. Since W is often small relative to other budget components, large percentage errors for W may result.

A variety of methods exist for calculating ET from climatic, soil and plant parameters. Most methods require, as a first step, the calculation of *potential evapotranspiration* ET_p. The potential evapotranspiration is the water-use rate for a reference crop that fully shades the ground and is not limited by water availability. Both short grasses and alfalfa have been used as the reference crop. The evaporation from a shallow lake is considered by many to closely approximate potential evapotranspiration.

The Jensen-Haise (1963) equation for potential evapotranspiration is

$$ET_p = C_T (T - T_x) R_s/\rho L_v \qquad (2\text{-}41)$$

where ET_p is potential evapotranspiration in centimeters per day, C_T is a temperature coefficient in $(°C)^{-1}$, T is mean air temperature for the period of interest in °C, T_x is a constant in °C , R_s is the solar radiation in langleys per day (calories per cm^2 per day) and L_v is the latent heat of vaporization for water. At 20°C, $L_v \cong 585$ cal/g . Eq. 2-41 was developed from data in the Western United States and is applicable to inland, arid and semi-arid regions. Equations are available from which C_T and T_x can be calculated based on vapor pressure and elevation (Jensen, 1973). When the relative humidity is less than 40% , $C_T = 0.025\ (°C)^{-1}$ and $T_x = -3°C$ can be used. The solar radiation R_s must be measured, a fact that sometimes limits the usefulness of the Jensen-Haise method because R_s has not been a widely measured parameter, historically.

Numerous attempts to relate the evaporation, E_{pan} from evaporation pans to ET_p have been made. This approach is attractive because of the ease with which pan evaporation can be measured. The relationship between ET_p and pan evaporation is a simple proportionality

$$ET_p = C_p\ E_{pan} \qquad (2\text{-}42)$$

where C_p is the pan coefficient. The pan coefficient depends upon such factors as size of pan, its location with respect to buildings and other obstructions, and the micro-climate of the pan. It has been suggested that great care is required to obtain accurate estimates of potential evapotranspiration from pan evaporation. Doorenbas and Pruitt (1974) suggest the pan coefficients presented in the following table. The coefficients in Table 2-3 apply to evaporation from a United States Weather Bureau, Class A pan.

The methods for estimating ET_p presented are but two of several methods that exist. The Jensen-Haise and the pan evaporation method were selected for presentation herein because of their simplicity and the relatively small data requirements. The Jensen-Haise method is regarded as one of the best methods in arid and semi-arid regions (Jensen, 1973) but other procedures are superior in humid climates. Also widely used is the Penman method (Penman, 1948) and modifications thereof. Calculation of potential evapotranspiration from pan evaporation is not regarded as a superior method, but the minimal data requirements, its simplicity and long history of use among hydrologists, seem to justify its inclusion.

The actual evapotranspiration by plants is less than or equal to the potential evapotranspiration, even when the crop is adequately watered. The actual evapotranspiration is

Table 2-3. Suggested Pan Coefficients for Relating Pan Evaporation to Evapotranspiration of Well-Watered, Grass Turf

Wind, km/day	Upwind fetch of green crop-m	Coefficient C_p Relative humidity-% 20-40	40-70	>70
Light >170 km/d	0	0.55	0.65	0.75
	10	0.65	0.75	0.85
	100	0.70	0.80	0.85
	1000	0.75	0.85	0.85
Moderate 170-425 km/d	0	0.50	0.60	0.65
	10	0.60	0.70	0.75
	100	0.65	0.75	0.80
	1000	0.70	0.80	0.80
Strong 425-700 km/d	0	0.45	0.50	0.60
	10	0.55	0.60	0.65
	100	0.60	0.65	0.70
	1000	0.65	0.70	0.75
Very strong >700 km/d	0	0.40	0.45	0.50
	10	0.45	0.55	0.60
	100	0.50	0.60	0.65
	1000	0.55	0.60	0.65

is computed from the potential by

$$ET = K_c \; ET_p \qquad (2\text{-}43)$$

where K_c is a coefficient that depends upon the type of crop, the degree to which the crop covers the soil (i.e., the growth stage) and the availability of water. When water is not limiting, $K_c = K_{co}$ where K_{co} is the crop coefficient under well-watered conditions. Values for K_{co} have been determined experimentally for many crops. Table 2-4 contains estimated values of K_{co} for several types of natural vegetation common in the Western United States. Wymore (1974) regards the coefficients in Table 2-4 as first estimates, subject to refinement.

Plants use water from *available water*, AW , stored in the root zone. Water existing in the root zone at volumetric water contents greater than the specific retention, S_r , is not usually considered available for plant use because this water drains by gravity from the root zone in a matter of a few days or less, provided the water table is sufficiently far below the root zone. On the other hand, plants cannot use water that

Table 2-4. Estimated Plant-Water-Use Coefficients K_{co} for Native Vegetation (From Wymore, 1974)

Vegetation	K_{co} Nov.-March	April	May	June	July
Sagebrush-grass	0.50	0.60	0.80	0.80	0.80
Pinyon-Juniper	0.65	0.70	0.80	0.80	0.80
Mixed Mountain shrub	0.60	0.67	0.81	0.85	0.82
Coniferous forest	0.70	0.71	0.80	0.80	0.80
Aspen forest	0.60	0.67	0.85	0.90	0.86
Rockland & Misc.	0.50	0.60	0.65	0.65	0.65
Phreatophytes	1.00	1.00	1.00	1.00	1.00
		Aug.	Sept.	Oct.	
Sagebrush-grass		0.71	0.53	0.50	
Pinyon-Juniper		0.80	0.69	0.65	
Mixed Mountain shrub		0.74	0.65	0.60	
Coniferous forest		0.79	0.75	0.71	
Aspen forest		0.75	0.65	0.60	
Rockland & Misc.		0.60	0.50	0.50	
Phreatophytes		1.00	1.00	1.00	

exists at apparent suctions greater than about 15 atmospheres. The volumetric-water content below which plants cannot use water is the *permanent wilting* water content, θ_{wp}. The maximum available water, AW_m, is therefore given by

$$AW_m = (S_r - \theta_{wp})\, D_r \tag{2-44}$$

where D_r is the rooting depth of the plant. Note that AW_m is a volume per unit area. The available water AW can vary between AW_m and zero, the latter condition occurring when the volumetric water content is θ_{wp} everywhere in the root zone.

Water which enters the root zone is assumed to add to the existing available water until AW_m is reached. Any water in addition to that required to bring the available water to its maximum value and not used as ET is assumed to leave the root zone as deep percolation W in a matter of a few days. Plants are believed to transpire at a rate which decreases as AW decreases. Jensen et al. (1970) suggests that

$$K_c = K_{co} \frac{\ln\left(\frac{AW}{AW_m} \times 100 + 1\right)}{\ln 101} \tag{2-45}$$

can be used to calculate the water use coefficient under circumstances of less than a fully adequate water supply in the root zone.

EXAMPLE 2-8

Estimate the evapotranspiration for June from a small watershed using the following data.

Mean solar radiation for June = 611 ly/d
Mean maximum air temperature = 27°C
Mean minimum air temperature = 9°C
Mean air temperature = 18°C
Percent of watershed in sagebrush-grass = 65
Percent of watershed in aspen forest = 20
Percent of watershed in phreatophytes = 15

Solution:

From Eq. 2-41,

$$ET_p = 0.025\ (18 + 3)\ 611/585 = 0.55\ \text{cm/d}$$

Sagebrush-grass, ET = (0.80)(0.55) = 0.44 cm/d
Aspen forest, ET = (0.90)(0.55) = 0.50 cm/d
Phreatophyte, ET = (1.00)(0.55) = 0.55 cm/d
Watershed ET = (0.65)(0.44) + (0.20)(0.50) + (0.15)(0.55) = 0.47 cm/day
Total evapotranspiration for June = (0.47)(30) = 14.2 cm .

The Ground-Water Budget

Previous sections of this chapter have dealt with methods by which deep percolation W and the phreatophyte transpiration components of the ground-water budget can be estimated. Usually a consideration of material balance in the surface and soil-water subsystems is required. Also presented in this chapter are procedures whereby observed changes of water level in wells and piezometers can be variously interpreted in terms of ground-water storage changes or as the result of external factors such as barometric pressure fluctuations. Components of the ground-water budget remaining to be discussed include pumping, artificial drainage, artificial recharge, and discharge to or recharge from streams and other surface-water bodies. Leakage to and from aquifers through adjacent aquitards must also be considered. Ground-water hydraulics provides the basis for an understanding of these components as well as methods by which they can be estimated. The remainder of this book is devoted to aspects of ground-water hydraulics.

EXAMPLE 2-9

A small Caribbean island with a surface area of 5400 km^2, receives a long term mean precipitation of 135 cm annually. The island is essentially flat and is composed of a highly permeable limestone. The regolith has a high infiltration capacity that prevents appreciable runoff. The underlying water-table aquifer discharges fresh water to the sea on the perimeter of the island. The island is undeveloped and no significant quantities of water are used by man for any purpose. The island is covered with a mixture of marsh land, phreatophytes, grass, and shrubs. The marsh land and phreatophytes comprise approximately 35 percent of the land cover, the remainder being grass and shrubs. The water-use coefficient for the grass and shrubs is estimated to be 0.60 on an annual basis. The evaporation from a USWB, Class A pan, located downwind of an approximately 10 m fetch of green vegetation, is 190.5 cm. The wind is moderate, and the relative humidity averages about 78 percent. Estimate the average annual ground-water discharge to the sea.

Solution:

Because the infiltration capacity of the limestone is high, it is assumed that precipitation is removed from the ground surface immediately, thus providing no opportunity for depression storage and subsequent evaporation. Evaporation from free-water surfaces is assumed to occur at the potential rate and is included as a contribution in the phreatophyte evapotranspiration. Under the above circumstances, the surface-water budget yields precipitation equal to infiltration.

The solution sought is for the long term, average annual condition. It is, therefore, reasonable to assume that the net change in soil-water storage over the average year is zero. The soil-water budget becomes

$$I - ET - W = 0$$

wherein all quantities are average annual values. From the surface-water budget, $I = 135$ cm/yr . The soil water lost by evapotranspiration results from the grass and shrubs. The potential evapotranspiration is $(0.75)(190.5) = 142.9$ cm . The pan coefficient of 0.75 is obtained from Table 2-3. Actual evapotranspiration is estimated to be $(0.60)(142.9)(0.65) =$ 55.7 cm , wherein 0.65 represents the fraction of the land covered by the grass and shrubs and 0.60 is the water-use coefficient. The soil-water budget yields

$$W = 135 - 55.7 = 79.3 \text{ cm} \quad .$$

Again because average annual conditions are under consideration, it is permissible to assume no annual change in ground water storage. Thus,

$$W - ET - q = 0$$

wherein ET is the evapotranspiration by phreatophytes and from the marsh land and q is the discharge of ground water to the sea. ET=(1.00)(142.9)(0.35)=50.0 cm , in which 0.35 is the fraction of the land area covered by phreatophytes and marsh land. From the ground-water budget

$$q = 79.3 - 50.0 = 29.3 \text{ cm} \quad .$$

The total discharge to the sea is the discharge per unit area multiplied by the area:

$$Q = 0.29 \text{ m/yr} \times 5.4 \times 10^9 \text{ m}^2 = 1.57 \times 10^9 \text{ m}^3/\text{yr} \quad .$$

In one sense, the above annual discharge to the sea is wasted water. However, the very existence of the fresh-water aquifer beneath the island depends upon a maintenance of some flow to the ocean, a fact that will become clear when the hydraulics of fresh water floating on more dense salt water are studied in Chapter IV.

REFERENCES

Bear, J., 1972. Dynamics of Fluids in Porous Media. American Elsevier Publishing Co., Inc., New York, London, Amsterdam, 764 p.

Cooper, H. H., Jr., 1966. The Equation of Groundwater Flow in Fixed and Deforming Coordinates. Journ. of Geophysical Research, V. 71, n. 20, p. 4785-4790.

Corapcioglu, M. Y., 1975. Prediction of Land Subsidence with a Rheological Model, Unpublished PhD Thesis, Cornell University, Ithaca, N. Y.

DeWeist, R. J. M., 1966. On the Storage Coefficient and Equation of Groundwater Flow. Journ. of Geophysical Research, V. 71, n. 4, p. 1117-1122.

Doorenbos, J. and Pruitt, W. O., 1974. Guidelines for Prediction of Crop Water Requirements. FAO, Irrig. and Drainage Paper No. 25, Rome.

Duke, H. R., 1972. Capillary Properties of Soils - Influence Upon Specific Yield. Transactions, ASAE, V. 15, n. 4.

Fatt, I., 1958. Pore Volume Compressibilities of Sandstone Reservoir Rocks. Journ. of Petroleum Tech., March.

Geertsma, J., 1957. The Effect of Fluid Pressure Decline on Volumetric Changes of Porous Rocks. Transactions, AIME, V. 210.

Hansen, B. L., 1977. Entrapped Gas in an Unconfined Aquifer. Unpublished PhD Dissertation, Colorado State University, Fort Collins, Colorado, 111 p.

Jacob, C. E., 1950. Chapter 5 in Engineering Hydraulics. H. Rouse, ed., John Wiley and Sons, New York. p. 321-386.

Jensen, M. E. and Haise, H. R., 1963. Estimating Evapotranspiration from Solar Radiation. Journ. of Irrig. and Drainage Div., ASCE, V. 89, p. 15-41.

Jensen, M. E., Robb, D. C. N., and Franzoy, C. E., 1970. Scheduling Irrigations Using Climate-Crop-Soil Data. Journ. of Irrig. and Drainage Div., ASCE, V. 96, p. 25-28.

Jensen, M. E., ed., 1973. Consumptive Use of Water and Irrigation Water Requirements. Technical Committee Report, Irrig. and Drainage Div., ASCE, 215 p.

McWhorter, D. B., 1971. Infiltration Affected by Flow of Air. Colorado State University Hydrology Paper No. 49, May, Fort Collins, Colorado, 43 p.

Meinzer, O. E., 1932. Outline of Methods for Estimating Ground-Water Supplies. U. S. Geological Survey Water Supply Paper 638-C.

Morris, D. A. and Johnson, A. I., 1967. Summary of Hydrologic and Physical Properties of Rock and Soil Materials as Analyzed by the Hydrologic Laboratory of the U. S. Geological Survey - 1948 - 1960. U. S. Geological Survey Water Supply Paper 1839-D.

Peck, A. J., 1960. The Water Table as Affected by Atmospheric Pressure. Journ. of Geophysical Research, V. 65, n. 8, p. 2303-2388.

Penman, H. L., 1948. Natural Evaporation from Open Water, Bare Soil, and Grass. Proc. Royal Soc. of London, A 193, p. 120-146.

Poland, J. F., Lofgren, B. E., Ireland, R. L. and Pugh, R. G., 1973. Land Subsidence in the San Joaquin Valley, California as of 1972. U. S. Geological Survey Professional Paper 437-H.

Robinson, T. W., 1950. Phreatophytes. U. S. Geological Survey Water Supply Paper 1423.

Wymore, I. F., 1974. Water Requirements for Stabilization of Spent Shale. Unpublished PhD Dissertation, Colorado State Univ., Fort Collins, Colorado, 137 p.

Youngs, E. G., 1969. Unconfined Aquifers and the Concept of the Specific Yield. Bull. Intern. Assoc. of Scientific Hydrology. V. 14, n. 2, p. 191-197.

PROBLEMS AND STUDY QUESTIONS

1. The cross-sectional area and length of a cylindrical sample of sandstone are 7.89 cm^2 and 6.10 cm, respectively. The sample is dried in a vacuum oven several hours at 105°C and then weighed. The weight is 9.24×10^4 dynes. The sample is then vacuum saturated by first placing it in a dry vacuum tank and subjecting it to a vacuum for about 30 minutes. With the vacuum pump remaining in operation, oil with a density of 0.755 g/cm^3 is allowed to enter the tank until the sample is completely submerged. The tank is then returned to atmospheric pressure. The sample is removed, the excess oil carefully wiped away, and the sample weight measured and found to be 1.01×10^5 dynes. Calculate the porosity and discuss possible sources of error. Answer: $\phi = 0.24$.

2. Derive a formula from which the porosity of a material can be calculated from the bulk density ρ_b and the particle density ρ_s. Calculate the porosity of a material for which $\rho_b = 1.42\ g/cm^3$ and $\rho_s = 2.68\ g/cm^3$. Answer: $\phi = 1 - \rho_b/\rho_s$, $\phi = 0.47$.

3. Using the capillary pressure-water content data of Example 2-3, plot the equilibrium distribution of water content above a water table located 70 cm below the soil surface. Suppose the water table should fall to 80 cm below the surface and remain stationary until equilibrium conditions again prevail. Calculate the volume of water removed per unit area by graphically integrating the area between the two equilibrium distributions of water content. Calculate the average apparent specific yield and compare with the specific yield. Answer: 1.28 cm^3/cm^2 , average $S_{ya} = 0.128$, $S_y = 0.19$.

4. Explain, using the results of problem 3 as a basis, why the apparent specific yield approaches zero when the water table is near the surface, regardless of the aquifer material.

5. Show explicitly that the volume of water removed per unit area in Fig. 2-9 is $(\phi - S_r)(z_2 - z_1)$ and, therefore, $S_{ya} = S_y$.

6. Consider a stratified aquifer with a thick gravel layer overlying a fine textured sand. Prior to development of the ground water, the water table was located a short

distance above the gravel-sand interface. Use of the water supply caused the water table to fall to a level a few centimeters below the gravel-sand interface. It was observed that the apparent specific yield was significantly smaller with the water table in the latter position than for the initial level. Further development caused the water table to fall several meters below the gravel-sand interface, at which time it was observed that S_{ya} was larger than at any previous water table level. Provide a possible explanation for these observations.

7. Discuss the role of interfacial tension at water-air interfaces in the retention of water above the water table.

8. Explain why the height of the capillary fringe is greater in fine textured than in coarse grained materials.

9. Explain why the coefficient of specific storage is usually much smaller than the apparent specific yield.

10. The bulk vertical compressibility is defined by

$$\alpha_b = \frac{-1}{\Delta z}\frac{d\Delta z}{d\sigma_z}$$

Show that $\alpha_b = \phi\alpha_p$ under the assumption that the solid grains in the aquifer are incompressible.

11. Water was treated as a compressible fluid for the derivation of Eq. 2-19. On the other hand, the relation $dp = \rho g dh_p$, which implies the water is incompressible, was also used. Derive a more exact relation between dp and dh_p in terms of water compressibility β and ρ. Rationalize the treatment of water as incompressible fluid when a change in pressure is converted to a change in pressure head. Answer:

$$dp = \frac{\rho g dh_p}{1-\beta\rho g h_p}$$

12. Calculate the barometric efficiency of an aquifer for which $\phi = 0.3$ and $S_s = 2\text{x}10^{-4}$ per meter. Assume that the aquifer is confined by soft, pliable strata that are incapable of supporting an applied load by a bridging or arching action. Repeat the computation for the same aquifer material confined by fairly rigid strata that transmit only 30 percent of the additional load dp_a^o. Answer: BE=0.07, BE=0.72.

13. Compute the anticipated change of water level in wells penetrating the two aquifers of problem 12 if the barometric pressure increases by 18.4 mm of mercury. Answer: $dh_p = -1.8$ cm, -18 cm.

14. Derive Eq. 2-26 from Eqs. 2-24 and 2-25.

15. Discuss, qualitatively, the effects of apparent specific yield and hydraulic conductivity upon the observed magnitude of diurnal fluctuations of the water table in response to transpiration by phreatophytes.

16. Show explicitly that the absolute pressure of the air in a bubble entrapped at depth h-z below the water table is $p_a + \rho g(h-z) + p_c$.

17. Explain why a restriction in the depth to the water table is required if the capillary pressure contribution p_c to the pressure of the entrapped air is to be neglected.

18. Carry out the mathematical steps required to obtain Eq. 2-35 from Eq. 2-34.

19. An undisturbed sample of aquifer is taken immediately below the water table and it is determined that the entrapped gas volume is 0.05 cm^3/cm^3. The barometric pressure head at the time of the sampling is 920 cm of water. The total depth of geologic material between the impermeable floor of the aquifer and the soil surface is 31 m. The water table is located 1.85 m below the soil surface, and the apparent specific yield is 0.28. Calculate the apparent barometric efficiency. Answer: 0.15.

20. Explain in detail why restricted flow of air between the atmosphere and a water-table aquifer can cause the water level in a piezometer to respond to changes in barometric pressure.

21. The following data were obtained for a watershed in the eastern United States. Compute the mean monthly groundwater discharges to the streams for each month, and prepare a mean stage-ground-water discharge curve.

    Water-table stages measured from mean-sea-level.
    $S_{ya} = 0.11$
    Mean stages represent average of 25 observation wells.
    Watershed area = 50.6 km^2

Date		Mean Water Table stage - m		W cm/mo.	Ground-Water ET cm/mo.
		Beginning	End		
1950	April	14.27	13.99	2.3	2.6
	May	13.99	14.09	5.0	1.4
	June	14.09	13.54	0.7	4.8
	July	13.54	13.48	3.7	3.0
	Aug.	13.48	13.11	0.7	3.6
	Sept.	13.11	13.14	3.0	1.8
	Oct.	13.14	12.93	0.0	1.5
	Nov.	12.93	13.23	5.0	0.9
	Dec.	13.23	13.66	7.2	0.9
1951	Jan.	13.66	13.72	2.7	0.3
	Feb.	13.72	13.95	4.2	0.0
	March	13.95	14.02	4.4	1.2
	April	14.02	13.83	1.5	1.4
	May	13.83	13.96	5.5	2.1
	June	13.96	13.96	7.4	4.7
	July	13.96	13.69	4.8	5.8
	Aug.	13.69	13.48	2.7	3.5
	Sept.	13.48	13.23	1.0	2.6
	Oct.	13.23	13.28	2.3	0.8
	Nov.	13.28	14.02	11.9	1.5
	Dec.	14.02	14.42	7.9	0.0
1952	Jan.	14.42	14.70	8.2	0.5
	Feb.	14.70	14.56	5.4	1.6
	March	14.56	14.66	10.7	3.0

Chapter III

DARCY'S LAW AND BASIC DIFFERENTIAL EQUATIONS

3.1 DARCY'S EMPIRICAL EQUATION

The concepts of pressure, density, and piezometric head are familiar to students of fluid mechanics and hydraulics. In hydrodynamics, these concepts apply to fluid elements which are large relative to molecular dimensions but sufficiently small to be considered "points" relative to the dimensions of the flow problem under consideration. The Navier-Stokes equations are the differential equations describing the motion of fluid elements.

Flow through the void spaces in aquifers is also described by the Navier-Stokes equations. Rigorous integration of these equations is virtually impossible, however, owing to the tremendous complexity of the geometry of the flow channels. Furthermore, the behavior of fluid elements in the voids of a porous solid cannot be measured by any practical means. These difficulties are circumvented by considering average values of hydraulic variables and media properties, applicable to representative volume elements as defined in Chapter II.

A mathematical point, p, is associated with each representative volume element. Average values of medium properties and hydraulic variables, determined in the volume element, are regarded as values existing at point p. Inasmuch as the location of p (and of the representative volume element) is a continuous function of the space coordinates, the average values of all parameters and variables are also continuous functions of the space coordinates. Thus, the actual geometry and hydrodynamics, that are so difficult to characterize on the "micro" scale, are replaced by a conceptual continuum of averages that can be measured and observed on a "macro" scale.

Pressure and density, as applied to fluids in a representative volume element, maintain their traditional meanings. The concept of ground-water velocity, however, is substantially different from the normal meaning of velocity; that being the time rate of displacement of a fluid element. Ground-water velocity, or *Darcy velocity*, is a volume flux defined as the volume of discharge per unit of bulk area (including both pore space and solids) per unit of time. The Darcy velocity is a macroscopic flux, defined on a representative element of bulk area and, in no case, is it equal to the displacement per unit time of fluid elements. The Darcy velocity can be visualized by reference to Fig. 3-1 in which water is shown flowing through a cylindrical pipe packed with sand. The volume of water collected per unit time is the discharge rate, Q, and the Darcy

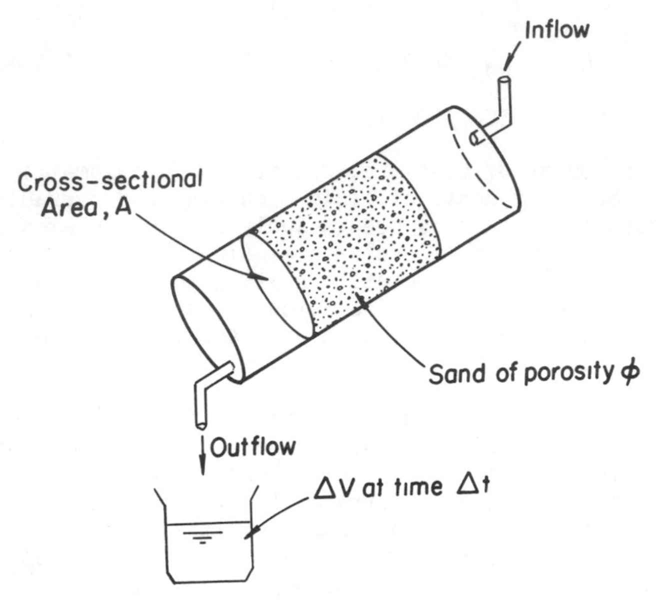

Figure 3-1. Flow in a pipe packed with sand.

velocity is

$$q = Q/A = \frac{\Delta V}{A\Delta t} \quad . \tag{3-1}$$

Another velocity, called the *seepage* velocity, v, is defined as the discharge rate per unit of pore area in the representative element of bulk area.

$$v = \frac{Q}{\phi A} = q/\phi \tag{3-2}$$

The seepage velocity is the average velocity of fluid elements through the voids. For example, suppose that the water being introduced into the sand column of Fig. 3-1 is abruptly changed to a water containing a concentration C_0 of radioactive elements that move, unretarded and unchanged, with the individual water elements. This is known as a *conservative tracer*. Fluid elements with the greater velocities transport the tracer farther into the column during any time period than do the slower fluid elements so that a distribution of tracer concentration C along the column results. The plane (normal to the axis of the column) on which $C/C_0 = 0.5$ moves through the column with a speed equal to v (Bear, 1972).

Forces On Fluids In Porous Solids

Ground water in a representative volume element is subjected to a surface force due to pressure and a body force due to gravity. These two forces are called *driving* forces because they are responsible for the motion of ground water. Ground water also experiences forces that resist the motion of the fluid. Resistance forces, F, act only when the water is in motion.

A balance of force components in the direction ℓ on the water flowing at a constant rate through the volume element shown in Fig. 3-2 leads to (Collins, 1961)

$$p\phi dA - (p + \frac{dp}{d\ell} d\ell)\phi dA = (\rho g\phi dA d\ell) \sin\delta + F \qquad (3\text{-}3)$$

which simplifies to

$$\frac{F}{\phi dA d\ell} = - \left(\frac{dp}{d\ell} + \rho g \frac{dz}{d\ell} \right) \qquad (3\text{-}4)$$

upon recognizing that $\sin \delta = dz/d\ell$. The first term on the right is the component, in the ℓ-direction, of the driving force per unit volume of fluid due to pressure difference and the second is the ℓ-direction component of the force per unit volume of fluid due to gravity.

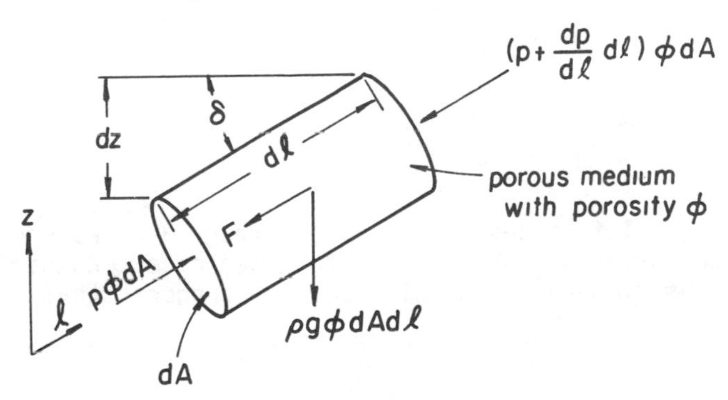

Figure 3-2. Forces on water flowing in a volume element of a porous solid. (Adapted from Hubbert, 1940)

The left side of Eq. 3-4 is the resistance force per unit volume of water. In the simplest case, F is dominated by the surface force due to shear resulting from the drag of the solid particles on the viscous fluid (Rumer, 1969). The force F is directly proportional to the Darcy velocity when the resistance is dominated by viscous shear. Large q results in resistance forces that are in addition to the shear forces. The additional resistance results from *convective acceleration* of fluid elements. Convective acceleration is caused by the change of magnitude and/or direction of the velocity of fluid elements as they pass through the void space of the porous solid. On a macroscopic basis, convective acceleration is zero for uniform rectilinear flow, but on a microscopic basis, convective acceleration distorts and crowds the flow path of fluid elements thereby changing the rate of shear (Hubbert, 1940). Inertial forces become important at values of q well below that required for the onset of turbulence.

Insight concerning the factors affecting the resistance to flow when shear forces predominate (i.e. small q) can be gained by examining exact solutions of the Navier-Stokes equations for three simple cases of viscous flow. The relationship between the driving forces and average velocity $\bar{v}$ in a cylindrical tube of small radius R is

$$\frac{8\mu}{R^2}\bar{v} = -\left(\frac{dp}{d\ell} + \rho g\frac{dz}{d\ell}\right) \quad , \qquad (3\text{-}5)$$

where μ is the dynamic viscosity of the flowing fluid and ℓ is the coordinate measured along the length of the tube. The corresponding equations for flow in a thin film of thickness d and between two flat plates spaced a distance b apart are, respectively,

$$\frac{3\mu}{d^2}\bar{v} = -\left(\frac{dp}{d\ell} + \rho g\frac{dz}{d\ell}\right) \qquad (3\text{-}6)$$

and

$$\frac{12\mu}{b^2}\bar{v} = -\left(\frac{dp}{d\ell} + \rho g\frac{dz}{d\ell}\right) \quad . \qquad (3\text{-}7)$$

The left sides of Eqs. 3-5 through 3-7 represent the resistance forces per unit volume for each case. Comparison of these viscous flow equations with Eq. 3-4 suggests that

$$\frac{F}{\phi dAd\ell} = \frac{C\mu}{\bar{d}^2}q \qquad (3\text{-}8)$$

where C is a dimensionless number that depends upon the shape of the flow channels and $\bar{d}^2$ is a characteristic dimension of the flow space. Note, however, that ℓ in Eqs. 3-5 through 3-7 is a coordinate length measured in the actual direction of motion of fluid elements while ℓ in Eq. 3-4 is a macroscopic coordinate distance measured in the macroscopic direction of flow. Fluid elements in a porous medium follow a tortuous path and travel a distance substantially greater than the macroscopic distance between two points. Therefore, the effects of the tortuous path traversed by fluid elements is a porous medium are included in the parameters $\bar{d}$ and C.

Putting Eq. 3-8 into Eq. 3-4 and solving for the Darcy velocity q yields

$$q = - \frac{\bar{d}^2}{C\mu} \left(\frac{dp}{d\ell} + \rho g \frac{dz}{d\ell} \right) \quad . \tag{3-9}$$

The parameter $\bar{d}^2/C$ is given the symbol k and is known as the *intrinsic permeability* of the porous solid. Note that k has dimensions of L^2. Eq. 3-9 becomes

$$q = - \frac{k}{\mu} \left(\frac{dp}{d\ell} + \rho g \frac{dz}{d\ell} \right) \quad , \tag{3-10}$$

which is one form of Darcy's Law.

The linear relationship between the Darcy velocity and the sum of the driving forces does not persist as convective acceleration on the microscale become significant. Instead of Eq. 3-8, one can write the Forchheimer equation as (Sunada, 1965)

$$\frac{F}{\phi dA d\ell} = \frac{\mu}{k} q + \frac{\rho}{\sqrt{k/C}} q^2 = - \left(\frac{dp}{d\ell} + \rho g \frac{dz}{d\ell} \right) \quad , \tag{3-11}$$

wherein the term in q^2 represents the inertial forces. Dividing by inertial forces yields

$$- \frac{\sqrt{k/C}}{\rho q^2} \left(\frac{dp}{d\ell} + \rho g \frac{dz}{d\ell} \right) = \frac{\mu}{\rho q \sqrt{kC}} + 1 \quad . \tag{3-12}$$

The first term on the right side of Eq. 3-12 is the ratio of viscous forces to inertial forces which is the inverse of the Reynolds' number, R_e,

$$R_e = \frac{\rho q \sqrt{kC}}{\mu} = \frac{\rho q \bar{d}}{\mu} \quad . \tag{3-13}$$

Defining the left side of Eq. 3-12 as the friction factor f and using Eq. 3-13, Eq. 3-12 becomes

$$f = 1/R_e + 1 \quad . \tag{3-14}$$

At small R_e, $f \approx 1/R_e$ and the driving forces are linearily related to q in accordance with Eq. 3-10. As R_e increases, the inertial effects become increasingly important and Darcy's

Law (Eq. 3-10) is no longer valid. Eq. 3-14 is in excellent agreement with experimental observation as shown in Fig. 3-3.

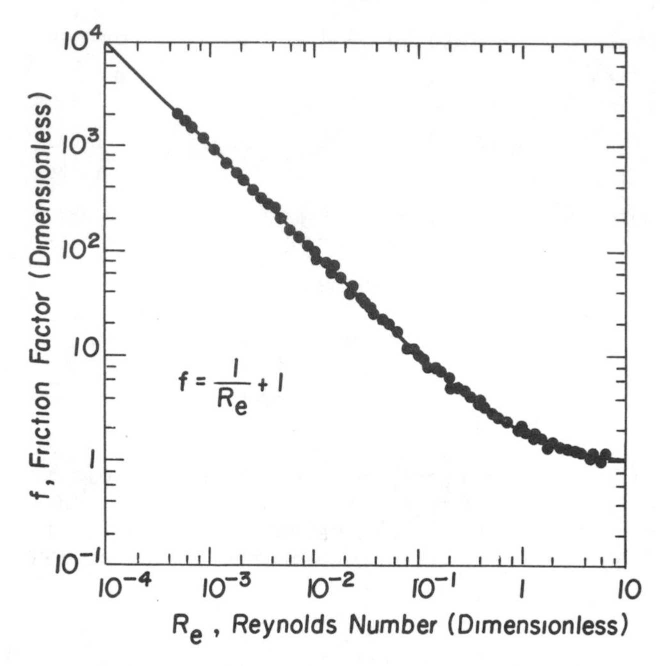

Figure 3-3. Graphical correlation of friction factor versus Reynolds number (Ahmed and Sunada, 1969).

Evidently Darcy's Law is extremely accurate up to $R_e = 0.02$ and is probably acceptable up to $R_e = 0.1$. The reader is cautioned that R_e for flow in porous media can and has been defined in which the characteristic length parameter is different from that used herein. The range of R_e over which Darcy's Law is quoted as being valid depends, of course, upon the definition of the characteristic length parameter. For example, average grain diameter is often used in lieu of $\bar{d}$ as defined above in which case Darcy's Law is valid for R_e up to approximately 1.0 (Todd, 1959; Rose, 1945).

A disadvantage of defining the characteristic length $\bar{d}$ as $\sqrt{kC}$ is that both k and C must be known to determine the Reynolds number for a particular flow condition. While it is possible to measure k from a single experiment by measuring q and the driving forces in Eq. 3-10, an entire set of experiments in which q is measured as a function of the sum of the driving forces (i.e. $dp/d\ell + \rho g\, dz/d\ell$) is required to determine C (Example 3-2). However, experimentally determined values of C range from 0.05 to 2.0, with the majority of the data in the range of

0.5 to 0.6 (Ward, 1964). Consequently, an estimate of $\bar{d}$ can be made by measuring k and using a value of 0.6 for C.

Equation 3-10, Darcy's Law, was obtained in the above paragraphs through a heuristic line of reasoning. The preceding developments do not constitute a derivation of Darcy's Law, however. More rigorous developments can be found in Hubbert (1956) and DeWiest (1965). The purpose of the presentation was to provide some insight into the physical meaning of the Darcy equation. It should be remembered that Eq. 3-10 is basically an empirical formula in which the intrinsic permeability must be measured, either directly by determining all other factors in Eq. 3-10 or indirectly from equations derived therefrom.

EXAMPLE 3-1

Suppose δ, in Fig. 3-2, is 30° and $dp/d\ell$ is -294 dynes/cm^3. Calculate the direction and magnitude of the resistance force per unit volume of water. In which direction is flow occurring?

Solution:

From Eq. 3-4,

Force per unit volume of water = $-\ (-294 + 980 \sin \delta)$

$= -\ 196$ dynes/cm^3 .

Thus, the magnitude of the resistance force is 196 dynes/cm^3. and the resistance force acts in the direction of increasing ℓ. The flow is downward in the direction of decreasing ℓ.

EXAMPLE 3-2

The following data relating the sum of the driving forces to the Darcy velocity were measured for a sample of porous medium. Determine k, C and $\bar{d}$ for the sample and compute the value of q and R_e at which a 10% departure from Darcy's Law can be expected. The properties of water are μ = 0.016 dyne-sec/cm^2 and ρ = 1.0 g/cm^3.

$-\frac{1}{\rho g}\left(\frac{dp}{d\ell} + \rho g \frac{dz}{d\ell}\right)$ dimensionless	q cm/s	$-\frac{1}{\rho g}\left(\frac{dp}{d\ell} + \rho g \frac{dz}{d\ell}\right)$ dimensionless	q cm/s
0.011	0.0049	0.44	0.19
0.017	0.0073	0.81	0.34
0.026	0.011	1.23	0.50
0.041	0.018	1.82	0.72
0.060	0.026	2.44	0.94
0.113	0.049	2.93	1.11
0.16	0.068	4.21	1.53
0.26	0.11	10.1	3.10
0.34	0.14	18.9	4.94

$-\frac{1}{\rho g}\left(\frac{dp}{d\ell}+\rho g\frac{dz}{d\ell}\right)$ dimensionless	q cm/s
26.1	6.20
38.4	8.05
52.7	9.89
70.1	11.8
103.0	15.0

Solution:

Note that Eq. 3-11 can be written as

$$|\nabla h| = Aq + Bq^2$$

where

$$|\nabla h| = -\frac{1}{\rho g}\left(\frac{dp}{d\ell}+\rho g\frac{dz}{d\ell}\right)$$

and

$$A = \frac{\mu}{\rho g k}$$

$$B = \frac{1}{g\sqrt{k/C}} \quad .$$

The problem reduces to determining the coefficients A and B by fitting the equation to the data provided by the statistical technique used in problem 6. The coefficients are found to be A = 2.3 s/cm and B = 0.31 $(s/cm)^2$ from which

$$k = \frac{\mu}{\rho g A} = \frac{(0.016)}{(1)(980)(2.3)} = 7.1\text{x}10^{-6}\ cm^2$$

and

$$C = B^2 g^2 k = (0.31)^2(980)^2(7.1\text{x}10^{-6}) = 0.66 \quad .$$

Using the definition of intrinsic permeability, the characteristic length parameter is

$$\bar{d} = \sqrt{Ck} = \sqrt{(0.66)(7.1\text{x}10^{-6}} = 2.2\text{x}10^{-3}\ cm \quad .$$

A 10% departure from Darcy's Law implies that

$$\frac{|\nabla h|}{Aq} = 1.1 = 1 + \frac{B}{A} q$$

or that

$$q = \frac{(0.1)A}{B} = \frac{(0.1)(2.3)}{0.31} = 0.74\ cm/s \quad .$$

The corresponding value of R_e is from Eq. 3-13:

$$R_e = \frac{(1)(0.74)(2.2\text{x}10^{-3})}{0.016} = 0.1 \quad .$$

EXAMPLE 3-3

The external radius (measured from the axis of the well bore) of the gravel pack around a water well is 35 cm (see Fig. 1-6). The length of well screen is 25 m and the well discharge is 0.15 m^3/s. The intrinsic permeability of the gravel is $4x10^{-5}$ cm^2. Assess the validity of Darcy's Law for this flow.

Solution:

To a close approximation, the entire discharge rate $Q = 0.15\ m^3/s$ must pass through the lateral surface area of the cylinder formed by the external surface of the gravel pack. This cylinder is of radius 35 cm and height 25 m. Therefore,

$$q = Q/A = \frac{0.15\ m^3/s}{(2\pi)(0.35)(25)m^2} = 2.73x10^{-3}\ m/s = 0.273\ cm/s \quad .$$

It is assumed that $\rho = 1\ g/cm^3$ and $\mu = 0.01$ poise (0.01 g/cm-s) for water and C = 0.6. From Eq. 3-13,

$$R_e = \frac{(1\ g/cm^3)\sqrt{(4x10^{-5}cm^2)(0.6)}(0.273\ cm/s)}{0.01\ g/cm\text{-}s} = 0.13 \quad .$$

This large value of R_e indicates that inertial forces are influencing the flow into the gravel pack and Darcy's Law is not, strictly speaking, applicable for this flow.

Note that the Darcy velocity in this example is 0.273 cm/s. This velocity would be considered, ordinarily, quite small in hydraulics, but is a very large velocity in ground-water hydraulics.

Force Potential, Velocity Potential, and Darcy's Law

Eq. 3-10 is a rather general form of Darcy's Law which applies for fluids with either constant or variable density contained in porous solids whose intrinsic permeability may depend upon both direction and location. For many applications in ground-water hydrology, it is unnecessarily general and alternate forms prove more convenient. It is often convenient to write the driving force as the gradient of a scalar *force potential*, when such a potential exists (Hubbert, 1940, 1953). Let the density in Eq. 3-10 be constant. Then

$$q = -\frac{k\rho g}{\mu}\frac{d}{d\ell}\left(\frac{p}{\rho g} + z\right) \qquad (3\text{-}15)$$

wherein the quantity in parentheses is the *piezometric head*, h, and is

$$h = \frac{p}{\rho g} + z = h_p + z \quad , \qquad (3\text{-}16)$$

a variable that is familiar to students of fluid mechanics and hydraulics. The piezometric head is an especially useful variable in practice because the depth of water in a piezometer is $p/\rho g$ and the elevation (with respect to an arbitrary datum) of the terminal point of the piezometer is z. Thus, the water level in piezometers is a direct indicator of the piezometric head.

Written for the three-dimensional case,

$$\vec{q} = -\frac{k\rho g}{\mu} \nabla h \qquad (3\text{-}17)$$

in which ∇ is the gradient operator

$$\nabla = \frac{\partial}{\partial x}\vec{i} + \frac{\partial}{\partial y}\vec{j} + \frac{\partial}{\partial z}\vec{k} \qquad (3\text{-}18)$$

where $\vec{i}$, $\vec{j}$, and $\vec{k}$ are the unit vectors in the x, y, and z coordinate directions, respectively.

Note that the piezometric head is a scalar, the negative gradient of which is a vector representing the force per unit weight acting on the fluid. Provided ρ is constant, piezometric head is a force potential with the physical significance of energy per unit weight of fluid. Equation 3-10 is more general than Eq. 3-17 because the latter is restricted to applications in which the fluid density is constant. The most common case in which the ground-water hydrologist must deal with variable density is that of flow in aquifers containing both saline and fresh water (i.e. a nonhomogeneous fluid). Even in this case, however, it is often possible to idealize the problem to a degree and, thereby, retain the use of Eq. 3-17.

The density of a homogeneous fluid can often be expressed as a function of pressure only. For example, the density of water with constant compressibility β can be written as a function of pressure. Fluids in which $\rho=\rho(p)$ are *barotropic* fluids. It is possible to write a force potential in terms of energy per unit mass for barotropic fluids. As will be demonstrated subsequently in this chapter, however, it is not necessary to account for the barotropic behavior of water except with respect to changes in storage, and the use of the piezometric head as a force potential is permissible.

The coefficient multiplying the driving force in Eq. 3-17 is called *hydraulic conductivity*, K and is,

$$K = \frac{k\rho g}{\mu} \quad . \tag{3-19}$$

The hydraulic conductivity is a hydro-geologic parameter, with dimensions L/T, that combines both fluid properties and the intrinsic permeability k. Using Eq. 3-19 in Eq. 3-17,

$$\vec{q} = - K\nabla h \tag{3-20}$$

which is probably the most common form of Darcy's Law used in the practice of ground-water hydrology. Factors affecting K and k and methods by which they can be measured in the laboratory are discussed in a following section of this chapter.

Suppose K is constant. Then it is permissible to define the scalar quantity

$$\Phi = Kh \tag{3-21}$$

from which

$$\vec{q} = - \nabla\Phi \ . \tag{3-22}$$

The scalar Φ is the *velocity potential*. The velocity potential is a useful mathematical artifice, but it should not be interpreted in terms of potential energy, and its gradient is not a force (Hubbert, 1940). Furthermore, the concept of the velocity potential is valid only when K is constant; otherwise Eq. 3-22 leads to the erroneous conclusion that flow is promoted by spatial changes of K.

EXAMPLE 3-4

Expand Eq. 3-20 and write $\vec{q}$ in terms of the orthogonal components q_x, q_y, and q_z.

Solution:

Expansion of the gradient operator in Eq. 3-20 yields

$$\vec{q} = - K \left(\frac{\partial h}{\partial x}\vec{i} + \frac{\partial h}{\partial y}\vec{j} + \frac{\partial h}{\partial z}\vec{k}\right) \quad .$$

But,

$$q_x = - K\frac{\partial h}{\partial x}, \quad q_y = - K\frac{\partial h}{\partial y}, \quad \text{and} \quad q_z = - K\frac{\partial h}{\partial z} \quad .$$

So

$$\vec{q} = q_x\vec{i} + q_y\vec{j} + q_z\vec{k} \quad .$$

EXAMPLE 3-5

The hydraulic conductivity was measured for a porous solid and found to be 4.8×10^{-4} cm/s for water with a density of 1 g/cm^3 and a viscosity of 1 centi-poise. Calculate the intrinsic permeability of the material and the hydraulic conductivity for oil with $\rho = 0.73$ g/cm^3 and $\mu = 1.8$ centi-poise.

Solution:

From Eq. 3-19,

$$k = \frac{\mu}{\rho g} K = \frac{(0.01 \text{ dynes-s/cm}^2)(4.8 \times 10^{-4} \text{ cm/s})}{980 \text{ dynes/cm}^3}$$

$$k = 4.9 \times 10^{-9} \text{ cm}^2 \quad .$$

Also

$$K_{oil} = \frac{(4.9 \times 10^{-9})(0.73)(980)}{0.018} = 1.95 \times 10^{-4} \text{ cm/s} \quad .$$

EXAMPLE 3-6

Figure 3-4 shows two piezometers, one terminating in a water-table aquifer and one in an underlying confined aquifer. The hydraulic conductivity of the layer between the two aquifers is 7.2×10^{-5} cm/s. Calculate the discharge rate Q per unit area through the layer between the aquifers. Assume steady flow.

Solution:

A simplified version of Eq. 3-20, appropriate for this computation, is obtained by noting that the flow is one-dimensional and parallel to the z coordinate. Hence, from Example 3-4,

$$\nabla h = \frac{dh}{dz} \vec{k}$$

and

$$q_z = - K \frac{dh}{dz} \quad .$$

Because the flow is steady, q_z is constant. Furthermore, K is constant. The gradient of piezometric head must also be constant, therefore, and

$$q_z = - K \frac{(h_2 - h_1)}{b}$$

where b is the thickness of the layer and the subscripts 1 and 2 denote the bottom and top of the layer, respectively.

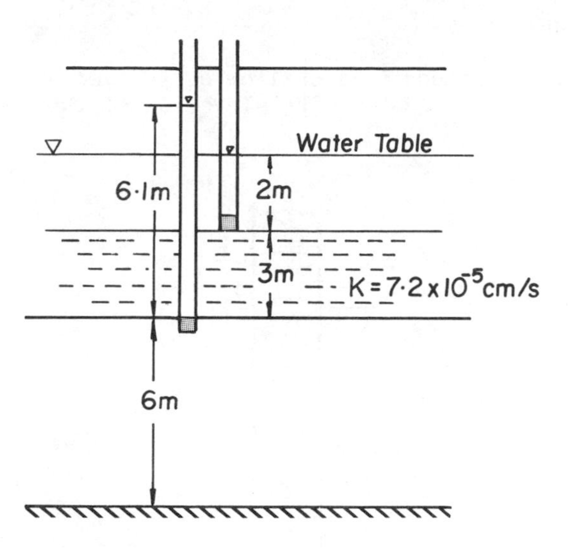

Figure 3-4. Upward seepage from a confined aquifer to a water table aquifer.

The impermeable floor of the aquifer is selected as the datum for calculating piezometric head. Thus, from Eq. 3-16,

$$h_1 = 6.1 + 6 = 12.1 \text{ m} \quad ,$$

and

$$h_2 = 9 + 2 = 11 \text{ m} \quad .$$

The discharge per unit area is the Darcy velocity and is

$$q_z = -\ 7.2\text{x}10^{-5} \text{ cm/s } \left(\frac{11 - 12.1}{3} \right)$$

$$q_z = 2.64\text{x}10^{-5} \text{ cm/s} \quad .$$

The fact that q_z is positive means the flow is in the direction of increasing z (i.e. upward).

The student should note that precisely the same value of q_z would be obtained, irrespective of the location of the datum. The values of piezometric head will differ, depending on the selection of the datum, but the difference between the values at the top and bottom of the layer will be the same and the Darcy velocity depends only upon the *difference* in piezometric head. The flow is always in the direction of decreasing head. Furthermore, it is to be noted that the difference in piezometric head is simply the difference in water-surface elevation in the piezometers. This illustrates the convenience of using the piezometric head as a force potential.

EXAMPLE 3-7

Water is ponded to a shallow depth over a column of aquifer material (Fig. 3-5). The elevation of the terminal end of

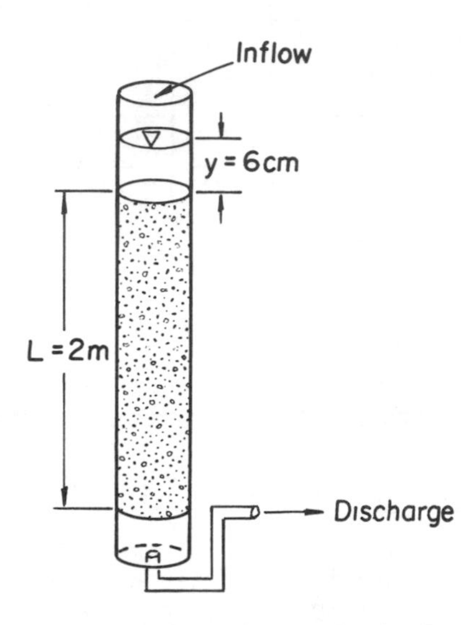

Figure 3-5. Steady downward flow in a vertical column.

the discharge tube is maintained equal to the elevation of the bottom of the column. Head loss through the discharge tube is negligible. Water is added at the top at the same rate it is discharged from the bottom. The hydraulic conductivity of the porous material is 1.1×10^{-3} cm/s. Calculate the Darcy velocity in the column. Also, derive a general formula for q in terms of K; the depth of ponded water, y; and the length of column, L.

Solution:

The equation of Example 3-5 applies to this case since, again, the flow is parallel to the z-coordinate, and both q and K are constants. The bottom of the porous material in the column is selected as the datum. The pressure head at the bottom of the porous solid is zero since the discharge tube is maintained at the same elevation as the bottom of the material. Also, z=0, at this location, so $h_1=0$.

The pressure head at the top of the porous solid is 6 cm, and the elevation is 2 m. Thus, $h_2=2.06$ m. The length of flow path is 2 m, and

$$q_z = -1.1 \times 10^{-3} \left(\frac{2.06 - 0}{2.0} \right) = -1.13 \times 10^{-3} \text{ cm/s} \quad .$$

The negative sign indicates that q is in the direction of decreasing z (i.e. downward).

In general, for any datum

$$h_i = 0 + z_1 = z_1$$

and

$$h_2 = y + z_1 + L.$$

Thus,

$$q_z = - K\left(\frac{y+L}{L} \right) = - K(1 + y/L) \quad .$$

Note that $y/L \to 0$ for either very shallow pond depths or large L, and $|q_z| \to K$. This is the condition of a unit gradient of piezometric head. When $dh/dz = 1$, $dp/dz = 0$ (i.e. the pressure is constant) and the downward flow results from the gravitational driving force only. This condition is sometimes approached by downward seepage from shallow lakes or recharge ponds, below which the water table depth is large relative to the pond depth.

Laboratory Determination of Hydraulic Conductivity

The most reliable method of determining the hydraulic conductivity and intrinsic permeability is from aquifer tests conducted in the field. These methods are discussed in Chapter V. Values of hydraulic conductivity, determined by carefully conducted experiments in the laboratory, are quite accurate and reproducible. By necessity, laboratory samples are extremely small compared to the aquifer as a whole, however. Furthermore, some degree of disturbance always accompanies the collection of the samples. For these reasons, it is extremely difficult to characterize the hydraulic conductivity of an aquifer, or even a small portion, by means of laboratory measurements.

Hydraulic conductivity and intrinsic permeability can be measured directly with *permeameters*. Two commonly used permeameters are shown in Fig. 3-6. Figure 3-6a depicts a constant-head permeameter in which steady upward flow through the sample is established. Darcy's equation can be applied directly in this case to compute K,

$$K = \frac{QL}{\Delta hA} \tag{3-23}$$

The total head loss through the permeameter is indicated by the difference in elevation between the inflow and outflow water levels. In a properly designed permeameter, the head loss through the retaining screens and the inflow and outflow

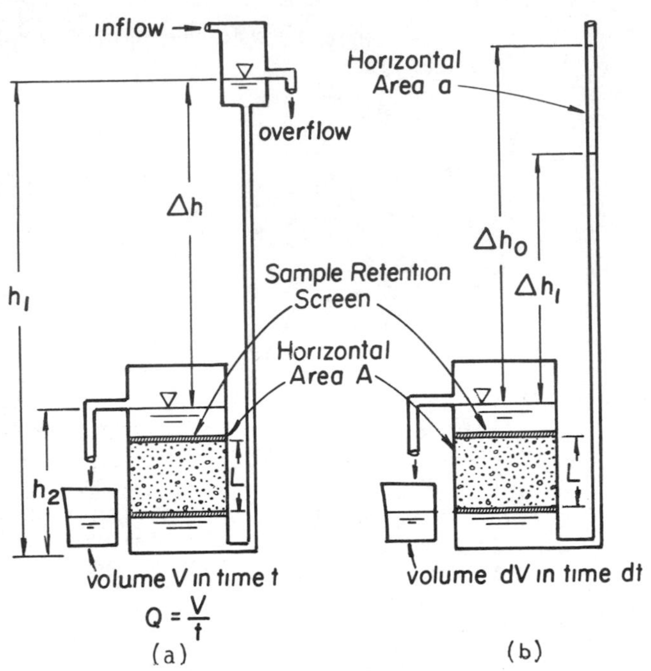

Figure 3-6. Constant head (a) and falling head (b) permeameters (After Todd, 1959).

plumbing is negligibly small and the head loss through the sample is very nearly equal to the difference in the inflow and outflow water levels. Some constant head permeameters are equipped with piezometer taps located in the test section and the difference in piezometric head is indicated by the difference in water level in the two piezometers. This difference reflects the head loss between the two piezometer taps, regardless of the head loss in the remainder of the system.

The falling head permeameter is shown in Fig. 3-6b. The rate of discharge through the sample decreases with time because the driving head decreases as the water level falls in the inflow standpipe. It is left to the student to show that the equation for hydraulic conductivity is

$$K = \frac{aL}{At} \ln \frac{\Delta h_o}{\Delta h(t)} \tag{3-24}$$

in which Δh_o is the initial head loss (at t=0) and $\Delta h(t)$ is the head loss at time t. Again, care must be taken to insure that the loss of head in the retaining screens and inflow-outflow plumbing is very small relative to that which occurs in the sample.

The intrinsic permeability can be estimated from grain-size distribution data like that shown in Fig. 3-7, and from the Fair-Hatch (1933) formula.

$$k = \frac{1}{A}\left[\frac{(1-\phi)^2}{\phi^3}\left(\frac{B}{100}\Sigma\frac{F}{d_m}\right)^2\right]^{-1} \quad . \qquad (3\text{-}25)$$

In Eq. 3-25, A is a dimensionless packing factor found to be approximately 5, and B is a particle shape factor equal to 6 for spherical particles and 7.7 for highly angular ones. The factor F is the percent by weight of the sample between two arbitrary particle sizes, and d_m is the geometric mean of the particle sizes corresponding to F.

The Fair-Hatch formula takes account of the distribution in grain sizes. Other equations require only a characteristic grain size d. For example,

$$k = (6.54 \times 10^{-4})\, d^2 \qquad (3\text{-}26)$$

yields the intrinsic permeability in cm^2 as a function of a characteristic particle diameter in cm (Harleman et al., 1963). Equation 3-26 is most nearly valid for materials of very uniform particle size and shape. The *effective* grain size, d_{90},

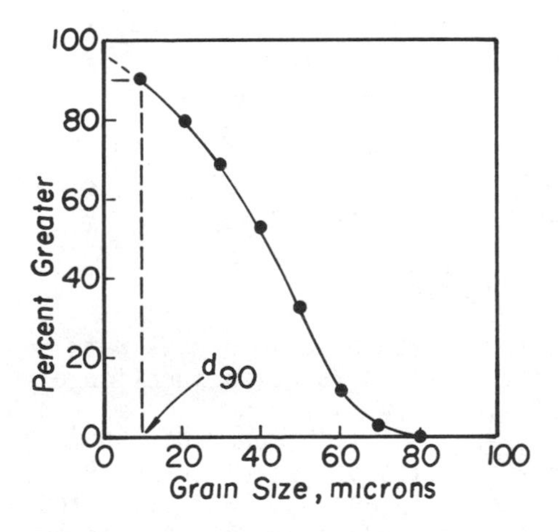

Figure 3-7. Typical grain-size distribution curve.

defined as the sieve size on which 90 percent of the sample is retained, is often used for nonuniform aquifer materials. Clearly, Eq. 3-26 makes no account of packing, porosity, or particle shape.

Table 3-1 contains values of hydraulic conductivity to water for several different porous materials. It is striking to note that the hydraulic conductivity of materials with which the hydrologist must deal range over 9 orders of magnitude.

Table 3-1. Hydraulic Conductivity of Porous Materials (Adapted from Morris & Johnson, 1967)

Material	No. of Analyses	Range cm/s	Arithmetic Mean cm/s
Igneous Rocks			
Weathered granite	7	$(3.3 - 52) \times 10^{-4}$	1.65×10^{-3}
Weathered gabbro	4	$(0.5 - 3.8) \times 10^{-4}$	1.89×10^{-4}
Basalt	93	$(0.2 - 4250) \times 10^{-8}$	9.45×10^{-6}
Sedimentary Materials			
Sandstone (fine)	20	$(0.5 - 2270) \times 10^{-6}$	3.31×10^{-4}
Siltstone	8	$(0.1 - 142) \times 10^{-8}$	1.9×10^{-7}
Sand (fine)	159	$(0.2 - 189) \times 10^{-4}$	2.88×10^{-3}
Sand (med.)	255	$(0.9 - 567) \times 10^{-4}$	1.42×10^{-2}
Sand (coarse)	158	$(0.9 - 6610) \times 10^{-4}$	5.20×10^{-2}
Gravel	40	$(0.3 - 31.2) \times 10^{-1}$	4.03×10^{-1}
Silt	39	$(0.09 - 7090) \times 10^{-7}$	2.83×10^{-5}
Clay	19	$(0.1 - 47) \times 10^{-8}$	9×10^{-8}
Metamorphic Rocks			
Schist	17	$(0.002 - 1130) \times 10^{-6}$	1.9×10^{-4}

3.2 NON-HOMOGENIETY AND ANISOTROPY

Non-homogeneous Aquifers

An aquifer is said to be *homogeneous* with respect to hydraulic conductivity if K is not a function of the space coordinates. If $K = K(x,y,z)$ the aquifer is *non-homogeneous*. Rarely is an aquifer actually homogeneous, but the practical and financial difficulties of establishing the function $K = K(x,y,z)$ in the field are substantial. Flow nets, a method of analysis to be discussed in Chapter IV, is one method by which non-homogeniety can be deduced quantitatively. Nevertheless, the hydrologist is often obliged to assign a constant value to K in full realization that doing so is only an approximation.

No modification of Darcy's Law is required for flow in non-homogeneous aquifers. Both Eqs. 3-10 and 3-20 apply because

$K = K(x,y,z)$ remains a scalar, and the vector $\vec{q}$ is colinear with, but opposite in sense to, the gradient of piezometric head just as in the case of constant K. Occasionally, the non-homogeniety that must be dealt with results from stratification of the aquifer. The hydraulic conductivity may be constant within individual stratum but different from layer to layer. Such an aquifer can be regarded as being block-wise homogeneous with discontinuities in K at the interfaces between adjacent layers.

Treatment of block-wise homogeneous aquifers can be accomplished by defining a piezometric head for each sub-region in which K is constant. For example, consider two sub-regions a and b in which the hydraulic conductivities are K_a and K_b and the piezometric heads are h_a and h_b. The interface between the two regions is denoted by curve C (Fig. 3-8). The Darcy velocity $\vec{q}_a$ at any point on C is decomposed into a component

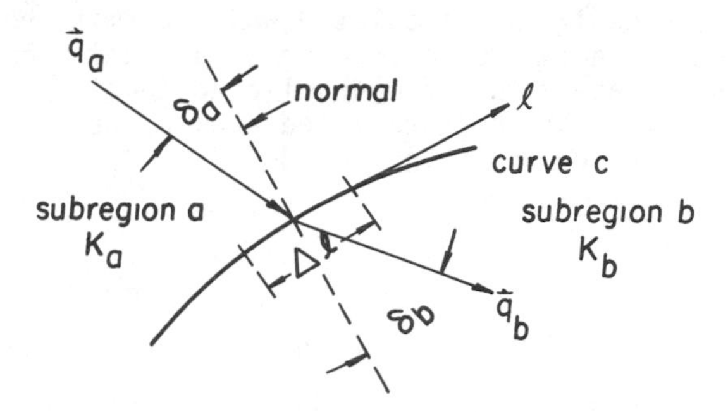

Figure 3-8. Refraction of the velocity vector at a boundary between different homogeneous media.

q_{an} normal to C and one tangent to C. The discharge per unit depth into the paper through a segment $\Delta\ell$ is $(q_{an})\Delta\ell$, where ℓ is the coordinate measured tangent to curve C. The vector $\vec{q}_b$ can be similarly decomposed. Since the flow through the segment $\Delta\ell$ must be the same regardless of which sub-region is used to calculate it, the conclusion is that the components of $\vec{q}_a$ and $\vec{q}_b$ normal to C must be equal. Therefore,

$$q_a \cos \delta_a = q_b \cos \delta_b \quad . \qquad (3\text{-}27)$$

The fact that $(p/\rho g)_a = (p/\rho g)_b$ and $z_a = z_b$ on curve C means that $h_a = h_b$ on C. Therefore,

$$\frac{\partial h_a}{\partial \ell} = \frac{\partial h_b}{\partial \ell} \tag{3-28}$$

on C. Equations 3-27 and 3-28 constitute the conditions that must be satisfied on the interfaces between adjacent subregions in a blockwise homogeneous aquifer. From Eq. 3-28,

$$\frac{q_a}{K_a} \sin \delta_a = \frac{q_b}{K_b} \sin \delta_b \tag{3-29}$$

which is combined with Eq. 3-27 to yield

$$\frac{K_a}{K_b} = \frac{\tan \delta_a}{\tan \delta_b} \quad . \tag{3-30}$$

Conditions are encountered in which the direction of flow is either parallel or normal to the stratifications. Average values of K can easily be calculated which permit the blockwise homogeneous medium to be treated as a single homogeneous mass with an average value of hydraulic conductivity $\overline{K}$. The discharge rate through the stratified aquifer per unit width normal to the section shown in Fig. 3-9 is

$$Q = - b\overline{K} \frac{(h_2 - h_1)}{L} \tag{3-31}$$

where $b = b_a + b_b$. The total discharge is also expressible as the sum of the discharges through each layer

$$Q = - b_a K_a \frac{(h_2 - h_1)}{L} - b_b K_b \frac{(h_2 - h_1)}{L} \quad . \tag{3-32}$$

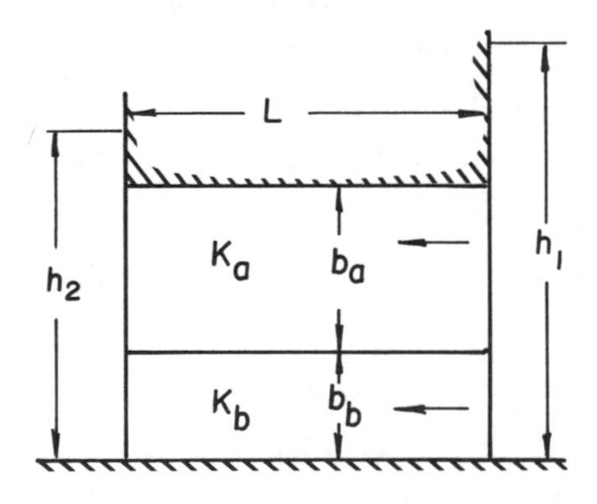

Figure 3-9. Flow parallel to the layers in a stratified aquifer.

Equating the quantities on the right sides of Eqs. 3-31 and 3-32 and simplification yields

$$\overline{K} = \frac{K_a b_a + K_b b_b}{b} \quad . \tag{3-33}$$

This result can be generalized for n layers:

$$\overline{K} = \frac{\sum_{i=1}^{n} b_i K_i}{\sum_{i=1}^{n} b_i} \quad . \tag{3-34}$$

A corresponding formula for flow through strata in series is obtained by recognizing that the Darcy velocity is the same in each layer (Fig. 3-10) and the total difference in piezometric head across the aquifer is the sum of the head losses across the individual strata.

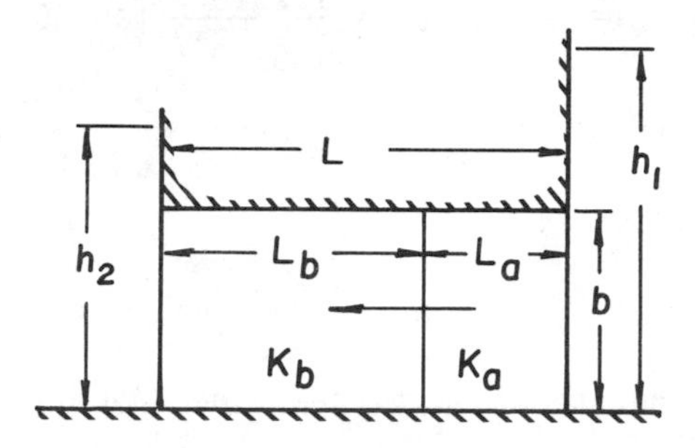

Figure 3-10. Flow through beds in series.

Referring to Fig. 3-10,

$$\frac{QL}{bK} = \frac{QL_a}{bK_a} + \frac{QL_b}{bK_b} \tag{3-35}$$

where $L = L_a + L_b$. From Eq. 3-35,

$$\overline{K} = \frac{L}{\frac{L_a}{K_a} + \frac{L_b}{K_b}} \tag{3-36}$$

which, for n layers, becomes

$$\overline{K} = \frac{\sum_{i=1}^{n} L_i}{\sum_{i=1}^{n} \frac{L_i}{K_i}} \quad . \qquad (3\text{-}37)$$

EXAMPLE 3-8

The Darcy velocity is incident on the interface between coarse and fine textured materials with hydraulic conductivities $K_a = 1.6 \times 10^{-3}$ cm/s and $K_b = 1.2 \times 10^{-4}$ cm/s. Flow occurs from the coarse material into the fine and makes an angle $\delta = 30°$ with the normal to the interface. Calculate the angle δ_b in the fine textured material.

Solution:

From Eq. 3-30

$$\tan \delta_b = \frac{K_b}{K_a} \tan \delta_a = \frac{1.2 \times 10^{-4}}{1.6 \times 10^{-3}} \tan 30°$$

$$= 0.0433 \quad .$$

Hence

$$\delta_b = 2.5° \quad .$$

It is observed that the flow in the fine layer is nearly normal to the interface.

EXAMPLE 3-9

One-dimensional, steady flow, between parallel channels (Fig. 3-10), occurs in a homogeneous aquifer of thickness 4 m. With a difference of head equal to 1.3 m, the discharge rate per unit length of channel is 1.82×10^{-5} m^3/m-s. The channels are 10 m apart. A layer of sediment is ultimately deposited on the inflow face. The sediment layer is 4 cm thick with a conductivity of 1.4×10^{-5} cm/s. Calculate the discharge rate per km between the two channels following the deposition of the sediment.

Solution:

$$q = \frac{Q}{b} = \frac{1.82 \times 10^{-5}}{4m} \, m^3/m\text{-}s = 4.55 \times 10^{-6} \, m/s \quad .$$

The conductivity of the aquifer is

$$K = \frac{qL}{\Delta h} = \frac{(4.55 \times 10^{-6}\ \text{m/s})(10)}{1.3\text{m}}$$

$$= 3.5 \times 10^{-5}\ \text{m/s} = 3.5 \times 10^{-3}\ \text{cm/s} \quad .$$

From Eq. 3-36

$$\overline{K} = \frac{10 + 0.04}{\dfrac{0.04}{1.4 \times 10^{-5}} + \dfrac{10}{3.5 \times 10^{-3}}} = 1.76 \times 10^{-3}\ \text{cm/s}$$

The discharge rate per km between the channels after deposition of the sediment layer is

$$Q = \frac{\overline{K}b\ (h_2 - h_1)}{L} = \frac{(1.76 \times 10^{-5})(4\text{m})(1.3\text{m})}{10.04\text{m}} \times 1000\ \text{m/km}$$

$$= 9.1 \times 10^{-3}\ \text{m}^3/\text{s - km} \quad .$$

Anisotropic Aquifers

The thicknesses of the strata discussed in the previous section are large relative to the characteristic dimension of the representative volume element. A stratification or bedding, observable only on a much smaller scale, is also quite common. When the thicknesses of the lamina are less than the characteristic dimension of the representative volume element, the stratification does not cause the aquifer to be non-homogeneous The characteristics of each representative volume element are identical, regardless of location in the material. The material must, therefore, be classified as homogeneous on the macroscopic scale.

The resistance to flow in directions normal to the microbedding planes is usually greater than for flow parallel to the lamina. In other words, the hydraulic conductivity takes on a direction property, being smaller in directions normal to the bedding than in directions parallel to the bedding. This is an example of *anisotropy*. Aquifers in which the resistance to flow through a representative volume element depends upon the direction of the flow through the element are said to be *anisotropic*; otherwise they are *isotropic*. Anisotropy can be the result of fracture patterns or solution channels with a preferred direction as well as from micro-scale bedding.

Description of flow in anisotropic porous solids requires a substantial generalization of the Darcy equation because K can no longer be considered a scalar. An effect of the directional properties of K is to cause the velocity vector $\vec{q}$ and the gradient vector ∇h to be non-colinear. This can be understood by considering flow in the hypothetical homogeneous but anisotropic medium depicted in Fig. 3-11. The anisotropy

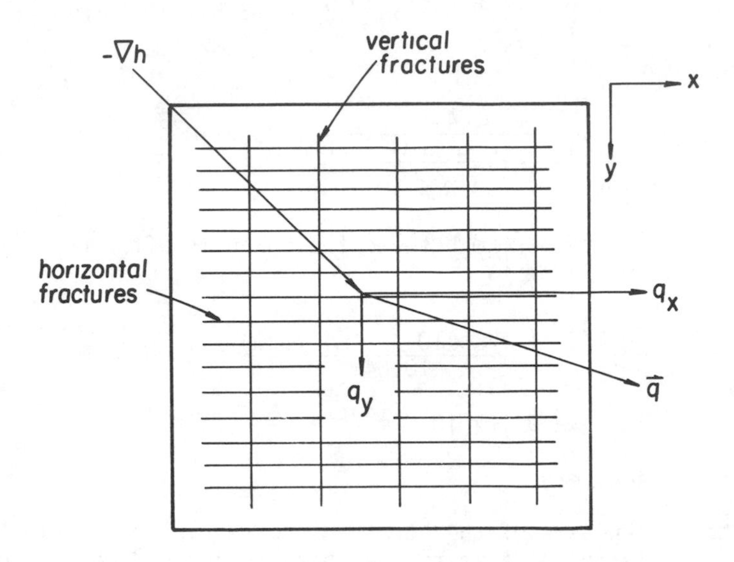

Figure 3-11. Representative volume element in an anisotropic aquifer.

in this material is caused by a greater fracture intensity in horizontal planes than in vertical planes. Thus, the hydraulic conductivity is greater in the horizontal than in the vertical direction. A gradient of piezometric head, oriented at a 45° angle to the horizontal results in the components of Darcy velocity as shown. Note that the horizontal component of velocity is greater in magnitude than the vertical component, yielding a resultant velocity vector favoring the horizontal direction.

In the general case of anisotropic porous solids, the hydraulic conductivity is regarded as a second rank tensor and the product of K and ∇h is calculated using well-defined procedures. The hydraulic conductivity tensor in three-dimensional space has nine components. It is always possible to find three orthogonal coordinate axes, called the *principal directions*, in which the 9-component matrix representing the K-tensor is reduced to a diagonal matrix.

Fortunately, the large fraction of anisotropic aquifers do not require such a general treatment. Usually, the hydraulic conductivity is a maximum and the same in all directions measured in planes parallel to the bedding. The minimum K is usually in the direction normal to the layering. The principal directions are parallel and normal to the bedding planes, therefore. In this case the components of Darcy velocity bear a simple relationship to the components of ∇h. Explicitly,

$$q_y = -K_y \frac{\partial h}{\partial y} \quad \text{and} \quad q_x = -K_x \frac{\partial h}{\partial x} \tag{3-38}$$

where y is the coordinate parallel to the direction of minimum K and x is the coordinate parallel to the direction of maximum K. The resultant velocity vector is co-linear with the gradient only when the gradient is parallel to one of the principal directions. If the aquifer is homogeneous, K_x and K_y are constants independent of the space coordinates. Otherwise, the aquifer is both anisotropic and non-homogeneous.

Measurement of hydraulic conductivity in anisotropic media presents an interesting problem (Marcus and Evenson, 1961). Consider, for example, an anisotropic sample whose thickness is small relative to the diameter of the permeameter (Fig. 3-12). Clearly, the gradient is directed vertically. Water will tend to take the path of least resistance through the sample and the velocity vector will not be co-linear with ∇h. According to

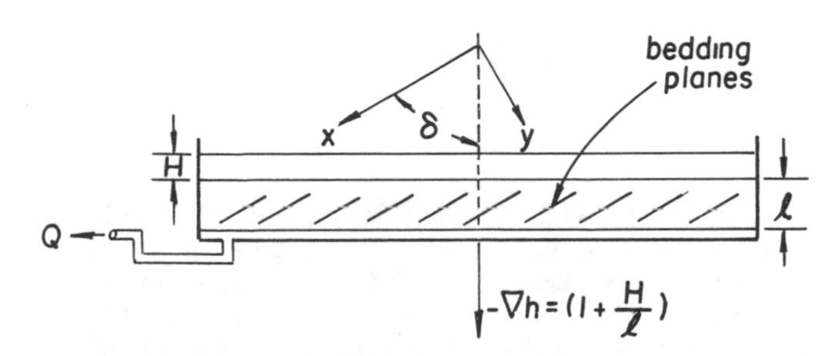

Figure 3-12. Definitive sketch for measuring conductivity in the direction of the gradient.

the results of Example 3-7, or Eq. 3-23, the apparent hydraulic conductivity in the direction of the gradient is

$$K_g = \frac{Q}{A(1 + H/\ell)} = \frac{-q_g}{|\nabla h|} \quad . \tag{3-39}$$

It will be seen, however, that K_g depends upon both the orientation of the principal directions with respect to the axis of the column and upon the thickness to diameter ratio of the sample.

From Fig. 3-12,

$$\frac{\partial h}{\partial y} = |\nabla h| \sin \delta$$

and (3-40)

$$\frac{\partial h}{\partial x} = |\nabla h| \cos \delta \quad .$$

Also, from Eq. 3-38, the components of the Darcy velocity in the principle directions are

$$q_x = -K_x \frac{\partial h}{\partial x} = -K_x |\nabla h| \cos \delta$$

and (3-41)

$$q_y = -K_y \frac{\partial h}{\partial y} = -K_y |\nabla h| \sin \delta \ .$$

The apparent velocity in the direction of the gradient is

$$q_g = q_x \cos \delta + q_y \sin \delta \tag{3-42}$$

which, when combined with Eq. 3-41, becomes

$$q_g = -K_x |\nabla h| \cos^2 \delta - K_y |\nabla h| \sin^2 \delta \quad . \tag{3-43}$$

Finally, since $q_g = -K_g |\nabla h|$ (see Eq. 3-39),

$$K_g = K_x \cos^2 \delta + K_y \sin^2 \delta \quad . \tag{3-44}$$

Equation 3-44 shows how the measured hydraulic conductivity depends upon the orientation of the sample in the permeameter. Now suppose the permeameter of Fig. 3-12 is very long, relative to the diameter. In this case the direction of flow through the sample is forced to be co-linear with the column axis. It is apparent that the value of K_g computed from Eq. 3-39 for this case would be less than for the situation shown in Fig. 3-12, because the water would not be free to take the path of least resistance (see problem 20).

3.3 FLOW IN CONFINED AQUIFERS

The differential equations of ground water are developed by combining the Darcy equation with the principle of mass balance. Mass balance involves consideration of inflow, outflow, and changes in ground water storage. The difference between the treatment here and that in Chapter II is that, now, the balance is applied to very small control volumes and a flux rate equation (Darcy's Law) is introduced into the balance. For reasons that will become apparent subsequently, the differential equations for confined and unconfined flow are derived separately.

A Differential Mass Balance

The control volume for which the mass balance is to be written, can be any shape. The control volume shown in Fig. 3-13 simplifies the analysis, however. The mass discharge across any plane of the control volume is the product of the water density ρ and the volume discharge Q. The difference between the outflow and inflow mass discharges across the two planes normal to the x-axis is

$$\text{Outflow rate} - \text{Inflow rate} = (\rho Q)_{\Delta x} - (\rho Q)_{ox} \quad . \quad (3\text{-}45)$$

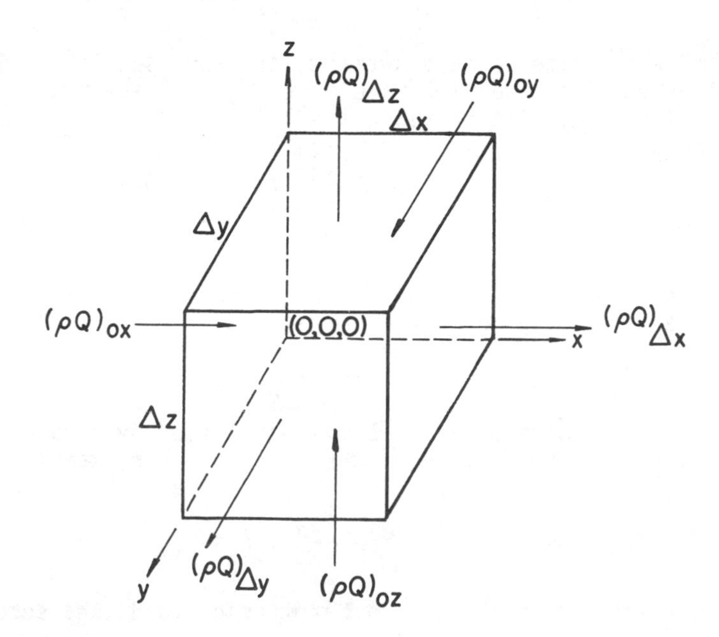

Figure 3-13. Control volume for mass balance calculation.

A similar expression can be written for the y and z directions. Provided the dimensions of the control volume are small, the discharge across the plane at $x=\Delta x$ can be related to that across the plane at $x=0$ by expansion in a Taylor series:

$$(\rho Q)_{\Delta x} = (\rho Q)_{ox} + \frac{\partial}{\partial x}(\rho Q)\Delta x + O(\Delta x^2 \ldots) \quad , \quad (3\text{-}46)$$

in which the last term on the right represents terms of order Δx^2, Δx^3 etc. For small Δx, these terms can be neglected and

Eq. 3-45 becomes

$$\text{Outflow rate} - \text{Inflow rate} = \frac{\partial}{\partial x}(\rho Q)\Delta x \quad . \qquad (3\text{-}47)$$

The volume discharge Q is the product of the Darcy velocity and the area normal to flow,

$$Q_x = q_x \Delta y \Delta z, \; Q_y = q_y \Delta x \Delta z, \; \text{and} \; Q_z = q_z \Delta x \Delta y \quad . (3\text{-}48)$$

Considering all planes on the control volume, the total outflow rate minus the total inflow rate is

$$\text{Outflow rate} - \text{Inflow rate} = \left[\frac{\partial}{\partial x}(\rho q_x) + \frac{\partial}{\partial y}(\rho q_y) + \frac{\partial}{\partial z}(\rho q_z) \right] \Delta x \Delta y \Delta z \quad . \qquad (3\text{-}49)$$

The net rate of mass outflow, as expressed in Eq. 3-49, must be equal to the time-rate of change of mass M within the control volume. Therefore,

$$\left[\frac{\partial}{\partial x}(\rho q_x) + \frac{\partial}{\partial y}(\rho q_y) + \frac{\partial}{\partial z}(\rho q_z) \right] \Delta x \Delta y \Delta z = -\frac{\partial M}{\partial t} \quad . \qquad (3\text{-}50)$$

The negative sign is included to make the net outflow positive when mass is being depleted from storage.

Differential Equations For Confined Flow

The left side of Eq. 3-50 is simplified by expanding the indicated differentiation of the product. For example,

$$\frac{\partial(\rho q_x)}{\partial x} = \rho \frac{\partial q_x}{\partial x} + q_x \frac{\partial \rho}{\partial x} \quad . \qquad (3\text{-}51)$$

Because water is considered a barotropic fluid, the second term on the right can be written

$$q_x \frac{\partial \rho}{\partial x} = q_x \frac{d\rho}{dp} \frac{\partial p}{\partial x} = q_x \, \rho \beta \frac{\partial p}{\partial x} \quad , \qquad (3\text{-}52)$$

the second equality following from Eq. 2-14. The expansion in Eq. 3-51 becomes

$$\frac{\partial}{\partial x}(\rho q_x) = \rho \frac{\partial q_x}{\partial x} + q_x \, \rho \beta \frac{\partial p}{\partial x} \approx \rho \frac{\partial q_x}{\partial x} \quad . \qquad (3\text{-}53)$$

The term which involves the compressibility of water is dropped because it is usually small compared to the one retained. The terms for the coordinates y and z in Eq. 3-50 can be similarly expanded and simplified, so that the mass balance equation is

$$\rho\left(\frac{\partial q_x}{\partial x} + \frac{\partial q_y}{\partial y} + \frac{\partial q_z}{\partial z} \right)\Delta x \Delta y \Delta z = - \frac{\partial M}{\partial t} \quad . \qquad (3\text{-}54)$$

The next step is to introduce the Darcy equation for flow in a non-homogeneous, anisotropic aquifer in which the principal directions of hydraulic conductivity coincide with the coordinate directions x, y, and z. Substitution from the Darcy equation into Eq. 3-54 and division by $\rho \Delta x \Delta y \Delta z$ gives

$$\frac{\partial}{\partial x}\left(K_x \frac{\partial h}{\partial x}\right) + \frac{\partial}{\partial y}\left(K_y \frac{\partial h}{\partial y}\right) + \frac{\partial}{\partial z}\left(K_z \frac{\partial h}{\partial z}\right) = \frac{1}{\rho \Delta x \Delta y \Delta z} \frac{\partial M}{\partial t} \quad . \qquad (3\text{-}55)$$

The change of water mass per unit volume was calculated in Chapter II and is given by Eq. 2-16. Therefore, the mass balance can be written as

$$\frac{\partial}{\partial x}\left(K_x \frac{\partial h}{\partial x}\right) + \frac{\partial}{\partial y}\left(K_y \frac{\partial h}{\partial y}\right) + \frac{\partial}{\partial z}\left(K_z \frac{\partial h}{\partial z}\right) = \phi(\alpha_p + \beta) \frac{\partial p}{\partial t} \quad . \qquad (3\text{-}56)$$

Finally, since $dp \approx \rho g \, dh$ (see Eqs. 2-18 and 3-16), and in view of Eq. 2-19,

$$\frac{\partial}{\partial x}\left(K_x \frac{\partial h}{\partial x}\right) + \frac{\partial}{\partial y}\left(K_y \frac{\partial h}{\partial y}\right) + \frac{\partial}{\partial z}\left(K_z \frac{\partial h}{\partial z}\right) = S_s \frac{\partial h}{\partial t} \quad . \qquad (3\text{-}57)$$

Equation 3-57 is a linear partial differential equation, solutions of which represent time and space distributions of piezometric head in non-homogeneous, anisotropic, confined aquifers. The coordinate system is not arbitrary in Eq. 3-57, but must be selected so that the coordinate axes are co-linear with the principal directions of hydraulic conductivity. Less general, but more tractable, forms apply for aquifers that meet additional restrictions with respect to hydraulic conductivity and geometry. For example,

$$K_x \frac{\partial^2 h}{\partial x^2} + K_y \frac{\partial^2 h}{\partial y^2} + K_z \frac{\partial^2 h}{\partial z^2} = S_s \frac{\partial h}{\partial t} \qquad (3\text{-}58)$$

applies when the aquifer is homogeneous but anisotropic. Further simplification is possible if $K_x=K_y=K_z=K$;

$$\frac{\partial^2 h}{\partial x^2} + \frac{\partial^2 h}{\partial y^2} + \frac{\partial^2 h}{\partial z^2} = \frac{S_s}{K}\frac{\partial h}{\partial t} \quad , \tag{3-59}$$

a form applicable for homogeneous and isotropic aquifers. In this case, the selection of the orientation of the coordinate axis is arbitrary. Equation 3-59 is a linear, parabolic, partial-differential equation that occurs in several branches of the physical and engineering sciences. Literature dealing with the conduction of heat in solids is a particularly prolific source for solutions to Eq. 3-59.

In many applications, the ground-water hydrologist can assume the aquifer to be of constant thickness b and the flow to be horizontal (in the x-y plane). Recalling that the product of the specific storage and thickness is the storage coefficient (Eq. 2-20), the differential equation becomes

$$\frac{\partial^2 h}{\partial x^2} + \frac{\partial^2 h}{\partial y^2} = \frac{S}{bK}\frac{\partial h}{\partial t} \quad . \tag{3-60}$$

The product bK is called *transmissivity*, T, and has the dimensions of L^2T^{-1}. The piezometric head does not vary with elevation in flows described by Eq. 3-60, and both piezometers and observation wells indicate the elevation of the piezometric surface. Piezometers must be used to indicate the piezometric head in flows with significant vertical components of velocity, and there does not exist a single piezometric surface that is indicative of the flow conditions in the aquifer as a whole.

Drawdown, s, is sometimes used as the dependent variable in the differential equations for ground water flow. The drawdown is

$$s = h_o - h \quad , \tag{3-61}$$

where h_o is a convenient reference value of piezometric head, usually taken as the initial or static value. Since $ds = -dh$, Eq. 3-60 becomes

$$\frac{\partial^2 s}{\partial x^2} + \frac{\partial^2 s}{\partial y^2} = \frac{S}{T}\frac{\partial s}{\partial t} \quad . \tag{3-62}$$

Ground-water flows sometimes exist for which the replenishment rate is equal to the outflow rate, at least approximately. No change in storage occurs under these conditions and piezometric heads do not change with time. Hence,

$$\frac{\partial^2 h}{\partial x^2} + \frac{\partial^2 h}{\partial y^2} + \frac{\partial^2 h}{\partial z^2} = 0 \quad , \qquad (3\text{-}63)$$

known as the *Laplace* equation, describes steady flow in confined aquifers. The Laplace equation has been the subject of study by mathematicians and physical scientists for many decades and powerful, elegant methods exist for its solution, particularly the two-dimensional form.

3.4 FLOW IN WATER-TABLE AQUIFERS

Recall from Chapter II that water is derived from storage in unconfined aquifers by drainage of the pores, by water expansion and by rock compaction; the contribution from the latter two processes usually being negligible relative to the first. Consider an aquifer in which the water table is initially horizontal and from which water is removed by a well. The pumped well causes the water table to draw down, forming a cone of depression as shown in Fig. 2-11. Most of the water supplied to the well is derived by the dewatering of the aquifer within the cone of depression (see Example 2-5). The water derived from water and aquifer expansion in the portion of the aquifer below the water table is negligible compared to that derived from the cone of depression. Thus, the change in aquifer storage can be adequately accounted for by determining the change in the volume of the cone of depression and multiplying by the apparent specific yield.

In principle, the location of the water table in time and space can be computed by solving Eq. 3-57 with $S_s=0$ which is

$$\frac{\partial}{\partial x}\left(K_x \frac{\partial h}{\partial x}\right) + \frac{\partial}{\partial y}\left(K_y \frac{\partial h}{\partial y}\right) + \frac{\partial}{\partial z}\left(K_z \frac{\partial h}{\partial z}\right) = 0 \quad . (3\text{-}64)$$

This equation reduces to the Laplace equation for homogeneous and isotropic aquifers. Unlike the case for confined flow, however, the right side of Eq. 3-64 is zero because $S_s \approx 0$ and not because the flow is steady. Furthermore, the flow domain for which solutions of Eq. 3-64 (or the Laplace equation) are sought is not constant because the water-table position changes with time. It is, in fact, the change of water-table position with time which accounts for the change in storage in the aquifer, as explained above.

A mathematically rigorous approach to flow in water-table aquifers is to solve Eq. 3-64 (or an appropriate less general form) for flow in the saturated zone. An outcome of the solution is the time and space distribution of the water table, from which the change in aquifer storage can be computed by the procedures of Example 2-5. The fact that the water-table position is an outcome of the solution, yet the water-table position is required (a priori) to define the flow domain in which Eq. 3-64 applies, makes it very difficult to obtain the required solutions. Even for steady flows, when the water table is fixed in time, calculation of its configuration via the above approach is not a simple task.

The Dupuit-Forchheimer Approximations

The difficulties attending the solution of Eq. 3-64, or appropriate less general forms, in water-table aquifers have led hydrologists to use a more practical, if less rigorous, approach. Consider a sloping water table above a horizontal impermeable boundary as shown in Fig. 3-14. The slope has been greatly exaggerated for clarity. The discharge, per unit width into the plane of the paper, across any vertical plane is

$$Q = \int_0^{z_f} q_x(x,z)\, dz \quad . \tag{3-65}$$

Evaluation of the integral in Eq. 3-65 requires that $q_x(x,z)$ be known. However, provided that the slope δ of the water table is small, q_x at the water table does not differ significantly from that on the impermeable boundary and $q_x(x,z) \approx q_x(x,z_f)$.

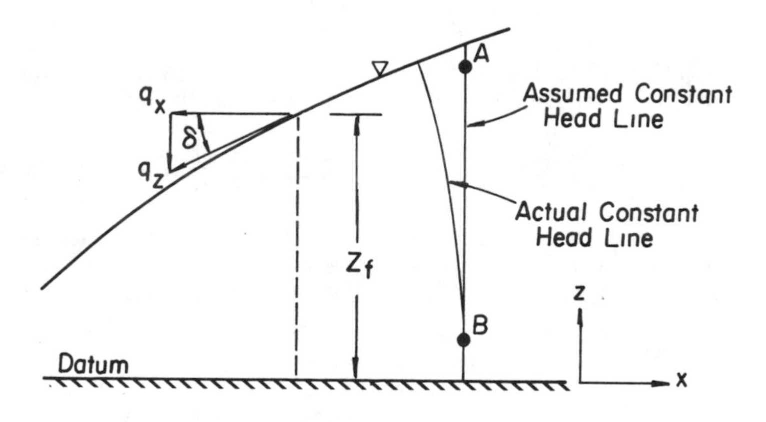

Figure 3-14. Flow in a water table aquifer.

In this case

$$Q = q_x(x,z_f)\, z_f = -K \frac{dh}{dx} z_f \quad , \tag{3-66}$$

where h is the piezometric head at the water table. By definition of a water table, the pressure head must be zero there, so $h=z_f$ and

$$Q = -Kh \frac{dh}{dx} \quad . \qquad (3\text{-}67)$$

In Eq. 3-67, h represents both the thickness of the flow and the piezometric head at the water table. The quantity dh/dx is the tangent of the angle the water table makes with the horizontal. Equation 3-67 actually implies that the flow is entirely horizontal, and that the pressure-head distribution along any vertical is hydrostatic. In other words, the piezometric head along any vertical is constant. It is emphasized that Eq. 3-67 is valid for situations in which the water-table slope is small. More explicitly

$$(dh/dx)^2 << 1 \qquad (3\text{-}68)$$

is the condition that must be satisfied (Bear, 1972).

The Boussinesq Equation

The Dupuit-Forchhiemer assumption of horizontal flow permits the use of a material-balance control volume that extends from the horizontal floor of the aquifer to the water table (Fig. 3-15). Because changes in water density are unimportant in unconfined aquifers, mass balance is assured by a volume balance. Following the procedures of the previous section, the rate of net outflow from the control volume is

$$\text{Net Outflow Rate} = \frac{\partial Q_x}{\partial x} \Delta x + \frac{\partial Q_y}{\partial y} \Delta y \quad . \qquad (3\text{-}69)$$

From Eq. 3-67, written for both the x and y directions in an isotropic aquifer,

$$\frac{\text{Net Outflow Rate}}{\Delta x \Delta y} = - \frac{\partial}{\partial x} (Kh \frac{\partial h}{\partial x}) - \frac{\partial}{\partial y} (Kh \frac{\partial h}{\partial y}) \quad . (3\text{-}70)$$

As before, the net rate of outflow must equal the negative time rate of reduction of stored water volume. The change in water volume associated with a change, dh, of water-table level follows from the definition of apparent specific yield discussed in Chapter II.

$$\frac{\partial V_w}{\partial t} = S_{ya} \frac{\partial h}{\partial t} \Delta x \Delta y \quad . \qquad (3\text{-}71)$$

Combining Eqs. 3-70 and 3-71 yields

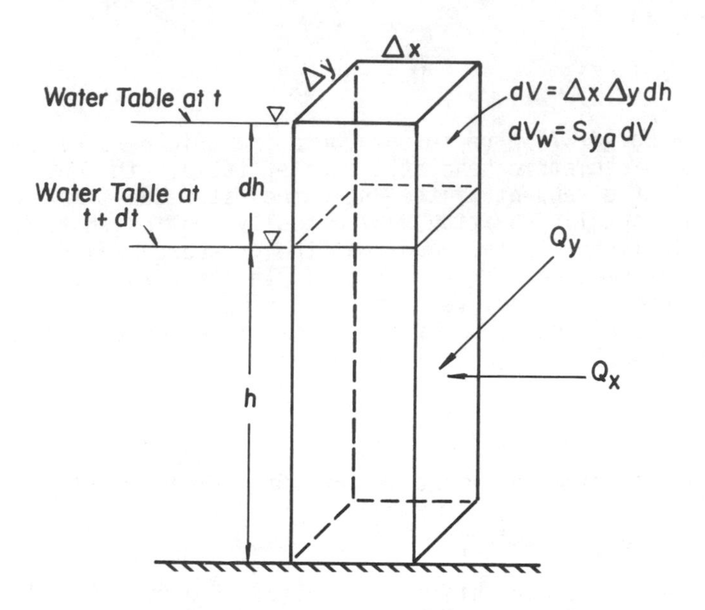

Figure 3-15. Control volume in an unconfined aquifer.

$$\frac{\partial}{\partial x}\left(Kh\frac{\partial h}{\partial x}\right)+\frac{\partial}{\partial y}\left(Kh\frac{\partial h}{\partial y}\right)=S_{ya}\frac{\partial h}{\partial t}\quad , \qquad (3\text{-}72)$$

and, if the aquifer is homogeneous,

$$\frac{\partial}{\partial x}\left(h\frac{\partial h}{\partial x}\right)+\frac{\partial}{\partial y}\left(h\frac{\partial h}{\partial y}\right)=\frac{S_{ya}}{K}\frac{\partial h}{\partial t}\quad . \qquad (3\text{-}73)$$

Equation 3-73 is the *non-linear Boussinesq* equation (Boussinesq, 1904). Because it is non-linear, Eq. 3-73 is difficult to solve by analytical methods, although several exact and approximate analytical solutions have been developed for particular boundary and initial conditions (Boussinesq, 1904; van Schilfgaarde, 1963; Brooks, 1961; McWhorter and Duke, 1976). Linearization of the Boussinesq equation is permissible when the spatial variation of h remains small relative to h. In this case it is possible to replace the variable saturated flow depth with an average thickness, b, and obtain

$$\frac{\partial^2 h}{\partial x^2}+\frac{\partial^2 h}{\partial y^2}=\frac{S_{ya}}{bK}\frac{\partial h}{\partial t}\quad , \qquad (3\text{-}74)$$

which is known as the *linearized Boussinesq* equation. Note that Eq. 3-74 is precisely the same form as Eq. 3-60. Other procedures have been used to linearize Eq. 3-73. Werner (1957), for example, puts $h = \sqrt{u}$ and writes the equation with $u = h^2$ as the dependent variable.

3.5 EQUATIONS FOR FLOW WITH VERTICAL ACCRETION

Seepage through underlying or overlying aquitards sometimes significantly affects the hydraulics in confined aquifers. In water-table aquifers, accretion to or depletion from the aquifers occur as the result of recharge or evapotranspiration, for example.

Flow in a confined homogeneous, isotropic aquifer, overlain by an aquitard through which seepage occurs, is considered. One approach is to apply Eq. 3-59 with the seepage through the aquitard taken into consideration as a boundary condition. A more practical approach is to regard the flow in the confined aquifer as entirely horizontal and to account for the seepage through the aquitard directly in the material balance on a control volume that extends from the floor of the aquifer to the aquitard (Fig. 3-16). Following the procedures used previously for establishing the material balance,

$$\frac{\partial Q_x}{\partial x} \Delta x + \frac{\partial Q_y}{\partial y} \Delta y - W\Delta x\Delta y = -S \frac{\partial h}{\partial t} \Delta x\Delta y \quad , \qquad (3\text{-}75)$$

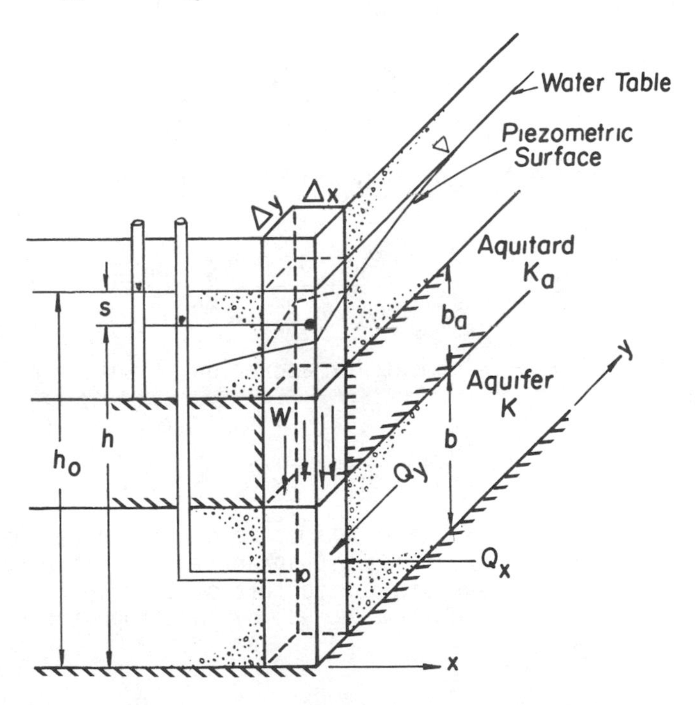

Figure 3-16. Flow in a leaky, confined aquifer.

or

$$\frac{\partial^2 h}{\partial x^2} + \frac{\partial^2 h}{\partial y^2} + \frac{W}{bK} = \frac{S}{bK}\frac{\partial h}{\partial t} \quad , \qquad (3\text{-}76)$$

since

$$Q_x = -\, Kb\, \Delta y\, \frac{\partial h}{\partial x} \quad \text{and} \quad Q_y = -\, Kb\, \Delta x\, \frac{\partial h}{\partial y} \quad . \qquad (3\text{-}77)$$

The vertical percolation rate W is a scalar discharge per unit area assumed to be positive under conditions of accretion. The magnitude of W can be computed directly from Darcy's Law, provided that changes in storage in the aquitard are neglected (see Examples 3-6 and 3-7):

$$W = K_a \frac{(h_o - h)}{b_a} \quad . \qquad (3\text{-}78)$$

Substitution of Eq. 3-78 into Eq. 3-76 gives

$$\frac{\partial^2 h}{\partial x^2} + \frac{\partial^2 h}{\partial y^2} + \frac{K_a}{K}\frac{(h_o - h)}{bb_a} = \frac{S}{bK}\frac{\partial h}{\partial t} \quad . \qquad (3\text{-}79)$$

A *leakage factor* $\mathcal{B}$, defined by

$$\mathcal{B} = \left(\frac{Kbb_a}{K_a} \right)^{\frac{1}{2}} \quad , \qquad (3\text{-}80)$$

is often introduced into Eq. 3-79 to yield the *leaky* aquifer equation

$$\frac{\partial^2 h}{\partial x^2} + \frac{\partial^2 h}{\partial y^2} + \frac{h_o - h}{\mathcal{B}^2} = \frac{S}{bK}\frac{\partial h}{\partial t} \quad . \qquad (3\text{-}81)$$

A similar development for accretion to water-table aquifers leads to

$$\frac{\partial}{\partial x}\left(h\frac{\partial h}{\partial x}\right) + \frac{\partial}{\partial y}\left(h\frac{\partial h}{\partial y}\right) + \frac{W}{K} = \frac{S_{ya}}{K}\frac{\partial h}{\partial t} \quad . \qquad (3\text{-}82)$$

The accretion rate W can sometimes be estimated by the water budget procedure discussed in Chapter II. In other cases, W follows from analyses similar to that of Example 3-7. An example of negative W is water use by phreatophytes.

REFERENCES

Ahmed, N. and Sunada, D. K., 1969. Nonlinear Flow in Porous Media. Journal of the Hydraulics Division, Am. Soc. Civil Engrs., Vol. 95, NO. HY 6, Proc. Paper 6883, pp. 1847-1857.

Bear, J., 1972. Dynamics of Fluids In Porous Media. American Elsevier Publishing Co., Inc., New York. 764 p.

Boussinesq, J., 1904. Recherches Theoriques sur l'e'coulement des Nappes d'eau Infiltrées dan le Sol et sur le Débit des Sources. Journal de Mathematiques Pure et Appl., Vol. 10, pp. 5-78.

Brooks, R. H., 1961. Unsteady Flow of Ground Water into Drain Tile. Journal of Irrig. and Drainage Div., Am. Soc. Civil Engrs., Vol. 87, No. IR 2, Proc. Paper 2836, pp. 27-37.

Collins, R. E., 1961. Flow of Fluids Through Porous Materials. Reinhold Publishing Co., New York.

DeWiest, R. J. M., 1965. Geohydrology. John Wiley and Sons, Inc. New York, 366 p.

Fair, G. M. and Hatch, L. P., 1933. Fundamental Factórs Governing the Streamline Flow of Water Through Sand. Journal of Am. Water Works Assoc., Vol. 25, p. 1551-1565.

Harleman, D. R. F., Mehlhorn, P. F., and Rumer, R. R. Jr., 1963. Dispersion-Permeability Correlation in Porous Media. Journal Hydraulics Div., Am. Soc. Civil Engrs., Vol. 89, No. 2.

Hubbert, M. K., 1940. The Theory of Ground-Water Motion. Journal of Geology, Vol. 48, No. 8, pp. 785-944.

Hubbert, M. K., 1953. Entrapment of Petroleum Under Hydrodynamic Conditions. Bull. Am. Assoc. Petroleum Geol., Vol. 37, No. 8, p. 1954-2026.

Hubbert, M. K., 1956. Darcy's Law and the Field Equations of the Flow of Underground Fluids. Journal of Petroleum Technology, October, Techn. Paper 4352.

Marcus, H. and Evenson, D. E., 1961. Directional Permeability in Anisotropic Porous Media. Water Resources Center, Contribution No. 31, Univ. of California, Berkeley.

McWhorter, D. B. and Duke, H. R., 1976. Transient Drainage with Nonlinearity and Capillarity. Journal of Irrig. and Drainage Div., Am. Soc. Civil Engrs., Vol. 102, No. IR 2, Proc. Paper 12185, pp. 193-204.

Morris, D. A. and Johnson, A. I., 1967. Summary of Hydrologic and Physical Properties of Rock and Soil Materials as Analyzed by the Hydrologic Laboratory of the U. S. Geological Survey - 1948-1960. U. S. Geological Survey Water Supply Paper 1839-D.

Rose, H. E., 1945. An Investigation Into the Laws of Flow of Fluids Through Beds of Granular Material. Proc. Instit. of Mech. Engr., Vol. 153, pp. 141-148.

Rumer, R. R. Jr., 1969. Chapter 3 in Flow Through Porous Media, R. J. M DeWiest, ed. Academic Press, New York and London, 530 p.

Sunada, D. K., 1965. Turbulent Flow Through Porous Media. Water Resources Center Contribution No. 103, Univ. of California, Berkeley.

Todd, D. K., 1959. Ground Water Hydrology. John Wiley and Sons, Inc., New York, 336 p.

van Schilfgaarde, J., 1963. Design of Tile Drainage For Falling Water Tables. Journal of Irrig. and Drainage Div., Am. Soc. Civil Engrs., Proc. Paper 3543, NO. IR 2, pp. 1-11.

Ward, J., 1964. Turbulent Flow in Porous Media. Journal Hydraulics Div., Am. Soc. Civil Engrs., Vol. 90, No. 5, pp. 1-12.

Werner, P. W., 1957. Some Problems in Non-artesian Groundwater Flow. Am. Geophysical Union, Trans., Vol. 38, No. 4, pp. 511-518.

PROBLEMS AND STUDY QUESTIONS

1. Explain what is meant by a representative volume element.

2. A conservative tracer is introduced into the flow entering a column of sand with uniform cross-section and porosity. The concentration of the tracer in the influent is C_o. The length of the column is 120 cm, the cross-section area is 11 cm^2, and the porosity is 0.45. Three hundred seconds after the introduction of the tracer, the concentration of tracer in the effluent was 0.5 C_o. Estimate the magnitudes of v, q, and Q in the column. Answer: v=0.4 cm/s, q=0.18 cm/s, Q=1.98 cm^3/s.

3. If the magnitude of the gradient of piezometric head is 2.6 in problem 2, what is the hydraulic conductivity? Answer: K=0.069 cm/s.

4. What is the fundamental difference between hydraulic conductivity and intrinsic permeability? Discuss some situations in which it might be advantageous to us k rather than K.

5. Show that the dimensions of K are L/T and k are L^2.

6. From the following data compute k, C, and $\bar{d}$ and the magnitude q of $\vec{q}$ that results in a 5% departure from Darcy's Law. Answer: $k=2.2\times10^{-5}$ cm^2, C=1.7, $\bar{d}=3.6\times10^{-3}$ cm. (Hint: Refer to Example 3-2 and compute A and B so that the residual R is minimized in

$$R = \sum_{i=1}^{n} \frac{[|\nabla h|_i - Aq_i - Bq_i^2]^2}{q_i^2} .$$

R is minimized when

$$A = \frac{\sum_{i=1}^{n} \left(\frac{|\nabla h|_i}{q_i}\right) \sum_{i=1}^{n} q_i^2 - \sum_{i=1}^{n} q_i \sum_{i=1}^{n} |\nabla h|_i}{n \sum_{i=1}^{n} q_i^2 - \sum_{i=1}^{n} q_i^2}$$

and

$$B = \frac{\sum_{i=1}^{n} |\nabla h|_i - \sum_{i=1}^{n} \frac{|\nabla h|_i}{q_i} \sum_{i=1}^{n} q_i}{n \sum_{i=1}^{n} q_i^2 - (\sum_{i=1}^{n} q_i)^2} .$$

$\|\nabla h\|$	q cm/s	$\|\nabla h\|$	q cm/s	$\|\nabla h\|$	q cm/s
0.0079	0.0112	0.252	0.337	4.82	3.70
0.0094	0.0137	0.294	0.388	6.07	4.32
0.0118	0.0176	0.339	0.443	7.45	4.94
0.0134	0.0196	0.387	0.499	9.00	5.58
0.0181	0.0264	0.435	0.554	10.64	6.20
0.0236	0.0342	0.484	0.610	12.31	6.79
0.0283	0.0401	0.534	0.665	14.26	7.43
0.0346	0.0489	0.585	0.721	16.27	8.05
0.0394	0.0562	0.637	0.776	18.52	8.70
0.0480	0.0684	0.691	0.832	20.68	9.29
0.0583	0.0826	0.745	0.887	22.99	9.89
0.0693	0.0973	0.799	0.942	25.59	10.53
0.0780	0.110	0.856	0.998	31.29	11.83
0.0866	0.121	0.910	1.05	39.84	13.58
0.1024	0.143	0.973	1.11	49.46	15.34
0.120	0.166	1.12	1.25	59.47	17.00
0.134	0.185	1.43	1.52	67.87	18.29
0.162	0.222	1.85	1.85	80.93	20.14
0.186	0.253	2.70	2.46	96.54	22.18
0.220	0.296	3.73	3.10	111.84	24.02
				124.96	25.50

7. Prove that the magnitude of the Darcy velocity during steady downward flow is equal to the hydraulic conductivity when the pressure is constant along the flow path. Discuss the magnitude and direction of the driving forces for this case.

8. A column of porous medium is oriented at an angle of 30° with the horizontal plane. The magnitude of the Darcy velocity is 0.005 cm/s. If the pressure is the same everywhere in the column, what is the hydraulic conductivity? Answer: K=0.01 cm/s.

9. 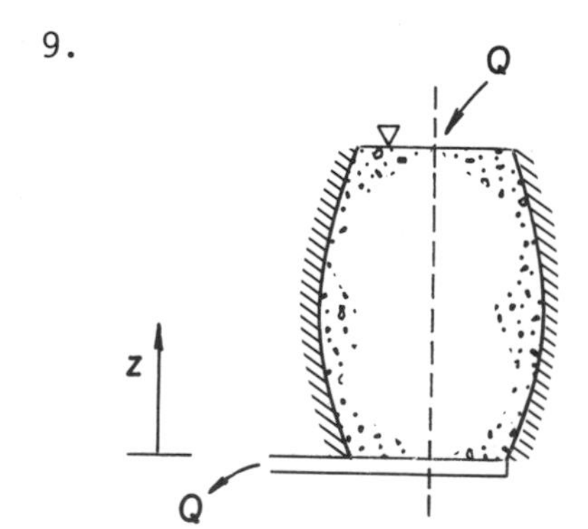

For the figure shown, carefully but schematically plot piezometric head, elevation head, and pressure head as a function of z along the axis. Also plot $\partial h/\partial z$ and $\partial(p/\rho g)/\partial z$. Note that the pressure head is zero on both ends of the sample.

10. An aquifer with K=1.0 cm/s and ϕ=0.25 is discharging into a stream. The flow in the aquifer is practically horizontal and the gradient of head is in a direction that makes a 45° angle with the stream (plan view). The magnitude is 0.01. A tracer is introduced into the aquifer at a point a perpendicular distance of 6 m from the stream. Supposing that dispersion and diffusion are negligible, estimate the time required for the tracer to appear in the stream. Answer: 2.12×10^4 seconds.

11. Derive Eq. 3-24.

12. Two cylindrical reservoirs are connected at the bottom with a sand filled pipe, 3 cm in diameter and 2 m long. The hydraulic conductivity of the sand in the connecting pipe is 9.1×10^{-4} cm/s. The cross-sectional areas of the large and small reservoirs are 1000 cm^2 and 250 cm^2, respectively. If the depth of water in the larger reservoir is 40 cm and 10 cm in the smaller at t=0, at what time will the water level in the larger reservoir be 35 cm? Answer: 128 days.

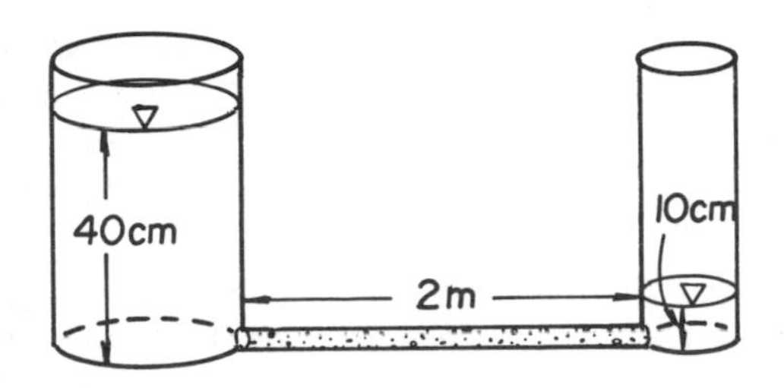

13. The following grain-size distributions were measured for two samples of sand with well rounded particles. The porosities of both samples was 0.31. Calculate the intrinsic permeabilities using the Fair-Hatch formula and using Eq. 3-26. Compare and discuss the results.
Answer: $k_A = 1.0 \times 10^{-9}$ cm^2, $k_B = 3.6 \times 10^{-10}$ cm^2 and from Eq. 3-26: $k_A = 0.65 \times 10^{-9}$ cm^2, $k_B = 1.4 \times 10^{-10}$ cm^2.

Sample A		Sample B	
Particle Size microns	Percent Greater	Particle Size microns	Percent Greater
80	0.0	80	0.0
70	2.5	70	2.5
60	11.0	60	11.0
50	32.0	50	32.0
40	53.0	40	40.0

Sample A		Sample B	
Particle Size microns	Percent Greater	Particle Size microns	Percent Greater
30	69.0	30	59.0
20	79.0	20	70.0
10	90.0	10	75.0
1	100.0	1	100.0

14. Two observation wells, 152 m apart, penetrate an artesian aquifer. The following information is obtained.

	Well A	Well B
Ground Surface Elev.	1579 m	1570 m
Top of Aquifer Elev.	1497 m	1478 m
Bottom of Aquifer Elev.	1466 m	1448 m

The static water level in Well B is 6.10 m below ground surface and in Well A the water pressure head measured at the ground surface is 3.05 m. If the hydraulic conductivity of the aquifer is 0.024 cm/s, calculate the total flow in a unit width of aquifer parallel to a line connecting the two wells. Answer: $8.6\text{x}10^{-4}$ m^3/s-m.

15. Horizontal flow occurs through a column of non-homogeneous porous medium in which the hydraulic conductivity varies continuously according to $K(x)=K_m \exp(-x/L)$ where K_m is the conductivity at x=0 and L is the length of the column. For K_m=0.019 cm/s, L=1.3 m, and a difference in pressure head across the interval L of 3 m, calculate the magnitude of the Darcy velocity. Plot the distribution of pressure head along the column. Answer: $|\vec{q}|=2.55\text{x}10^{-2}$ cm/s.

16. Discuss the relationship between the dimensions of the representative volume element and the concepts of homogeniety and isotropy in media with large scale stratification and in media with small scale stratification.

17. Show that Eq. 3-22 leads to physically absurd results for media in which K=K(x,y,z) and explain why $-\nabla\Phi$ is not a force.

18. An aquifer is homogeneous, but anisotropic with K_z=0.001 cm/s and K_x=K_y=0.01 cm/s where x, y, and z are the principle directions. The component of ∇h in the z-direction is -0.92, in the x-direction is 0.17, and zero in the y-direction. Calculate and plot (to scale) the vectors ∇h and $\vec{q}$. What is the magnitude of $\vec{q}$? What is the direction

and magnitude of $\vec{q}$ if the porous solid is isotropic with K=0.01 cm/s? Answer: Anisotropic case; $\vec{q}=-1.7\times10^{-3}\,\vec{i} + 9.2\times10^{-4}\,\vec{k}$, $|\vec{q}|=1.92\times10^{-3}$ cm/s : isotropic case; $\vec{q}=-1.7\times10^{-3}\,\vec{i} + 9.2\times10^{-3}\,\vec{k}$, $|\vec{q}|=9.4\times10^{-3}$ cm/s.

19. Downward seepage occurs from a shallow pond (30 cm depth) to a water table 30 m below the bottom of the pond as shown below. A conservative tracer is introduced into the flow at point A. Calculate the location of tracer appearance at the water table, assuming no dispersion or diffusion and that both strata remain fully saturated. Also compute the travel time of the tracer between point A and the water table. Discuss the variation of pressure head with elevation and the assumption of fully saturated flow. Answer: Tracer appears at the water table at x=18 m after a travel time of 2.9 days.

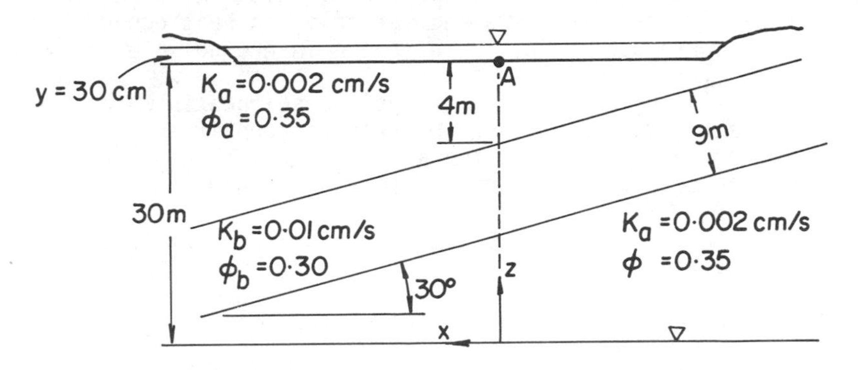

20. The hydraulic conductivity in the direction of the gradient is given by Eq. 3-44 for a permeameter with a large diameter to length ratio. Show that the hydraulic conductivity in the direction of $\vec{q}$ is given by

$$K_q = \frac{K_x K_y}{K_x \sin^2\beta + K_y \cos^2\beta}$$

where β is the angle between the x-coordinate direction and the vector $\vec{q}$ for a permeameter with a small diameter to length ratio (i.e., $\vec{q}$ is parallel to the axis of the permeameter).

21. By letting $x=r\cos\beta$, $y=r\sin\beta$, where $r=|\vec{q}|$ in the result from problem 20, derive the ellipse

$$\frac{1}{K_q} = \frac{(x/r)^2}{K_x} + \frac{(y/r)^2}{K_y}$$

for hydraulic conductivity in the direction of flow.

22. Repeat problem 21 using Eq. 3-44 to derive the ellipse for hydraulic conductivity in the direction of the gradient:

$$K_g = K_x(x/r)^2 + K_y(y/r)^2 \text{ where } r=|\nabla h| \ .$$

23. What are the values of K_q and K_g for the anisotropic situation in problem 18. Answer: K_q=3.3x10^{-3} cm/s, K_g=1.3x10^{-3} cm/s.

24. By reference to a text on heat conduction in solids, identify the parameters and variables in heat conduction that correspond to K, S_s, h, and q in ground-water flow.

25. Identify three branches of physical science in which the Laplace equation arises. Discuss the conditions under which the Lapalce equation arises in relation to those in which it arises in a ground-water hydrology.

26. Show that W in Eq. 3-78 approaches K_a for very thick aquitards.

27. Derive Eq. 3-82 from first principles.

28. Show that the discharge through the lateral surface of a cylinder of radius r and height h is

$$Q = -2\pi rKh \frac{dh}{dr}$$

for radially symmetric unconfined flow.

29. Using the control volume shown below, develop

$$\frac{1}{r}\frac{\partial}{\partial r}\left(r \frac{\partial h}{\partial r}\right) = \frac{\partial^2 h}{\partial r^2} + \frac{1}{r}\frac{\partial h}{\partial r} = \frac{S}{T}\frac{\partial h}{\partial t}$$

for radially symmetric flow in a confined aquifer.

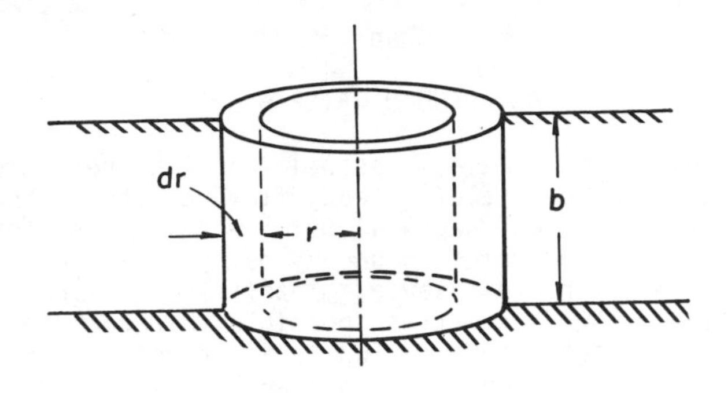

30. Repeat problem 29 for radially symmetric flow in an unconfined aquifer.

31. The data presented in problem 6 was obtained for a uniform sized sand of diameter equal to 0.254 cm and porosity equal to 0.41. Compute the permeability by Eqs. 3-25 and 3-26 and compare your results with the measured permeability given in problem 6.

Chapter IV

STEADY GROUND-WATER HYDRAULICS

A finite flow domain is associated with every ground-water flow problem in the field. Steady flow results in the domain of interest only when the total outflow equals the total inflow, all segments of the domain boundary being taken into consideration. In the strictest sense, steady flow is a rare occurrence in the field. It is sometimes possible, however, to derive a great deal of useful information by treatment of field problems as steady state. In any case, all analyses of ground-water flow are approximate, whether they be based on analytical developments, sophisticated simulation models, laboratory information or field observations because of the limited detail with which the geologic and hydrologic parameters can be characterized. The practicality of steady-state analysis in field problems depends to a significant degree upon the mathematical skill and physical insight possessed by the hydrologist.

The presentation of the material in this chapter begins with a discussion of the methods by which solutions to the two-dimensional Laplace equation are displayed in terms of equipotential and streamlines. The student is encouraged to practice visualizing and displaying the solutions for the subsequent elementary flows in terms of the appropriate network of equipotential and stream lines. The experience gained by so doing will substantially enhance the ability to obtain approximate solutions for more complex problems by sketching flow nets; a method that is introduced in section 4.4. The remainder of the chapter is devoted to a discussion of confined and unconfined flow situations which, in principle, are described by solutions of the Laplace equation but are more conveniently treated by writing approximate differential equations.

4.1 EQUIPOTENTIAL CONTOURS AND STREAMLINES

The material balance equations in Chapter III become the Laplace equation for steady flow in homogeneous, isotropic aquifers. Three-dimensional flow problems are beyond the scope of this text and attention is limited to two-dimensional flows in the subsequent discussions. The two-dimensional Laplace equation, written in terms of the velocity potential ($\Phi \equiv Kh$) is

$$\frac{\partial^2 \Phi}{\partial x^2} + \frac{\partial^2 \Phi}{\partial y^2} = 0 \quad . \qquad (4\text{-}1)$$

Recall that the use of velocity potential is permissible for flow in homogeneous aquifers only. Functions $\Phi(x,y)$ that satisfy Eq. 4-1 are called *harmonic* functions (Churchill, 1960).

Curves in the x-y plane for which $\Phi(x,y)$ = constant are called *equipotential* or equihead contours. An equipotential contour is the locus of points at which the piezometric head is a particular constant value. Equipotential contours are analogous to curves connecting points of equal elevation on a topographic map. Equipotential contours in the horizontal plane for water table aquifers actually represent the topographic configuration of the water-table surface just as a topographic map depicts the configuration of the ground surface. Equipotential contours in the horizontal plane for confined aquifers represent the configuration of the piezometric surface. Equation 4-1 can be also written for vertical (x-z or y-z) planes, in which case the equipotential contours are traced in vertical planes. The component of Darcy velocity in a homogeneous and isotropic aquifer in any direction ℓ can be computed by differentiating Φ in the direction ℓ:

$$q_\ell = - \frac{\partial \Phi}{\partial \ell} \quad . \tag{4-2}$$

Suppose the direction ℓ is oriented parallel to an equipotential contour at a particular point, P. By definition, the function $\Phi(x,y)$ is a constant on the contour. Thus, it must be concluded that the Darcy velocity $\vec{q}$ at P has no component in directions tangent to the equipotential contour. The fact that the Darcy velocity $\vec{q}$ is normal to the equipotential lines has great utility in visualizing flow patterns and in the solution of problems by flow nets.

It is often convenient to describe flow patterns by a family of curves that are everywhere tangent (rather than normal) to the velocity $\vec{q}$. Functions $\psi(x,y)$ that are everywhere tangent to $\vec{q}$ are called *stream functions*. The locus of points for which $\psi(x,y)$ = constant is called a *streamline*; different constants representing different streamlines. The relationship between the stream function and the velocity components can be deduced by considering the streamline shown in Fig. 4-1. The velocity $\vec{q}$ at point P has components q_x and q_y as shown. The slope of the streamline is $(dy/dx)_\psi$ which is equal to q_y/q_x. Thus,

$$\frac{q_y}{q_x} = \left(\frac{dy}{dx} \right)_\psi \quad ,$$

or (4-3)

$$q_y dx - q_x dy = 0 \quad .$$

For $\psi(x,y) = \psi_1$, where ψ_1 is the constant representing the particular streamline through P, we have

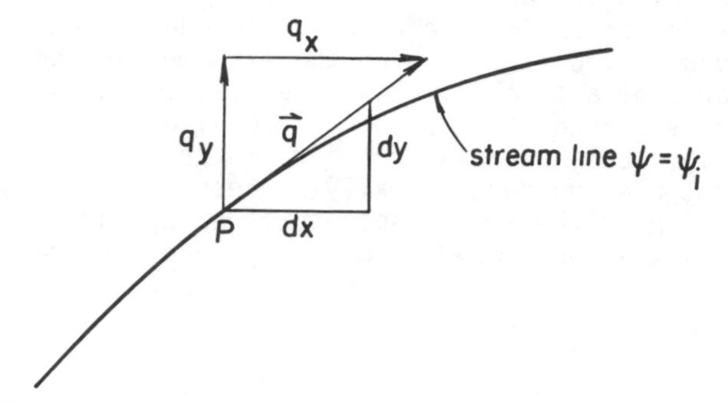

Figure 4-1. Relationship between velocity vector and stream line.

$$d\psi_1 = \frac{\partial \psi}{\partial x} dx + \frac{\partial \psi}{\partial y} dy = 0 \quad . \qquad (4\text{-}4)$$

The coefficients of dx and dy in Eqs. 4-3 and 4-4 must be equal since dx and dy are the same in both equations. Therefore,

$$q_x = - \frac{\partial \psi}{\partial y} \quad , \qquad (4\text{-}5)$$

and

$$q_y = \frac{\partial \psi}{\partial x} \quad . \qquad (4\text{-}6)$$

It can now be easily proven that streamlines and equipotential lines intersect at right angles. At point P on an equipotential line $\Phi(x,y) = \Phi_1$,

$$d\Phi_1 = \frac{\partial \Phi}{\partial x} dx + \frac{\partial \Phi}{\partial y} dy = 0 \quad , \qquad (4\text{-}7)$$

so that

$$\left(\frac{dy}{dx} \right)_\Phi = - \frac{\partial \Phi / \partial x}{\partial \Phi / \partial y} = - \frac{q_x}{q_y} \qquad (4\text{-}8)$$

where $(dy/dx)_\Phi$ is the slope of the equipotential line. Notice that Eqs. 4-3 and 4-8 show that the slopes of the streamline and the equipotential line are negative reciprocals of each other. This is the condition for orthogonality. It is left to the student to show that streamlines and contours of equal piezometric head are orthogonal at points of intersection in nonhomogeneous aquifers. The anisotropic case is discussed in Section 4.4.

It is not difficult to show, using Eqs. 4-5 and 4-6 and their counterparts in terms of Φ, that the ψ functions are also harmonic and, therefore, flow problems can be solved directly for the stream function. It is important to note that the functions ψ have the dimensions of L^2/T or L^3/LT; the numerical difference between any two adjacent streamlines $\Delta\psi = \psi_2 - \psi_1$ being equal to the discharge Q between ψ_2 and ψ_1 per unit of length normal to the x-y plane. Because flow does not cross streamlines, the discharge Q between adjacent streamlines is a constant throughout the flow domain.

Equipotential And Stream Surfaces In Relation To Physical Boundaries

The piezometric head on the wetted surface beneath bodies of static surface water is constant. Such surfaces are known as equipotential (constant head) boundary segments and correspond to mathematical surfaces of constant Φ. If the water levels in the surface water bodies change with time, the piezometric head on segments of aquifer-surface water intersection is regarded as being everywhere the same at each instant but changing with time. In practice, aquifer surfaces exposed beneath the water surface in streams and canals of constant stage are often treated as equipotential surfaces by regarding the slightly sloping water surface as being horizontal. The surface of an aquifer exposed in a well bore below the water level is another example of an equipotential boundary segment. The head loss along the axis of a pumped well is usually so small that the piezometric head below the water level in the well is practically constant.

Shales, clays, basement rock, and many man-made structures with impervious surfaces (concrete abuttments, sheet piling, grouted areas, etc.) often form boundary segments of aquifers. These boundary segments are regarded as stream surfaces since the only velocity component of significant magnitude must be directed tangent to the surface. The traces of the intersection of impermeable surfaces with the plane in which the flow is depicted are streamlines.

EXAMPLE 4-1

Plot several streamlines and equipotential lines in the upper right quadrant for a flow whose stream function is $\psi=xy$.

Solution:

The streamlines are calculated from $xy=\psi_i$ where the ψ_i represent individual streamlines. From $\psi_i=0$, it is seen that the entire x and y axes are streamlines. Putting $\psi_i=1=xy$ yields the curve labeled with $\psi=1$ in Fig. 4-2, and so on for other values of ψ_i.

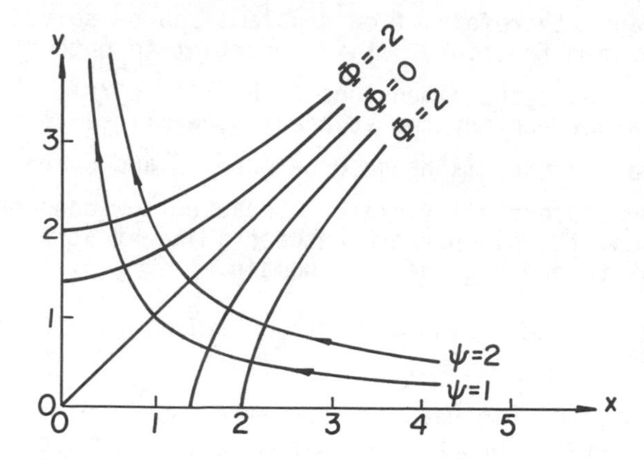

Figure 4-2. Streamlines and equipotential lines for the flow with ψ=xy.

The equipotential function is derived from the ψ-function by noting that

$$q_x = -\frac{\partial\Phi}{\partial x} \quad , \quad q_y = -\frac{\partial\Phi}{\partial y}$$

and, therefore, from Eqs. 4-5 and 4-6

$$\frac{\partial\Phi}{\partial x} = x \quad , \quad \frac{\partial\Phi}{\partial y} = -y \quad .$$

Integration gives

$$\Phi = \frac{x^2}{2} + f(y) \quad , \quad \Phi = \frac{-y^2}{2} + g(x) \quad .$$

Differentiating the above with respect to y results in

$$\frac{\partial\Phi}{\partial y} = \frac{df}{dy} = -y \quad .$$

Therefore, $f = -y^2/2$ + constant and

$$\Phi = \frac{1}{2}\,(x^2-y^2) + \text{constant}$$

is the equation for the velocity potential. The constant is arbitrary and can be set equal to zero. Curves representing different values of Φ are plotted in Fig. 4-2. It is readily verified that both ψ and Φ satisfy the Laplace equation.

4.2 ELEMENTARY SOLUTIONS FOR CONFINED FLOW

One-Dimensional Flow

The Laplace equation for one-dimensional flow in the x-direction is

$$\frac{d^2\Phi}{dx^2} = 0 \quad . \qquad (4\text{-}9)$$

Integration yields

$$\frac{d\Phi}{dx} = \text{constant} \quad , \qquad (4\text{-}10)$$

or

$$\frac{d\Phi}{dx} = -q_x \qquad (4\text{-}11)$$

from the definiton of velocity potential. A second integration results in

$$\Phi = -q_x x + \text{constant} \quad . \qquad (4\text{-}12)$$

Equation 4-12 is the velocity-potential function for one-dimensional uniform flow. The flow is shown schematically in Fig. 4-3.

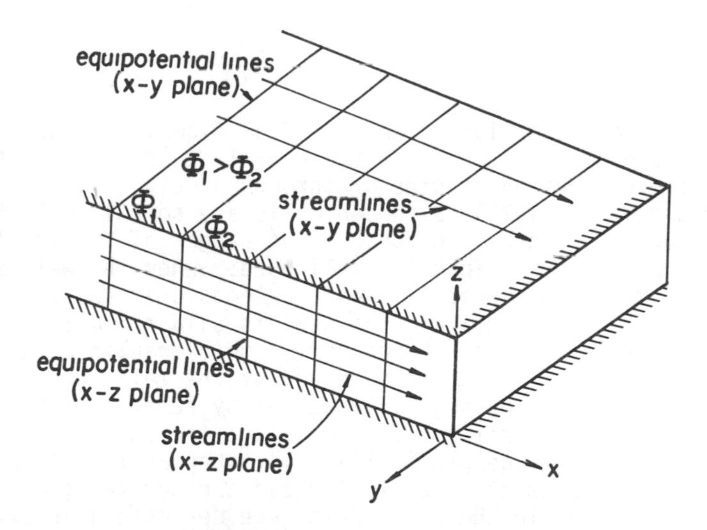

Figure 4-3. One-dimensional, uniform flow in a confined aquifer.

It is obvious that the streamlines are a family of horizontal lines in the x-z plane. This may be proven, however, by integrating Eqs. 4-6 and 4-7:

$$\psi = -\int q_x dz + f(x) = \int q_z dx + g(z) \quad , \qquad (4\text{-}13)$$

where f and g are functions that must be determined. Since it is known that q_z is zero, then

$$\psi = g(z) \qquad (4\text{-}14)$$

from Eq. 4-13. Therefore, f(x) is no more than a constant that can be selected as zero and

$$\psi = -\int q_x dz + f(x) = -q_x z \quad . \qquad (4\text{-}15)$$

The streamlines corresponding to $\psi=\psi_i$ are horizontal lines, parallel to the x-axis. It is clear that the streamline $\psi=0$ is the lower boundary of the aquifer, and any other value, ψ_i at z_i, represents the negative discharge rate between the streamlines $\psi=\psi_i$ and $\psi=0$, per unit of aquifer width measured along y.

It is equally pertinent to write Eqs. 4-13 through 4-15 for the x-y plane instead of the x-z plane. In this case the streamlines are again parallel to the x-axis, but in the x-y plane. The difference between any two values of the stream function is the discharge rate per unit of aquifer thickness measured along the z-axis.

Radial Flow

A common situation in which radial flow is encountered is flow toward a pumped well. Water usually enters a well through a perforated section of pipe or through a well screen (Ground Water and Wells, 1972), although wells are sometimes left uncased and unscreened when there is no danger of the aquifer collapsing or caving into the well bore. When the well screen, perforated pipe, or open well bore extends over the entire thickness b of the aquifer, the well is said to *fully penetrate* the aquifer (Fig. 4-4). A zone of large hydraulic conductivity, immediately adjacent to the exterior of the well screen, is sometimes provided by a gravel pack or by removal of the fines from the natural aquifer. In such cases the exterior radius of this zone of large permeability is taken as the effective well radius. Loss of piezometric head through the gravel pack and the openings in the well screen are designated as *well losses*. In a properly constructed well, the well losses are small, and in the following treatment they are assumed to be

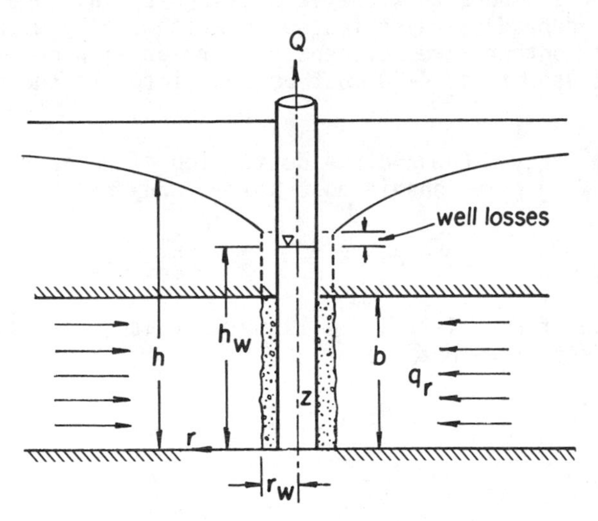

Figure 4-4. Radial flow toward a pumped well.

zero so that h_w is the piezometric head at $r=r_w$.

Flow toward a fully penetrating well in a homogeneous and isotropic aquifer is radially symmetric. In other words, the distributions of Φ on every vertical plane passing through the well axis are identical. The Laplace equation in the radial coordinate r is

$$\frac{d^2\Phi}{dr^2} + \frac{1}{r}\frac{d\Phi}{dr} = 0 \quad . \tag{4-16}$$

Putting

$$\Phi' = \frac{d\Phi}{dr} \tag{4-17}$$

and substituting into Eq. 4-16 gives

$$\frac{d\Phi'}{dr} + \frac{\Phi'}{r} = 0 \tag{4-18}$$

which is integrated to obtain

$$r\Phi' = C_1 = \text{constant} \quad . \tag{4-19}$$

Reintroducing the definition of Φ' and integrating a second time yields

$$\Phi = C_1 \ln r + C_2 \quad . \tag{4-20}$$

Equation 4-20 shows that the velocity potential (and therefore piezometric head) is distributed logarithmically with r. Equipotential contours are, evidently, circles with centers at r=0. Note that use of Eq. 4-20 must be restricted to the region $r \geq r_w > 0$.

From Darcy's Law and the definition of velocity potential, the following relationship must hold at any r:

$$Q = 2\pi rbK \frac{dh}{dr} = 2\pi rb \frac{d\Phi}{dr} \quad . \tag{4-21}$$

Comparison of Eqs. 4-21 and 4-19 shows that Q is a constant (independent of r) because

$$r \frac{d\Phi}{dr} = C_1 = \frac{Q}{2\pi b} \quad , \tag{4-22}$$

and Eq. 4-20 becomes

$$\Phi = \frac{Q}{2\pi b} \ln r + C_2 \quad . \tag{4-23}$$

In the polar coordinates r and δ, the velocity potential and the stream function are related by

$$q_r = - \frac{\partial \Phi}{\partial r} = - \frac{1}{r} \frac{\partial \psi}{\partial \delta}$$

and

$$q_\delta = - \frac{1}{r} \frac{\partial \Phi}{\partial \delta} = \frac{\partial \psi}{\partial r} \quad , \tag{4-24}$$

from which

$$\psi = \int r \frac{d\Phi}{dr} d\delta + \text{constant} \quad , \tag{4-25}$$

for the case at hand in which Φ depends only on r. Substituting into Eq. 4-25 from Eq. 4-22 yields

$$\psi = \frac{Q}{2\pi b} \delta + \text{constant} \tag{4-26}$$

for the stream function. The streamlines are radial lines emanating from r=0. Putting ψ=0 when δ is zero makes the arbitrary constant zero. Notice that when $\delta=2\pi$, $\psi_{2\pi}-\psi_o=Q/b$, the discharge per unit thickness of aquifer.

Writing Eq. 4-23 in terms of piezometric head results in

$$h = \frac{Q}{2\pi T} \ln r + \text{constant} \quad . \tag{4-27}$$

The constant of integration can be evaluated from the fact that $h=h_w$ at $r=r_w$ to obtain

$$h = \frac{Q}{2\pi T} \ln r/r_w + h_w \quad . \tag{4-28}$$

Equation 4-28 predicts that the piezometric head increases without bound as r becomes very large, a condition that is unacceptable on physical grounds. Strictly speaking, Eq. 4-28 is valid only for steady flow in an aquifer with the geometry of a cylinder of height b and finite radius r_e; a condition that is essentially non-existent in the field. However, when an isolated well is pumped for a sufficiently long period of time in a large aquifer, a psuedo-steady state condition is developed in a vicinity of the well and Eq. 4-28 yields reliable results. Exterior to the region of psuedo-steady state, the piezometric head increases less rapidly than predicted by Eq. 4-28 as shown in Fig. 2-11. Actually, the most practical use of the foregoing theory is made when heads h_1 and h_2 are measured at r_1 and r_2 in the region of psuedo-steady state and Eq. 4-28 is put in the form

$$h_2 - h_1 = \frac{Q}{2\pi T} \ln r_2/r_1 \quad . \tag{4-29}$$

The region around a pumped well in which a psuedo-steady state is developed expands with time and the region in which Eqs. 4-28 and 4-29 apply becomes increasingly larger. A detailed accounting of the time and space restrictions imposed on the use of the foregoing theory in a large aquifer with no recharge is given in Chapter V when transient flow toward a pumped well is discussed.

A few concluding remarks concerning the material in this subsection are in order. Radial flow was presented in the context of flow toward a pumped well because of the importance of the pumped well in ground-water hydrology and because of the relative ease with which radial flow can be visualized for this case. In a more theoretical context, the velocity potential and stream function for radial flow in a plane could have been presented as solutions to the Laplace equation for the case of a mathematical sink (or source) as is often done in hydrodynamics (Shames, 1962; Vallentine, 1959), the quantity Q/b being the sink (source) strength. In particular, the reader should note that by writing -Q in place of Q the analyses of this subsection apply for a *recharge well* through which water is added to the aquifer at rate Q. The results of this subsection also apply without modification for flow in an infinite vertical plane toward a drain tube with horizontal axis oriented normal to the vertical plane.

EXAMPLE 4-2

The data of Example 2-5 were collected for a water table aquifer. Assume, however, that the drawdown is everywhere small relative to the initial saturated thickness, making it permissible to use the equations for confined flow to analyze the flow in this water table aquifer. Using the water level and discharge data for Example 2-5, calculate the transmissivity T. Assume a psuedo-steady state.

Solution:

A plot of the water level (relative to an arbitrary datum) in the observation wells as a function of radial distance from the well is shown in Fig. 2-11. Indeed, the piezometric head apparently increases linearly with log r as predicted by the above theory for $r<20$ m. Using the data for observation wells HF1 and HF2 in Eq. 4-29, the transmissivity is

$$T = \frac{Q}{2\pi(h_2-h_1)} \ln r_2/r_1 = \frac{(0.0312 \text{ m}^3/\text{s})}{(2\pi)(95.58-95.14)} \ln \left(\frac{10.4}{4.6} \right)$$

$$= 0.0092 \text{ m}^2/\text{s} \quad .$$

Average T can be computed directly from the slope of the straight line through the data in Fig. 2-11 as follows. Notice that the ratio

$$\frac{\ln r_2/r_1}{h_2-h_1} = \frac{2.303 \log r_2/r_1}{h_2-h_1} = \frac{2\pi T}{Q} = \text{constant} \quad .$$

Thus,

$$T = \frac{2.303\ Q}{2\pi\Delta h}$$

where Δh is the difference in piezometric head per log cycle. From Fig. 2-11,

$$\Delta h = 1.22 \text{ m/log cycle}$$

and

$$T = \frac{(2.303)(0.0312)}{(2\pi)(1.22\text{m})} = 0.0094 \text{ m}^2/\text{s} \quad .$$

4.3 SUPERPOSITION OF ELEMENTARY SOLUTIONS

Superposition is the method in which linear combinations of elementary solutions are formed to provide additional solutions. The method is valid for linear, homogeneous, partial differential equations and greatly expands the range of problems for which analytical solutions can be obtained. Superposition of solutions in the time domain is discussed in Chapter V. Here, superposition in space is used to obtain solutions for several problems of interest in ground-water hydrology.

Drawdown In A Confined Aquifer Due To A Well Field

Ground water is commonly extracted from a group of wells located in a relatively small area. The drawdown at any particular point in the well field is influenced by all of the pumping wells in the field and can be calculated by adding the drawdowns produced by each individual pumped well. The drawdown at any point due to the ith well is

$$s_i = \frac{Q_i}{2\pi T} \ln r_e/r_i \quad , \quad r_i<r_e \tag{4-30}$$

where s_i is the drawdown at the point of interest, r_e is the distance from the ith well to a point where the drawdown can be considered negligible for the particular problem under consideration, and r_i is the distance from the ith well to the point of interest. The total drawdown at the point of interest produced by n wells is given by

$$s = \sum_{i=1}^{n} \frac{Q_i}{2\pi T} \ln r_e/r_i \quad , \quad r_i<r_e \tag{4-31}$$

which is simply the sum of the drawdowns from each of the wells.

When a well field consists of a large number of wells, computation of the drawdown by Eq. 4-31 becomes tedious. If the drawdown in the vicinity of individual wells is not important, it may be sufficiently accurate to conceptually replace the n individual withdrawal points by a uniformly distributed withdrawal rate W (discharge per unit area) which applies to the area of the well field. The uniform withdrawal rate is computed from

$$W = \frac{\sum_{i=1}^{n} Q_i}{A} \quad , \tag{4-32}$$

where A is the area of the well field idealized as a circle of radius R. Thus, the n discrete withdrawal points are replaced by a circular area over which the total withdrawal $\pi r^2 W$ is equal to the total pumping rate from the well field. A plan view and a section view through the center of the well field is shown in Fig. 4-5.

The differential drawdown at the center (r=0) of the well field produced by withdrawal at rate W over the annular differential strip of area $2\pi r dr$ is

$$ds = \frac{2\pi r dr W}{2\pi T} \ln r_e/r$$

$$= \frac{r dr W}{T} \ln r_e/r \ . \quad (4\text{-}33)$$

Equation 4-33 follows from Eq. 4-30 in which Q_i has been replaced by $dQ = 2\pi r dr W$, and s_i has become ds. The drawdown s_o at the center of the well field is the sum of the differential drawdowns:

$$s_o = \frac{W}{T}\int_0^R r \ln(r_e/r) dr$$

$$= \frac{WR^2}{2T}\left[\ln r_e/R + 1/2\right] \ . \quad (4\text{-}34)$$

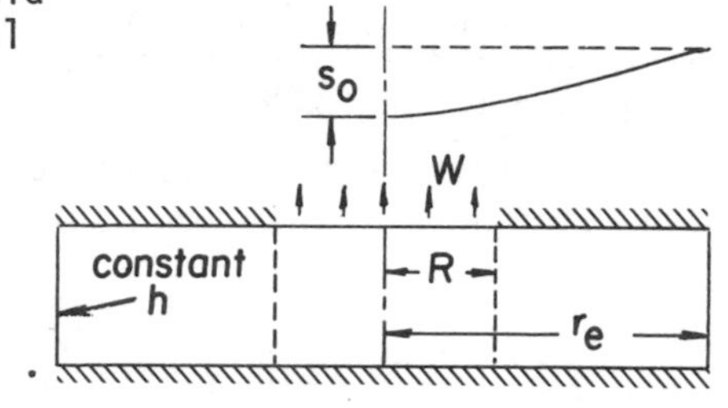

Figure 4-5. Idealization of a well field as a circular area of uniform withdrawal.

The integration performed in Equation 4-34 is analogous to the discrete summation indicated in Eq. 4-31. It is left to the student to show that Eq. 4-34 closely predicts the drawdown at a single isolated well when R becomes very small relative to r_e.

EXAMPLE 4-3

Collector wells consist of horizontal, perforated pipe driven radially from a common shaft (Fig. 4-6). Rigorous mathematical description of the hydraulic performance of collector wells is very difficult to accomplish. To some degree the performance of a collector well is approximated by a hypothetical, ordinary well with a very large radius (Mikels and

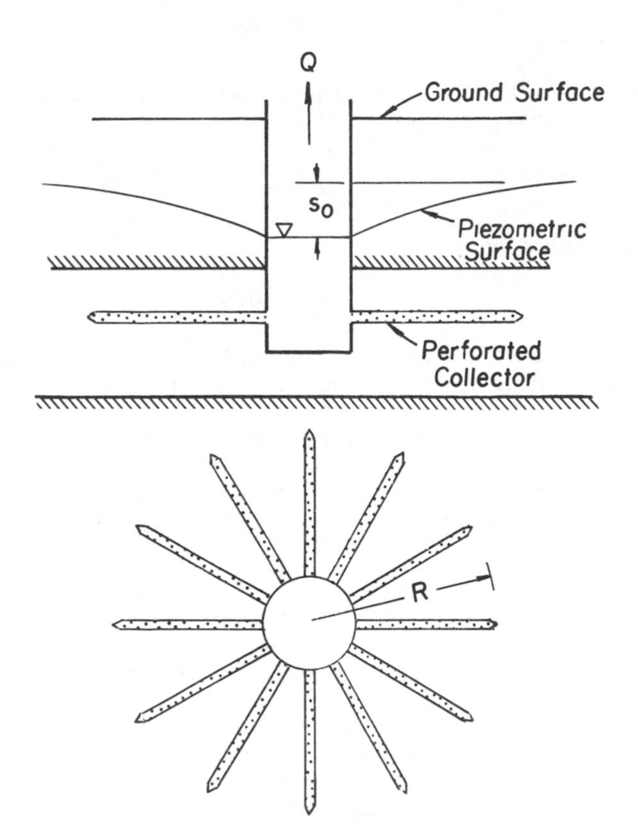

Figure 4-6. Schematic of a collector well - (a) vertical section, (b) plan view.

Klaer, 1956). Provided that the radius of the hypothetical equivalent well can be determined, the hypothetical well can be used to estimate the performance of the collector well. Use Eq. 4-34 to make a first estimate of the radius of the equivalent well for a collector well with a large number of collectors of average length R, measured from the center of the vertical shaft.

Solution:

As a first estimation, the withdrawal of water by the collector well is assumed to approximate a uniform withdrawal at rate W over the circular area of radius R. Thus, the discharge of the collector well is $Q=\pi R^2 w$, and Eq. 4-34 becomes

$$s_o = \frac{Q}{2\pi T}\left(\ell n\ r_e/R + 1/2\right) \quad .$$

The drawdown in a well of radius r_w and discharge Q is

$$s_w = \frac{Q}{2\pi T} \ln r_e/r_w$$

from Eq. 4-28. Equating the drawdowns in the collector and the ordinary well, yields

$$\ln r_e/r_w = \ln r_e/R + 1/2 \quad ,$$

from which

$$r_w = 0.61R \quad .$$

The above calculation suggests that the collector well can be represented by a hypothetical ordinary well with a radius of 0.61R. The approximation does not account accurately for the actual flow pattern into the collectors, of course. Experimental data (Mikels and Klaer, 1956) indicate that the collector well is somewhat more efficient than predicted by the analysis of this example, r_w being about 75-85 percent of the average length of the laterals.

Pumping Near Hydro-Geologic Boundaries

Consider a completely penetrating well pumping near a fully penetrating stream which intersects the confined aquifer along a straight line as shown in Fig. 4-7. The water level in the stream is assumed constant so that the aquifer face exposed in the stream is an equipotential boundary. In a strict sense, the following analysis applies for a well and stream that both completely penetrate a confined aquifer. The results, however, apply approximately to the more common situation in which a stream partially penetrates a water-table aquifer. Implicit in the analysis is the assumption that water supplied to the aquifer from the stream is not substantially retarded by a layer of sediment that sometimes exists on the wetted surface of the streambed. Should a layer of sediment (with hydraulic conductivity much less than that of the aquifer) be deposited on the bed of a partially penetrating stream, flow can be retarded to the extent that seepage from the stream is not sufficient to meet the demands of the pumped well. In such a case, the steady state assumed in this analysis will not develop.

Flow to the well is not radially symmetric because the source of the pumped water is the stream, idealized as an infinitely long straight line on which the piezometric head is constant. The solution is obtained by replacing the semi-infinite aquifer to the left of the stream in Fig. 4-7 and the constant head boundary by the mathematically equivalent system composed of an infinite aquifer with a pumped well (discharge rate Q) located at $x=a$, $y=0$ and with a recharge well (recharge rate $-Q$) located at $x=-a$, $y=0$. The drawdown anywhere in the semi-infinite

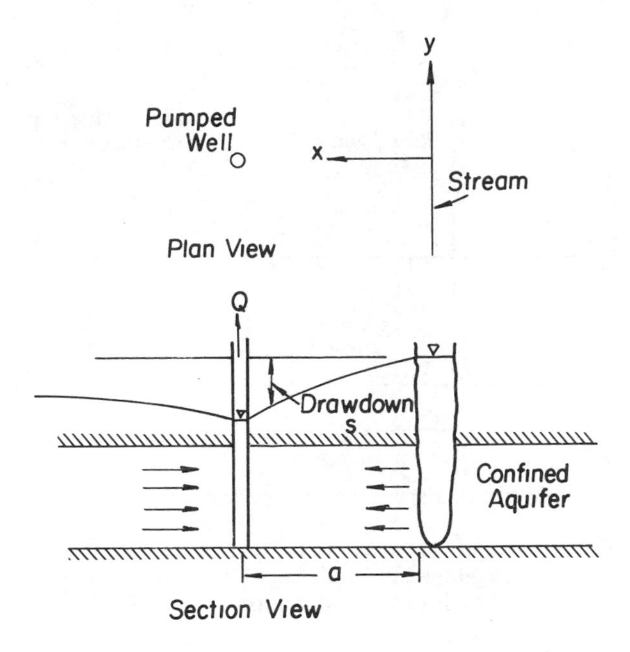

Figure 4-7. A pumped well near a stream with constant water level.

plane $x \geq 0$ is computed by adding the drawdown due to the real pumped well to the buildup due to the *image* recharge well. The result is

$$s = \frac{Q}{2\pi T} \ln r_i/r = \frac{Q}{2\pi T} \ln \left\{ \frac{(a+x)^2+y^2}{(a-x)^2+y^2} \right\}^{\frac{1}{2}} \quad , \quad (4\text{-}35)$$

where r and r_i are the distances from the pumped well and the recharge well, respectively, to the point at which the drawdown is s. The superposition is illustrated in section view in Fig. 4-8. Notice that Eq. 4-35 yields $s=0$ at all points on the line $x=0$ and, therefore, properly simulates the constant potential (i.e., $h=h_0$) boundary formed by the stream. To assist in understanding this, the reader can imagine the recharge well supplying water to the vertical surface represented by the line $x=0$ at the same rate at which the pumped well removes water from the surface. The drawdown in the pumped well is obtained by putting $r=r_w$ and $r_i=2a-r_w$ in Eq. 4-35 to yield

$$s_w = \frac{Q}{2\pi T} \ln\left(\frac{2a-r_w}{r_w} \right) \simeq \frac{Q}{2\pi T} \ln \frac{2a}{r_w} \quad . \qquad (4\text{-}36)$$

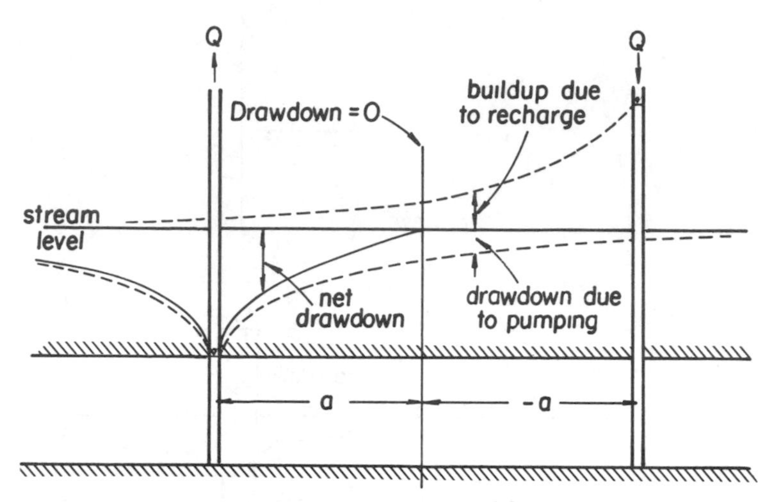

Figure 4-8. Section view of superposition of the drawdown from a pumped well and the buildup from a recharge well (Adapted from Ferris et al., 1962).

The above procedure is an example of the *method of images*. The recharge well is called an image well. Simulation of an infinitely long impermeable boundary located a perpendicular distance a from a pumped well is accomplished by replacing the semi-infinite aquifer with an infinite aquifer and by locating an image pumping well at x=-a, y=0.

The reason a pumping image well is required for impermeable boundaries can be understood by imagining that water located on the plane represented by x=0 does not flow because it has equal tendencies to move toward the real and image wells. Mathematically, the x-direction component of the gradient of piezometric head (or drawdown) is zero at all points on the line x=0 that perpendicularly besects the line joining the centers of the two wells pumping at the same rate. The student should verify this statement to his own satisfaction.

Other, more complicated patterns of image wells can be used to simulate flow in aquifers with a variety of different geometries imposed by impermeable and constant potential boundaries (Ferris et al., 1962). For example, the pattern of image wells required to simulate the flow toward a pumped well located between an impermeable and a constant-potential boundary is shown in Fig. 4-9. The indicated pattern repeats to infinity and the summation of drawdowns is an infinite series. The effects of additional image wells rapidly decreases, however, as their location becomes more remote from the real well. The pattern shown in Fig. 4-9 is useful for analyzing flow to a well located

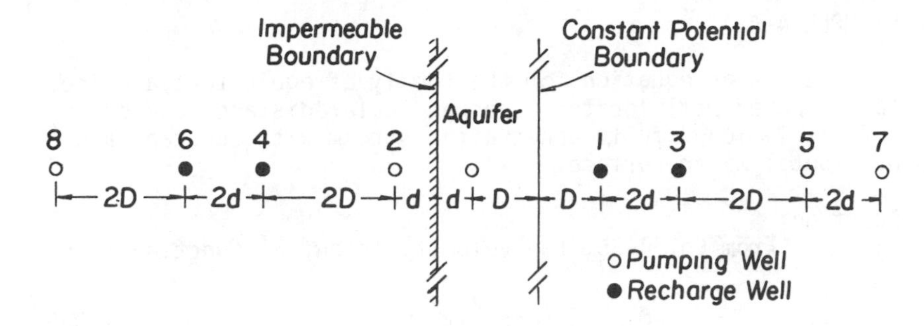

Figure 4-9. Pattern of images used to replace the impermeable and constant potential boundaries in an infinite strip aquifer (Adapted from Ferris et al., 1962).

in an alluvium filled valley between the stream and the edge of the valley, but the analysis is not limited to such an application. For example, consider flow in a vertical plane toward a horizontal drain located a distance D below the soil surface on which water is ponded as shown in Fig. 4-10. A

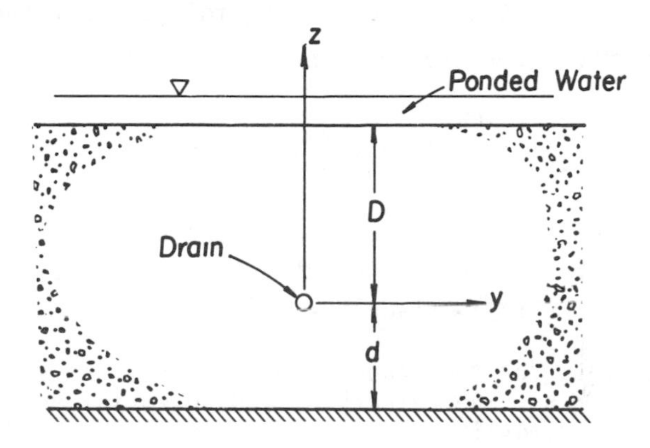

Figure 4-10. A horizontal drain below a constant head boundary.

horizontal, impermeable substratum is located at distance d below the drain. The pattern of image drains, both discharging and recharging, is that shown in Fig. 4-9 for the wells. Should the impermeable substratum be at a large distance below the drain, Eqs. 4-35 and 4-36 apply directly with a replaced by D, r_w by the radius of the drain, x by z, and b is the length of the drain measured along the axis of the drain.

EXAMPLE 4-4

Derive an equation for the family of equipotential lines for a pumped well located a perpendicular distance a from an infinitely long, fully penetrating stream with constant and horizontal water surface.

Solution:

From Eq. 4-35, the velocity potential function is

$$\Phi = \frac{-Q}{2\pi b} \ell n \left\{ \frac{(a+x)^2+y^2}{(a-x)^2+y^2} \right\}^{\frac{1}{2}} + \text{constant} \quad .$$

Putting the arbitrary constant equal to zero, noting that $\Phi=\Phi_i$ on an equipotential contour, and exponentiating yields

$$\left\{ \frac{x^2+y^2+2ax+a^2}{x^2+y^2-2ax+a^2} \right\}^{-1} = \exp\left(\frac{4\pi b\Phi_i}{Q} \right) = c_i \quad ,$$

where c_i is a constant corresponding to Φ_i. Rearranging and completing the square results in

$$\left(\frac{x}{a} - \frac{1+c_i}{1-c_i} \right)^2 + \left(\frac{y}{a} \right)^2 = \left(\frac{1+c_i}{1-c_i} \right)^2 - 1 \quad ,$$

which represents a family of circles in the dimensionless coordinates (x/a, y/a) with the centers at $x/a = (1+c_i)/(1-c_i)$, y=0 and radii $\{(1+c_i)^2/(1-c_i)^2 - 1\}^{\frac{1}{2}}$.

It is left for the student to show that the streamlines are also circles. The network of equipotential and streamlines is shown in Fig. 4-11.

EXAMPLE 4-5

Verify that the total discharge from the stream to the aquifer is equal to the well discharge for the steady flow depicted in Figs. 4-7 and 4-8.

Solution:

The discharge dQ_s from the stream into the aquifer on a reach of length dy follows from Darcy's law:

$$dQ_s = q_x b dy = T \left. \frac{\partial s}{\partial x} \right|_{x=0} dy \quad .$$

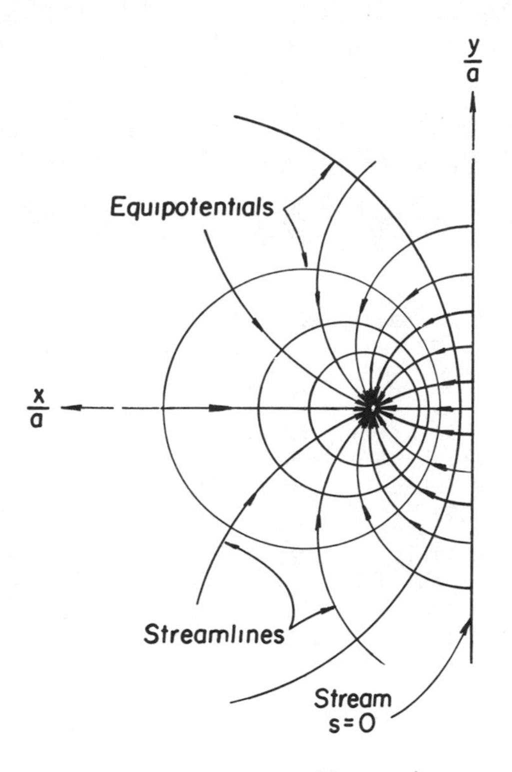

Figure 4-11. The flow net for a well pumping near a stream.

The drawdown is

$$s = \frac{Q}{2\pi T} \ell n \frac{r_i}{r} = \frac{Q}{4\pi T} \ell n\{ \frac{(x+a)^2+y^2}{(a-x)^2+y^2} \} \quad ,$$

which follows from Eq. 4-35 and the coordinate system indicated in Fig. 4-7. Calculation of the partial derivative of s with respect to x yields

$$\frac{\partial s}{\partial x} = \frac{Q}{4\pi T} \frac{(a-x)^2+y^2}{(x+a)^2+y^2} [\frac{2\{(a-x)^2+y^2\}(x+a) + 2\{(x+a)^2+y^2\}(a-x)}{\{(a-x)^2+y^2\}^2}] \quad .$$

At x=0, the result is

$$\left.\frac{\partial s}{\partial x}\right|_{x=0} = \frac{Q}{\pi T} (\frac{a}{a^2+y^2}) \quad ,$$

and the differential discharge from reach dy becomes

$$dQ_s = \frac{Q}{\pi} \frac{a}{a^2+y^2} dy \quad .$$

The discharge from the stream to the aquifer is obtained by integrating over the reach of stream extending from $-\infty$ to ∞:

$$Q_s = \frac{Qa}{\pi} \int_{-\infty}^{\infty} \frac{dy}{a^2+y^2} = \frac{2Q}{\pi} \tan^{-1} \frac{y}{a} \Big]_0^{\infty} \quad ,$$

from which Q_s=Q as required.

EXAMPLE 4-6

The center of a long, approximately horizontal, drain with a radius of cross-section equal to 15 cm is 2 m below the ground surface on which water is ponded to a depth of 20 cm. The depth to the impermeable floor of the aquifer is very great and the hydraulic conductivity of the aquifer is 4×10^{-4} cm/s. Assume that the drain runs full and that the water pressure at the top of the drain is equal to zero. Calculate the discharge of the drain per meter of drain tube.

Solution:

Eq. 4-36 is rearranged to give

$$Q = \frac{2\pi K b s_w}{\ell n\left(\frac{2a-r_w}{r_w} \right)}$$

where b=1 m, a=D=2 m, and r_w=0.15 m. The drawdown is the difference in piezometric head on the ground surface and on the surface of the drain tube. The piezometric head on the ground surface is $h_s = P/\rho g + z = 0.20 + 2 = 2.2$ m since the ground surface is 2 m above the datum (i.e., the center of the drain tube) and the pressure head is equal to the depth of water ponded on the surface. On the surface of the drain tube the piezometric head is constant and can be computed at any point at which the pressure head is known. Thus, the piezometric head at the drain is $h_d = P/\rho g + z = 0 + 0.15 = 0.15$ m since the pressure head is zero at the top of the drain. The drain discharge per meter of drain length is

$$Q = \frac{2\pi (4 \times 10^{-6})(1)(2.2-0.15)}{\ell n\left(\frac{4-0.15}{0.15} \right)} = 1.6 \times 10^{-5} \text{ m}^2/\text{s-m} \quad .$$

A Pumped Well In Uniform Flow

Superposition can be used to determine the flow pattern for a well placed in an aquifer in which one-dimensional uniform flow occurs. The equations for the streamlines are obtained by adding the ψ functions for uniform and radial flow. The stream function for one-dimensional, uniform flow in the x-direction is

$$\psi = -q_x y + \text{constant} = Kiy + \text{constant} \quad , \quad (4\text{-}37)$$

where $i=dh/dx$ has been introduced for simplicity in notation. Note that i may be positive or negative depending upon the coordinate system selected. Adding ψ from Eq. 4-37 and 4-26 yields

$$\psi = Kiy + \frac{Q}{2\pi b}\,\delta + \text{constant} \quad (4\text{-}38)$$

which is the stream function for a well in uniform flow. The flow pattern in the upper-half plane is shown in Fig. 4-12. The flow is symmetrical about the x-axis.

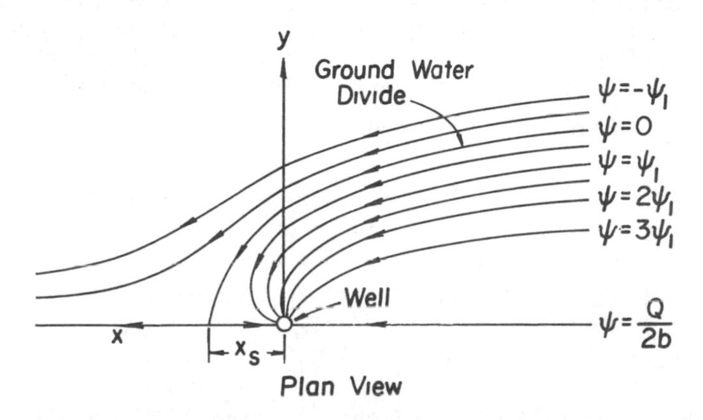

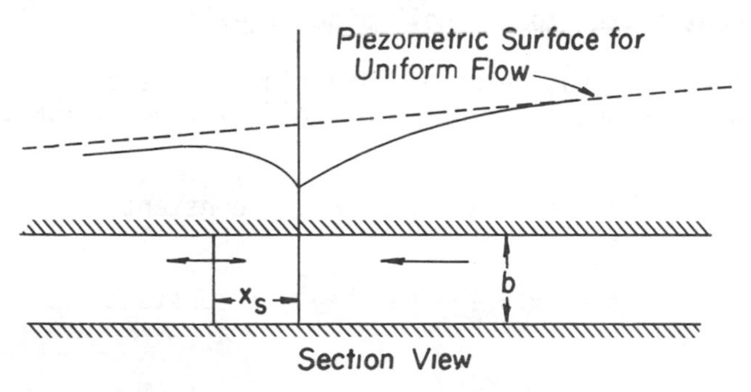

Figure 4-12. A pumped well in uniform flow.

Recall that the numerical difference between any two values of constant ψ is equal to the discharge per unit of aquifer thickness between the two streamlines. Selecting the arbitrary constant in Eq. 4-38 equal to zero, it is evident that ψ is Q/2b when y=0 and $\delta=\pi$. Thus, the entire negative x-axis is a streamline for which $\psi=Q/2b$. It follows that one-half of the well discharge must flow between the streamlines $\psi=Q/2b$ and $\psi=0$. Furthermore, the streamline for which $\psi=0$ must separate flow that eventually contributes to well discharge from flow that bypasses the well. The streamline $\psi=0$ is called the ground-water divide (Fig. 4-12). The relationship between the x and y coordinates of all points on the ground-water divide is

$$Kiy = -\frac{Q}{2\pi b}\,\delta \tag{4-39}$$

where $\delta=\tan^{-1}(y/x)$ for $x\geq 0$ and $\delta=\pi-\tan^{-1}|y/x|$ for $x\leq 0$. For the coordinate system shown in Fig. 4-12, the slope i of the piezometric surface in uniform flow is negative.

The point at which the ground-water divide streamline crosses the x-axis downstream of the well is called a stagnation point because the Darcy velocity is zero there. Water on the x-axis upstream of the stagnation point moves toward the well, while water on the x-axis downstream of the stagnation point moves away from the well. The coordinate x_s of the stagnation point is obtained from

$$x_s = \lim_{y\to 0}\left[\frac{y}{\tan\left(\frac{-2\pi Tiy}{Q}\right)}\right] = -\frac{Q}{2\pi Ti} \quad . \tag{4-40}$$

Far upstream of the well,(as $x\to-\infty$ and $\delta\to\pm\pi$) Eq. 4-39 yields

$$y = \pm\frac{Q}{2Ti} \tag{4-41}$$

which is the half-width of that portion of the aquifer in which flow contributes to the well discharge.

The distribution of piezometric head is calculated by adding the heads for radial and uniform flow. From Eqs. 4-12 and 4-23

$$\Phi = -q_x x + \frac{Q}{2\pi b}\,\ell n\, r + \text{constant} \tag{4-42}$$

or

$$h = ix + \frac{Q}{4\pi T}\,\ell n(x^2+y^2) + \text{constant} \quad . \tag{4-43}$$

EXAMPLE 4-7

Contaminated water from several waste-holding ponds seeps into an aquifer in which uniform flow exists. A well is to be constructed 300 m down gradient from the ponds in an attempt to intercept the contaminated ground water. The slope of the piezometric surface is -0.022 and the transmissivity is 0.013 m^2/s. The width of contaminant source area is 200 m, measured perpendicular to the direction of ground water flow (Fig. 4-13). Compute the well discharge required to intercept the contaminated ground water, assuming no dispersion. Assume further, that the discharge of contaminated water into the aquifer is small compared to the discharge through the aquifer.

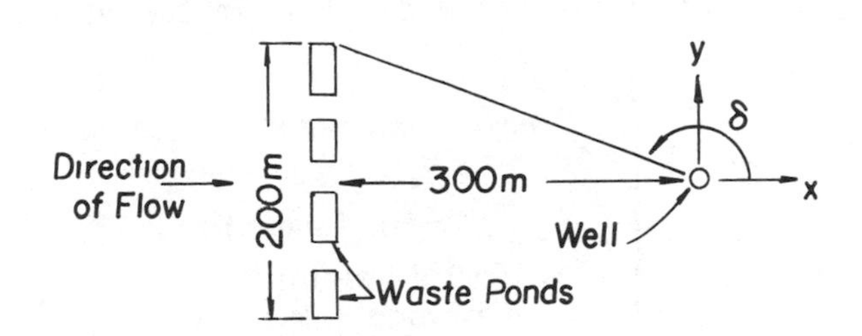

Figure 4-13. An interceptor well downstream of a source of ground-water contamination.

Solution:

The significance of the second assumption in the problem is that the discharge of contaminated water is too small to significantly influence the pattern of flow and that the quantity of contaminated water added does not appreciably alter the discharge per unit area in the aquifer. Provided lateral dispersion is negligible, the contaminated water will be intercepted if the ψ=0 streamline passes through the extremeties of the line of waste ponds. The coordinates of one end of the line on which contamination occurs are (-300,100). From Eq. 4-39,

$$Q = -\frac{2\pi Tiy}{\delta} ,$$

wherein δ is the angle indicated in Fig. 4-13. Hence,

$$\delta = \pi - \tan^{-1}\left(\frac{100}{300}\right)$$

$$= 2.82 \text{ radians.}$$

The required discharge is

$$Q = \frac{-2\pi(0.013)(-0.022)(100)}{2.82} = 0.064 \ m^3/s \quad .$$

Partially Penetrating Wells

The formulas for flow toward a well that have been discussed to this point apply for fully penetrating wells. In practice, wells are sometimes constructed so that water enters the well bore over a length which is less than the aquifer thickness. Flow toward a partially penetrating well experiences convergence that is in addition to the convergence in flow toward a fully penetrating well. Partial penetration causes the flow to have a vertical component in the vicinity of the well as shown in Fig. 4-14. The water level in piezometers depends not only upon r but also upon the vertical coordinate z of the terminal point of the piezometer. Thus, two piezometers located at equal r will exhibit different water levels if they terminate at different elevations in the aquifer. It

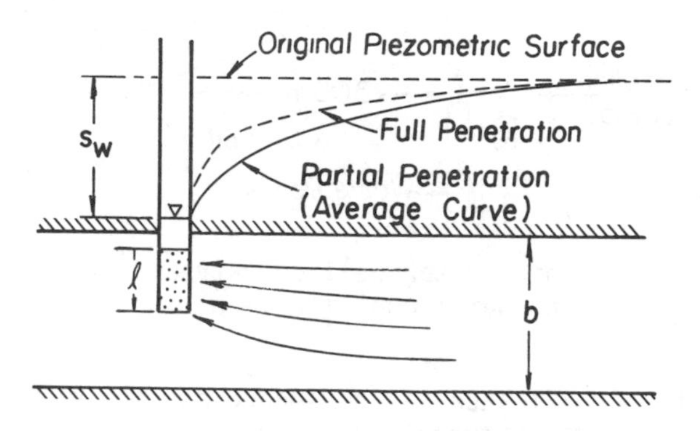

Figure 4-14. Flow toward a partially penetrating well.

follows that there does not exist a single drawdown vs. r curve or a single cone of depression that entirely characterizes the flow in the aquifer. Observation wells, screened in the same interval as the production well, yield the average drawdown curve shown in Fig. 4-14.

The vertical components of flow are negligible for distances from the well that exceed about 1.5 b. All of the foregoing equations for purely radial flow apply for the partially penetrating case, provided that $r \geq 1.5$ b, therefore. The additional convergence of the flow for $r < 1.5$ b causes the discharge per unit of drawdown to be smaller than if the well is fully penetrating. Muskat (1946) used a continuous superposition of sinks to represent the partially penetrating well to derive an approximate relationship between the discharge of a partially penetrating well and that for a fully penetrating well. The details of Muskat's solution are beyond the scope of this basic text, but the useful result is

$$\left(\frac{Q}{s_w} \right)_p = \left(\frac{Q}{s_w} \right) \left[\frac{\ell}{b} \left\{ 1+7 \left(\frac{r_w}{2\ell} \right)^{\frac{1}{2}} \cos \frac{\pi\ell}{2b} \right\} \right] , \quad (4\text{-}44)$$

where ℓ is the length of well bore over which water enters the well and the subscript p denotes the partially penetrating case. Equation 4-44 was derived for the case in which the screened interval is located at the top of the aquifer. In practice it is used to estimate the effects of partial penetration regardless of where the open interval (intervals) is located (Walton, 1970).

The discharge per unit drawdown that appears in Eq. 4-44 is called *specific capacity* and is widely used to characterize the discharge capacity of pumped wells. For steady state, the specific capacity of a well is a constant that depends upon well radius, degree of penetration, and aquifer transmissivity. In nonsteady flow, the specific capacity is also a function of time.

EXAMPLE 4-8

A well with an effective well radius of 0.2 m exhibits a specific capacity of 0.021 $m^3/s \cdot m$ in an aquifer for which b is 42 m. The length of screened well bore is 10 m. If the length of well screen is increased to 20 m, estimate the increase in specific capacity.

Solution:

From Eq. 4-44 the specific capacity for full penetration is

$$Q/s_w = \frac{(Q/s_w)_p}{\frac{\ell}{b} \left\{ 1+7 \left(\frac{r_w}{2\ell} \right)^{\frac{1}{2}} \cos \frac{\pi\ell}{2b} \right\}}$$

$$Q/s_w = \frac{0.021}{\frac{10}{42} \left\{ 1+7 \left(\frac{0.2}{20} \right)^{\frac{1}{2}} \cos \left(\frac{10\pi}{84} \right) \right\}} = 0.053 \ m^2/s \quad .$$

The specific capacity with ℓ=20 m is

$$(Q/s_w)_p = (0.053) \left[\frac{20}{42} \left\{ 1+7 \left(\frac{0.2}{40} \right)^{\frac{1}{2}} \cos \frac{20\pi}{84} \right\} \right] = 0.034 \ m^2/s \quad .$$

Increasing the length of well screen to 20 m will change the specific capacity from 0.021 to 0.034 m^2/s, an increase of 62%. This calculation assumes a homogeneous aquifer, of course, and would have little validity in a horizontally stratified aquifer.

Concluding Remarks On Superposition of Solutions

The reader should, by now, recognize that superposition of solutions is a very useful technique for extending analytical tools to progressively more complex problems. The few examples discussed in the preceding pages are not exhaustive by any means. In particular, occasions arise when an estimate must be made of the flow pattern in an aquifer subsequent to some man-made change such as construction of a well or drain. It is not necessary in such cases to express the pre-construction flow pattern mathematically in order to use superposition to deduce the post-construction flow pattern. All that is required is that the pre-construction flow pattern be known and displayed as a map of equipotentials or streamlines. Such knowledge is often deduced from water levels measured in a network of observation wells or piezometers.

When the pre-construction flow pattern is displayed in the form of a piezometric map, the procedure is to calculate or otherwise estimate the equipotentials that pertain to the proposed construction and to superimpose these equipotential lines on the map of pre-construction equipotential lines. The post-construction distribution of piezometric head is deduced by adding the values of head from both sets of equipotential lines at all points of intersection. The result is an array of points at which the post-construction piezometric head is known (up to an additive constant). Finally points of equal piezometric head are connected, yielding the desired equipotential contours from which streamlines can be drawn if required.

4.4 FLOW NETS

Thus far, analytic methods for obtaining solutions to the Laplace equation have been presented. It is apparent from the previous discussions that boundary conditions must be relatively simple in order to obtain analytic solutions even using superposition. Another solution technique which may be used is a graphical method known as the flow net.

Flow Nets In Homogeneous Aquifers

A flow net consists of a network of equipotential lines and the corresponding orthogonal streamlines (Fig. 4-11). From a properly constructed flow net one may obtain distribution of heads, discharges, areas of high (or low) velocities, and the general flow pattern. In addition, flow nets offer an excellent means of gaining insight into the general characteristics of ground water flow.

Referring to Fig. 4-15, the flow channel between adjacent streamlines is called a *streamtube*. The discharge in the streamtube per unit width perpendicular to the plane of the figure is

$$\Delta Q = \psi_2 - \psi_1 = \Delta\psi \quad (4\text{-}45)$$

and the Darcy velocity is

$$q = \Delta Q/\Delta B = \Delta\psi/\Delta B \quad .(4\text{-}46)$$

Darcy's Law gives, for this figure,

$$q = -\frac{K\Delta h}{\Delta \ell} = \frac{\Phi_2 - \Phi_1}{\Delta \ell} = \frac{\Delta\Phi}{\Delta \ell} \quad (4\text{-}47)$$

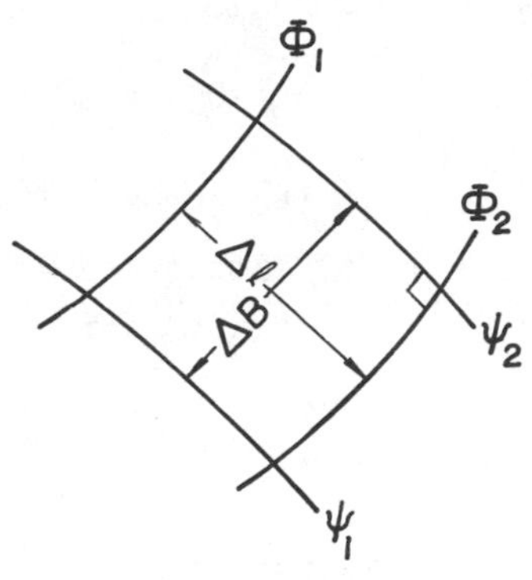

Figure 4-15. A flow net element for two-dimensional flow.

where $\Phi = Kh$. If the flow net is drawn so that $\Delta\ell$ is equal to ΔB, then the flow net reduces to a network of "squares" which can be easily identified by visual means. In addition, since continuity must be satisfied, then

$$q = \Delta\psi/\Delta B = \Delta\Phi/\Delta\ell \qquad (4\text{-}48)$$

and the increment $\Delta\psi$ must be equal to $\Delta\Phi$.

The first step in the solution of a flow problem by flow net construction is to draw the two-dimensional flow domain to scale so that all boundary locations, wells, etc. are in the correct positions, relative to one another. A trial and error procedure is used to sketch the flow net. The following rules aid in minimizing the number of trials needed to construct a proper flow net:

a) Equipotential contours and streamlines are orthogonal at points of intersection.

b) The flow net is constructed so that the spacing ΔB between streamlines is equal to the spacing $\Delta\ell$ between equipotential lines for each element of the net. In this way the flow net is constructed as a set of curvilinear squares of appropriate sizes.

c) All boundary conditions must be satisfied by the net, e.g., constant head boundaries or impermeable boundaries.

d) Take advantage of apparent symmetry (if any) by beginning the flow net construction around lines of symmetry.

e) Use no more than 4 or 5 streamtubes in the first trials.

The total discharge through the aquifer is (by Eqs. 4-45, 4-46, and 4-47)

$$Q = n_s \Delta Q = n_s \Delta\psi = n_s \Delta\Phi = n_s K \Delta h \qquad (4\text{-}49)$$

where n_s is the number of streamtubes and Δh is the head loss along the streamtube for one element. The head loss along a streamtube for each element is given by

$$\Delta h = \frac{H_t}{n_\ell} \qquad (4\text{-}50)$$

where H_t is the total head loss determined from the boundary conditions for a streamtube and n_ℓ is the number of equipotential drops determined from the flow net. From Eqs. 4-49 and 4-50, the discharge is

$$Q = K \frac{n_s}{n_\ell} H_t \quad . \qquad (4\text{-}51)$$

For a particular problem the ratio n_s/n_ℓ is a constant. Only by coincidence is n_ℓ an integer when an integer number of streamtubes is used at the outset.

The above procedures and equations permit one to determine discharges and potential distribution in cases in which the aquifer is homogeneous. Flow beneath impermeable dams and under cut-off walls are common cases that are analyzed by these methods.

Flow nets may also be used for anisotropic aquifers provided that the directional permeabilities are known. Steady flow in the x-y plane in a homogeneous but anisotropic aquifer is described by (from Eq. 3-58),

$$K_x \frac{\partial^2 h}{\partial x^2} + K_y \frac{\partial^2 h}{\partial y^2} = 0 \quad , \qquad (4\text{-}52)$$

where the coordinate axes are selected parallel to the direction of the principle permeabilities. This may be rearranged as

$$\frac{\partial^2 h}{\partial \dot{x}^2} + \frac{\partial^2 h}{\partial y^2} = 0 \quad , \qquad (4\text{-}53)$$

where $\dot{x} = x \sqrt{K_y/K_x}$. Equation 4-53 is the Laplace equation for a homogeneous isotropic aquifer in the distorted $\dot{x}$-y coordinate system. A set of curvilinear squares may now be drawn for the aquifer redrawn in the transformed $\dot{x}$-y plane and the results transferred back to the original plane. It is left for the reader to show that the discharge is given by

$$Q = \frac{n_s}{n_\ell} \sqrt{K_x K_y}\, H_t \quad .$$

EXAMPLE 4-9

Using the flow net method, estimate the discharge under the sheet piling shown in Fig. 4-16 for $K=2.5\times10^{-5}$ m/s and $H_t=2$ m. Note that the sheet piling penetrates to 1/2 of the aquifer thickness.

Solution:

Because of symmetry, the flow net construction is started with a vertical equipotential line, directly below the terminal point of the sheet piling. Because the velocity is expected to be greatest near the terminal point of the piling, the upper streamline is located quite near the end of the sheet piling. The spacing between the remaining streamlines is increased with depth because it is anticipated that the Darcy velocity will decrease with depth. The next step is to sketch the equipotential immediately adjacent to the vertical one that was placed at the start. The location and shape of this second equipotential is determined by the condition of orthogonality that must be satisfied at points of intersection and by the condition than each element form a square. Some readjustment of both the streamlines and the equipotential line is usually required. Once the streamlines and the equipotential lines are correctly drawn in the vicinity of the terminal point of the sheet pile, the next step is to extend the streamlines and repeat the procedure.

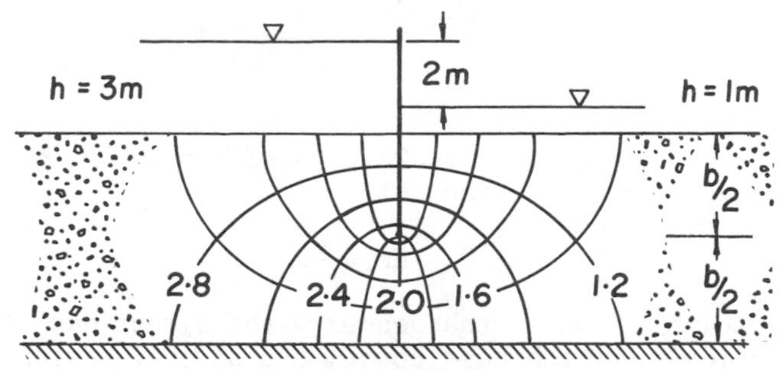

Figure 4-16. Flow net for seepage under a sheet pile (Adapted from Polubarinova-Kochina, 1962).

The number of streamtubes is 5 from Fig. 4-16 and the number of equipotential drops is 10. Thus, the discharge per meter of sheet pile length is

$$Q = K \frac{n_s}{n_\ell} H_t = 2.5\text{x}10^{-5}\left(\frac{5}{10} \right)(2) = 2.5\text{x}10^{-5} \; m^2/s \quad .$$

EXAMPLE 4-10

Calculate the seepage discharge beneath the dam and cut-off wall shown in Fig. 4-17. The hydraulic conductivity is 15 m/d and H_t=10 m.

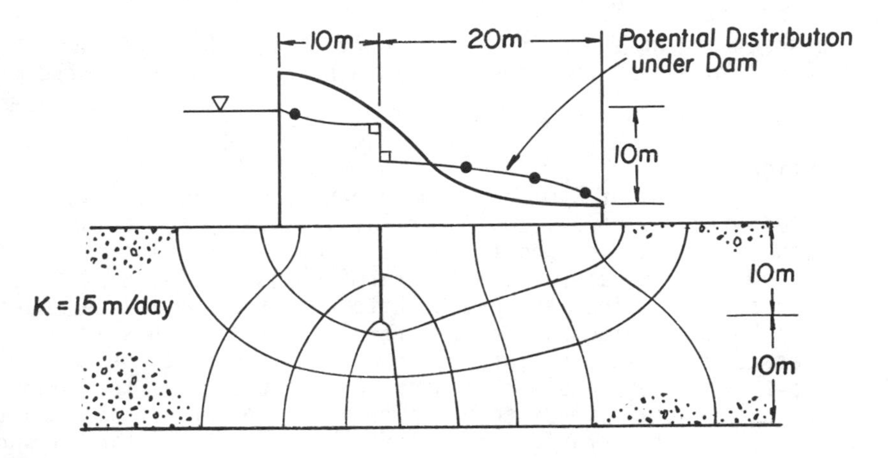

Figure 4-17. Flow net for seepage beneath a dam with a cut-off wall.

Solution:

A flow net is also shown with 9 equipotential drops and 3 stream channels. The discharge per unit width perpendicular to the plane of the cross-section is

$$Q = \left(\frac{n_s}{n_\ell} \right) KH_t = \left(\frac{3}{9} \right)(15)(10) = 50 \; \frac{m^3}{m \cdot day} \quad .$$

Polubarinova-Kochina (1962) presents an analytic solution of the above case in the form of

$$Q = KH_t c$$

where c is a function of the geometry and dimensions of the cross-section. For the case above, c=0.33 (Polubarinova-Kochina, 1962, Fig. 85) and

$$Q = (15)(10)(0.33) = 50 \frac{m^3}{m \cdot day} \quad .$$

It should be noted that one would have to be extremely careless in constructing a flow net for the above case to obtain n_ℓ less than 8 or more than 10. Consequently, the probable error for this case would be ±1 potential drop or an error for discharge of ±11%. Thus the solution in terms of discharges is, in general, not too sensitive to the quality of the flow net.

Areas of large velocity occur where the distance between streamlines is small. From the figure, large Darcy velocities would occur at the bottom of the cut-off wall and the up-stream and down-stream corner of the dam. For this case the down-stream corner of the dam is of special interest because high velocities could cause piping and subsequent failure of the dam.

By plotting the potential along the base of the dam as shown in the figure, one can determine the uplift forces due to the aquifer water pressures. It can be seen that the cut-off wall accounts for approximately 5 m of head loss and is very effective in increasing the length of flow path and, thereby, causing a large potential drop over a short horizontal distance.

Flow Nets In Non-homogeneous Aquifers

Aquifers can rarely be considered homogeneous over large areas and the student might well question the usefulness of the above procedures on a large scale. The hydrologist, however, normally makes use of measured values of piezometric head to guide him in the construction of flow nets. For example, it is very common to measure values of piezometric head in wells in the area of interest and the location of these wells and their corresponding values of head are plotted on a map. Equipotential contours are sketched by connecting points of equal head, using considerable interpolation and judgment. The flow net is constructed by sketching in streamlines normal to the equipotential lines. The non-homogeniety of the aquifer causes the elements of flow net to be distorted and irregular in regions of rapidly varying hydraulic conductivity and the elements will not be curvilinear squares.

Applying Darcy's Law to the flow in a streamtube, shown in Fig. 4-18, for a nonhomogeneous aquifer gives

$$\Delta Q = Kb\left(\frac{\Delta h}{\Delta \ell} \right) \Delta B \quad . \qquad (4\text{-}54)$$

Since ΔQ is a constant in the streamtube and Δh is also constant, one obtains

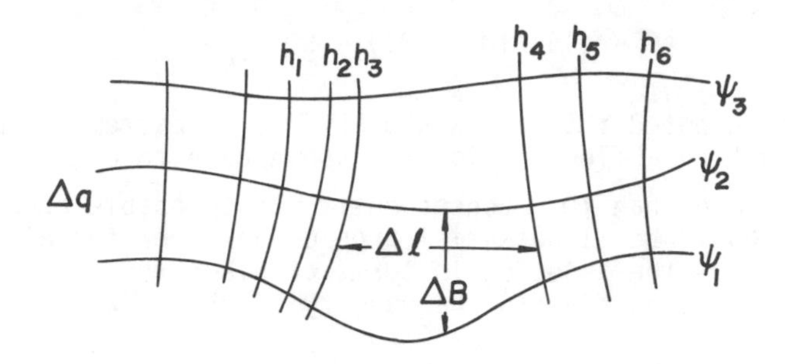

Figure 4-18. Plan view of streamlines and orthogonal equipotential lines for a non-homogeneous aquifer.

$$Kb \frac{\Delta B}{\Delta \ell} = \frac{\Delta Q}{\Delta h} = \text{constant} \quad . \qquad (4\text{-}55)$$

Equation 4-55 must hold anywhere along a selected streamtube. The ratio $\Delta B/\Delta \ell$ is determined for each element of the flow net. Provided b is known, the distribution of relative values K along the streamtube can be computed from Eq. 4-55. If b is unknown, then relative values of T along a streamtube may be evaluated. Locations favorable for well siting can be determined this way.

Another important application of flow nets constructed from measured water levels is to identify sources of recharge or discharge of ground water from the aquifer. In some cases, the quantities of recharge or discharge may be estimated from the flow net. It should be noted here that the water table is a streamline when the flow is steady. Thus, the equal head contours, as determined from observation well measurements, may actually represent the topographical configuration of the phreatic surface relative to an arbitrary datum.

4.5 FLOW IN CONFINED AQUIFERS WITH VERTICAL ACCRETION

Steady flow is sometimes sustained by vertical percolation through an over or under lying aquitard. Equations 3-76 and 3-81 are the appropriate differential equations from which the analyses are made. In a strict sense, the analyses in this section are applicable to confined flows only, but they provide approximate solutions for unconfined flows also, provided that the saturated aquifer thickness is large relative to the maximum drawdown.

One-Dimensional Flow In A Leaky Aquifer

Consider steady one-dimensional flow in the x-direction in a confined aquifer into which seepage occurs at a constant rate W from one of the adjacent geologic strata. Because the flow is steady, Eq. 3-76 reduces to

$$\frac{d^2h}{dx^2} + \frac{W}{T} = 0 \quad . \tag{4-56}$$

Two integrations of Eq. 4-56 yield

$$h = \frac{-W}{2T} x^2 + C_1 x + C_2 \quad , \tag{4-57}$$

in which the constants of integration, C_1 and C_2, are evaluated using the boundary conditions appropriate for the particular problem at hand.

Radial Flow In A Leaky Aquifer

The distribution of piezometric head (or drawdown) caused by pumping from an aquifer which experiences a constant uniform accretion at rate W is derived by solving Eq. 3-76 in the form

$$\frac{1}{r}\frac{d}{dr}\left(r\frac{dh}{dr}\right) + \frac{W}{T} = 0 \quad . \tag{4-58}$$

Integration of Eq. 4-58 subject to $h=h_o$ and $dh/dr=0$ at $r=r_e$ yields

$$h = \frac{W}{4T}(r_e^2 - r^2) + \frac{Wr_e^2}{2T} \ln r/r_e + h_o \quad . \tag{4-59}$$

The drawdown h_o-h in a well of radius r_w is

$$s_w = \frac{Wr_e^2}{2T} \ln r_e/r_w - \frac{W}{4T}(r_e^2 - r_w^2) \quad , \tag{4-60}$$

which becomes

$$s_w = \frac{Wr_e^2}{2T}\left\{\ln\ r_e/r_w - \frac{1}{2}\right\} \tag{4-61}$$

upon recognition that r_w^2 is very small compared to r_e^2. Because the flow is steady, the inflow from leakage must equal outflow. Hence,

$$Q = \pi(r_e^2 - r_w^2)W \simeq \pi r_e^2 W \quad , \tag{4-62}$$

which permits Eq. 4-61 to be written as

$$s_w = \frac{Q}{2\pi T} \{\ell n \; r_e/r_w - \frac{1}{2}\} \quad . \tag{4-63}$$

Equation 4-63 predicts the drawdown in a well pumping from an aquifer in which the recharge is vertical through the end of a cylinder of radius r_e. Equation 4-63 should be compared with Eq. 4-28 (rewritten in terms of drawdown) which applies when the recharge occurs as horizontal flow through the lateral surface area of a cylinder of radius r_e.

The methods of this section can be used to derive

$$s = \frac{Q}{2\pi T} \ell n \; r_e/R + \frac{W}{4T}(R^2 - r^2) \quad , \quad r \leq R \tag{4-64}$$

for the situation depicted in Fig. 4-5. Equation 4-64 is obtained by solving

$$\frac{1}{r}\frac{d}{dr}(r\frac{dh}{dr}) = 0 \quad , \quad R \leq r \leq r_e \tag{4-65}$$

and

$$\frac{1}{r}\frac{d}{dr}(r\frac{dh}{dr}) - \frac{W}{T} = 0 \quad , \quad r \leq R \tag{4-66}$$

with the condition that h is continuous at r=R. Equation 4-66 is Eq. 4-58 with a sign change that reflects that water is being withdrawn rather than recharged to the aquifer. Putting r equal to zero in Eq. 4-64 yields the drawdown at the center of the well field, derived previously by the method of continuous superposition of elementary solutions (Eq. 4-34).

Recall that Eq. 3-81 was obtained for the case in which the seepage through the aquitard is proportional to the difference in piezometric head between the supplying and main aquifer. For steady flow toward a well, Eq. 3-81 is rewritten as

$$\frac{\partial^2 s}{\partial r^2} + \frac{1}{r}\frac{\partial s}{\partial r} - \frac{s}{B^2} = 0 \tag{4-67}$$

with the boundary contitions

$$s(\infty) = 0 \tag{4-68}$$

and

$$\lim_{r \to 0} r\frac{\partial s}{\partial r} = -\frac{Q}{2\pi T} \tag{4-69}$$

The solution is (Hantush, 1956)

$$s = \frac{Q}{2\pi T} K_o(r/B) \qquad (4\text{-}70)$$

where K_o is the modified Bessel function of the second kind. Values of K_o are listed in Table 4-1.

Table 4-1. Values of $K_o(r/B)$, the Modified Bessel Function of the Second Kind and Order Zero. (After Hantush, 1956).

N	r/B-Nx10^{-3}	Nx10^{-2}	Nx10^{-1}	N
1.0	7.0237	4.7212	2.4271	0.4210
1.5	6.6182	4.3159	2.0300	0.2138
2.0	6.3305	4.0285	1.7527	0.1139
2.5	6.1074	3.8056	1.5415	0.0623
3.0	5.9251	3.6235	1.3725	0.0347
3.5	5.7709	3.4697	1.2327	0.0196
4.0	5.6374	3.3365	1.1145	0.0112
4.5	5.5196	3.2192	1.0129	0.0064
5.0	5.4143	3.1142	0.9244	0.0037
5.5	5.3190	3.0195	0.8466	
6.0	5.2320	2.9329	0.7775	0.0012
6.5	5.1520	2.8534	0.7159	
7.0	5.0779	2.7798	0.6605	0.0004
7.5	5.0089	2.7114	0.6106	
8.0	4.9443	2.6475	0.5653	
8.5	4.8837	2.5875	0.5242	
9.0	4.8266	2.5310	0.4867	
9.5	4.7725	2.4776	0.4524	

The boundary condition expressed in Eq. 4-69 is that of a line sink which is used to simulate the well.

EXAMPLE 4-11

Calculate the steady-state drawdown in a well of radius 0.2 m discharging 0.063 m^3/s from an aquifer experiencing a constant vertical recharge of 2.6×10^{-8} m/s. The transmissivity of the aquifer is 0.018 m^2/s.

Solution:

The radius of influence (the radius of the catchment area) is computed by

$$r_e = (Q/\pi W)^{\frac{1}{2}} = \{ \frac{0.063}{(\pi)(2.6\times10^{-8})} \}^{\frac{1}{2}} = 878 \text{ m} \quad .$$

Drawdown in the well is computed from either of Eqs. 4-61 or 4-63.

$$s_w = \frac{0.063}{2\pi(0.018)} \{\ell n(\frac{878}{0.2}) - \frac{1}{2} \} = 4.4 \text{ m} \quad .$$

4.6 STEADY FLOW IN UNCONFINED AQUIFERS

It has been mentioned previously that flow in water table aquifers can sometimes be treated as confined flow provided that the drawdown is small relative to the saturated thickness. This fact is especially important in the analysis of unsteady flow in unconfined aquifers which is treated in the following chapter. Steady-state flows in water-table aquifers can, however, be treated without linearizing the differential equation.

The right hand side of Eq. 3-73 is zero for steady flow and the differential equation for steady flow in homogeneous unconfined aquifers is

$$\frac{\partial}{\partial x} (h \frac{\partial h}{\partial x}) + \frac{\partial}{\partial y} (h \frac{\partial h}{\partial y}) = 0 \quad , \tag{4-71}$$

which can be rewritten as

$$\frac{\partial^2 h^2}{\partial x^2} + \frac{\partial^2 h^2}{\partial y^2} = 0 \quad . \tag{4-72}$$

Equation 4-72 is the linear Laplace equation in the variable h^2. It should be evident, therefore, that solutions for one-dimensional and radial flow in water table aquifers will bear a close resemblance to the corresponding equations for confined flow. Superposition of elementary solutions for unconfined flow must be accomplished by adding solutions in the form $h_i^2(x,y)$, however, rather than $h_i(x,y)$ as was done for confined flow.

One-Dimensional Unconfined Flow

Steady one-dimensional flow between two parallel channels in which the water levels are not equal is depicted in cross-section in Fig. 4-19. An equation for the water table elevation can be derived by integration of Eq. 4-72, with the constants of integration evaluated from the indicated boundary conditions. Perhaps a better understanding of the flow can be gained by deriving the equation for the discharge between the channels

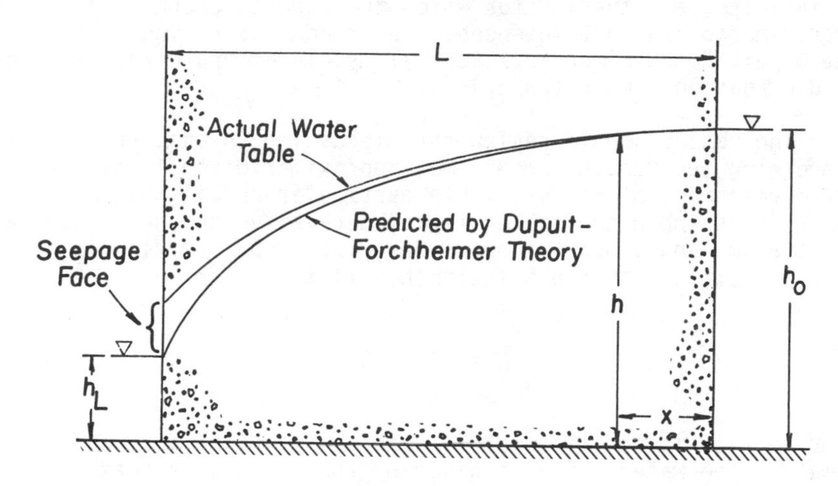

Figure 4-19. Steady flow between parallel channels with unequal water levels.

directly from the Dupuit-Forchheimer discharge equation (Eq. 3-67).

The fact that the flow is steady requires that the discharge through all vertical planes be the same. In other words, Q is not dependent upon x. Integration of Eq. 3-67 yields

$$Q = \frac{K}{2L}\,(h_o^{\,2}-h_L^{\,2}) \quad , \qquad (4\text{-}73)$$

from which the discharge between the channels can be computed.

Bear (1972) has shown that the discharge predicted by Eq. 4-73 is exact, even though it was derived by means of the Dupuit-Forchheimer approximations. However, the water-table elevation predicted by the Dupuit-Forchheimer approach is not exact. The water-table elevation is obtained by integrating Eq. 3-67 from 0 to an arbitrary coordinate value x:

$$h^2 = h_o^{\,2} - \frac{2Q}{K}\,x = h_o^{\,2} - \left(\frac{h_o^{\,2}-h_L^{\,2}}{L} \right)x \quad . \qquad (4\text{-}74)$$

Equation 4-74 predicts that the water table in profile is a parabola. The actual water-table elevation is somewhat greater than that calculated from Eq. 4-74 because the vertical components of flow have been neglected in the latter case. Notice that no mathematical difficulties are encountered by putting $h_L=0$ in Eqs. 4-73 and 4-74. However, putting $h_L=0$ implies that the area through which flow is occurring is zero and dh/dx is infinite, a situation for which the Dupuit-Forchheimer approximation was not intended. The reader is reminded that the Dupuit-Forchheimer approach will yield accurate water-table predictions only when $(dh/dx)^2 \ll 1$.

The reader should review the discussion in Chapter III concerning the Dupuit-Forchheimer approximation that leads to the development of Eq. 4-74. In particular recall that h(x) in Eq. 4-74 is the water-table elevation relative to the horizontal substratum, and a more rigorous description of the flow in Fig. 4-19 is obtained by the solution h(x,z) to

$$\frac{\partial^2 h}{\partial x^2} + \frac{\partial^2 h}{\partial z^2} = 0 \quad , \qquad (4\text{-}75)$$

where h(x,z) is the piezometric head at any point in the flow domain. The water table is given by the locus of points for which h=z. The solution can be displayed by drawing streamlines $\psi(x,z)$ and equipotential lines h(x,z) (or $\Phi(x,z) = Kh$). One of the streamlines is the water table, since the flow is steady and there is no vertical recharge. Also, that part of the aquifer exposed in the channels below the water level in the channels is an equipotential line. Recall that equipotential lines and streamlines must intersect at right angles. In Fig. 4-19 it is clear that the water table predicted by the approximate Dupuit-Forchheimer theory does not satisfy this requirement. The actual water table intersects the wall of the left channel somewhat above the water level in the channel. Between the water level in the channel and the point at which the water table intersects the channel is an interval called the *seepage face*, on which the water pressure is zero and the piezometric head increases linearly with elevation. Because the piezometric head is not constant on the seepage face, there is no requirement that the water table be normal to the channel wall at the point of intersection. It is emphasized that Eq. 4-73 is an exact discharge formula in spite of the fact that the theory by which it was derived is approximate.

One-Dimensional Unconfined Flow With Vertical Accretion

One-dimensional flow, sustained by constant, uniform recharge between parallel channels, is shown in Fig. 4-20. This flow can be analyzed by making appropriate simplifications of Eq. 3-82 and integrating the result. However, the student should recognize that

$$Q = Wx = -Kh \frac{dh}{dx} \qquad (4\text{-}76)$$

on any vertical plane between the channels. Integration of Eq. 4-76 over the interval x=0, L/2 yields

$$h^2 = h_o^2 - \frac{W}{K}\{x^2 - (L/2)^2\} \quad , \tag{4-77}$$

which is known as the ellipse equation because the predicted water table profile is a section of an ellipse. The height of the water table is a maximum midway between the channels and is obtained by putting x=0 in Eq. 4-77:

$$h_{max}^2 = \frac{WL^2}{4K} + h_o^2 \quad . \tag{4-78}$$

Equations 4-77 and 4-78 are useful for analyzing flow toward subsurface agricultural drains in humid regions where W is approximately constant. In particular, Eq. 4-78 can be used to estimate the drain spacing L required to insure that h_{max}

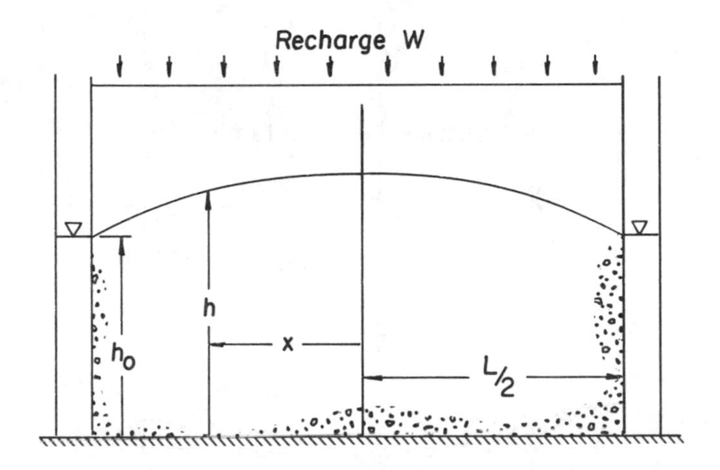

Figure 4-20. Steady unconfined flow with vertical recharge between parallel channels.

does not exceed a value that has been determined to be satisfactory for crop growth. It is not required that the channels in Fig. 4-20 be drains, of course. The channels can be natural streams, for example. Furthermore, since the plane located at x=0 is a ground-water divide, this plane can be regarded as impermeable. Thus, either half of Fig. 4-20 can be regarded as depicting the flow to a stream in an alluvial valley bounded by an impermeable formation at a distance L/2 from the stream. It is left to the student to derive equations corresponding to Eqs. 4-77 and 4-78 for the case in which the water levels in the channels are not equal.

EXAMPLE 4-12

A stream flows through an alluvial valley bounded by impermeable rocks as shown in Fig. 4-21. The lands adjacent to the stream are to be developed for irrigated agriculture. Recharge to the unconfined alluvial aquifer is expected to increase upon the start of irrigation because of deep percolation beneath the cropped fields and seepage from the diversion canals and distribution system. It is possible that the increase in recharge will cause the water table to become too close to the land surface unless artificial drainage is practiced. This possibility can be analyzed by estimating the

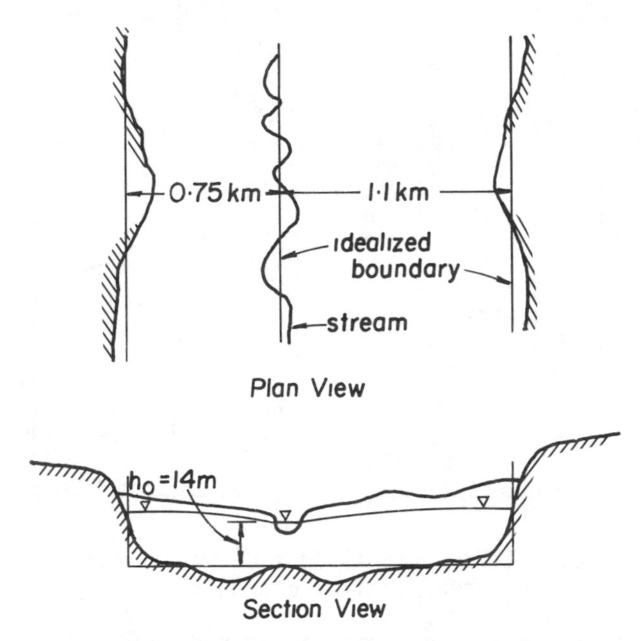

Figure 4-21. Definition sketch for example problem 4-12.

water-table elevation that is expected during the irrigation season and comparing the estimated water table levels with the land surface elevation. If $W=8.2 \times 10^{-4}$ m/d during the irrigation season and K=4.41 m/d, estimate the water table elevation as a function of distance from the stream relative to the water level in the stream. Other required information is given in Fig. 4-21. Assume that a steady state is achieved during the irrigation season.

Solution:

Equations 4-77 and 4-78 can be adapted for use in this problem. Because the water table elevation is to be estimated relative to the water level in the stream, Eq. 4-77 is rewritten as

$$H = [h_o^2 + \frac{W}{K}\{(\frac{L}{2})^2 - x^2\}]^{\frac{1}{2}} - h_o \quad ,$$

in which H is the water-table elevation above the water level in the stream. This result was obtained by writing $h=h_o+H$.

Consider, first, the right-hand side of the stream. The impermeable boundary is simulated by the ground-water divide at x=0 in Fig. 4-20. Thus, the origin of the x-coordinate is on the impermeable boundary and L/2 is 1.1 km. Substituting into the above equation yields

$$H = [(14)^2 + \frac{8.2 \times 10^{-4}}{4.41}\{(1100)^2 - x^2\}]^{\frac{1}{2}} - 14$$

from which the required water-table elevation can be computed at any desired distance from the stream. The maximum value of H is 6.5 m, obtained at x=0. The same calculation can be made for the left-hand side of the stream using L/2 = 750 m. The calculated water-table elevations can be compared with the land surface elevations to determine if natural drainage to the stream is adequate, or if it is probable that artificial drainage will be required. The actual case will be that of a water-table elevation that fluctuates about the average value computed in this example. Methods by which the fluctuation of the water table can be calculated are developed in Chapter V.

Radial Unconfined Flow

Steady radial flow toward a well (Fig. 4-22), according to the Dupuit-Forchheimer approximation, is described by

$$Q = K(2\pi rh)\frac{dh}{dr} \quad , \qquad (4\text{-}79)$$

which can be integrated to obtain

$$Q = \frac{\pi K(h_e^2 - h_w^2)}{\ell n\; r_e/r_w} \quad . \qquad (4\text{-}80)$$

Equation 4-80 is the formula for discharge of a well of radius r_w in which the water level is h_w. As before, well losses are assumed to be negligible. The discharge predicted by Eq. 4-80 is mathematically exact (Hantush, 1962; Hunt, 1970), just as

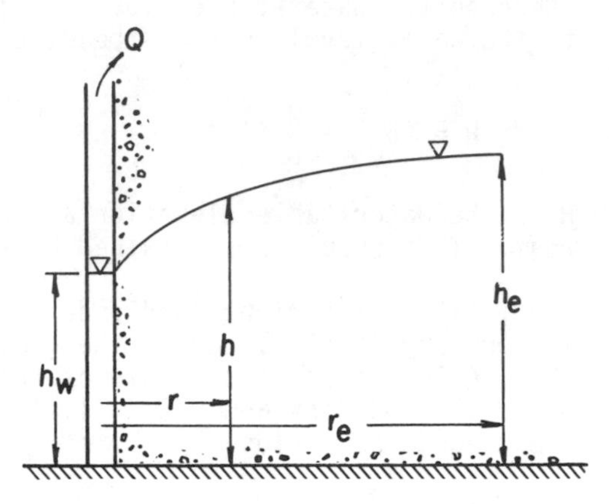

Figure 4-22. Radial flow in an unconfined aquifer.

in the case of one-dimensional flow. The variation of water-table elevation with distance from the well is given by

$$h^2 = h_e^2 - \frac{Q}{\pi K} \ln r_e/r = h_e^2 - (h_e^2 - h_w^2) \frac{\ln(r_e/r)}{\ln(r_e/r_w)} \quad , \tag{4-81}$$

a result that is accurate for $r \geq 1.5\ h_e$ even with $h_w=0$ (Hantush, 1962). Equation 4-81 is satisfactorily accurate for all r provided that $(h_e - h_w)/h_e << 1.0$.

It has been stated repeatedly that the confined flow equations can be used to describe flow in water table aquifers provided the spatial change of h is small compared to h. The following development shows that use of an average saturated thickness in the equation for confined flow to a well yields a mathematically exact prediction of well discharge. Expanding the quantity $h_e^2 - h_w^2$ gives

$$h_e^2 - h_w^2 = 2\left(\frac{h_e + h_w}{2} \right)(h_e - h_w) = 2b_{avg}(h_e - h_w) \quad . \tag{4-82}$$

Thus, Eq. 4-80 can be written in the form

$$Q = 2\pi T_{avg} \frac{(h_e - h_w)}{\ln r_e/r_w} \quad , \tag{4-83}$$

which is the equation for confined flow with an average transmissivity.

The discharge can also be written in terms of drawdown,

$$Q = \frac{\pi K}{\ell n\ r_e/r_w}\ s_w(2h_e - s_w) \tag{4-84}$$

from which it is clear that the maximum discharge is obtained when the well penetrates to the impermeable substratum and the water level is drawn down to the bottom of the well. In addition to the fact that it is impractical to draw the water level down to the bottom of the well, it is also undesirable to do so from an ecomonic point of view. Figure 4-23 shows how the well discharge, expressed as a fraction of the maximum obtainable

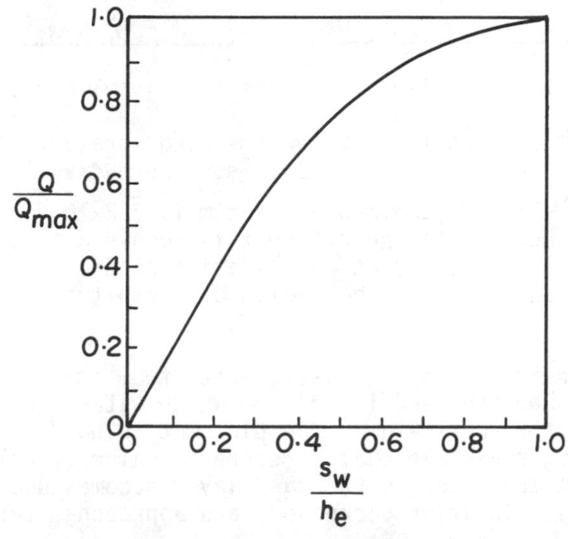

Figure 4-23. The dependence of well discharge on drawdown in a water-table aquifer.

discharge, varies with the drawdown. It is evident that the incremental increase in discharge associated with a given increase in drawdown becomes small for s_w greater than about 0.7 h_e. In other words, for $s_w > 0.7\ h_e$, the reduction of the area of inflow into the well, caused by increasing s_w further, becomes increasingly important relative to the increase in gradient. Thus, there is little to be gained by drawing the water level down beyond about two-thirds of the saturated thickness.

EXAMPLE 4-13

A cylindrical pit penetrates a horizontal layer of sand bounded on the top by an impermeable stratum and on the bottom by a slightly permeable layer (Fig. 4-24). The regional water

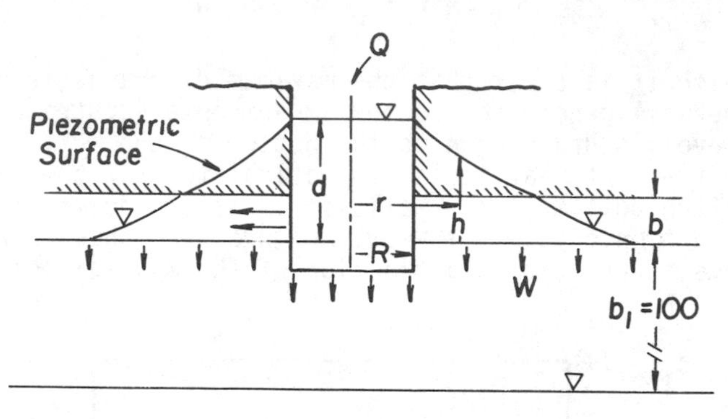

Figure 4-24. Section view of seepage from a cylindrical pit.

table is located about 100 m below the sand stratum. The sand layer is 1.6 m thick with K=0.052 cm/s. The hydraulic conductivity of the slightly permeable stratum is 2.2×10^{-5} cm/s. Estimate the recharge to the pit that is required to maintain the water level in the pit at an elevation of d=3 m above the bottom of the sand layer. The radius of the pit is 5 m.

Solution:

As indicated in Fig. 4-24, water from the pit flows laterally through the sand layer from which water seeps vertically through the substratum. The piezometric head decreases with increasing r and eventually reaches a value of h=1.6 m, at which point the water in the sand layer becomes unconfined. The water-table elevation decreases, and approaches zero at $r=r_e$.

The first step in the solution is to estimate W. Darcy's Law for vertical flow has been written previously in Example 3-6 and becomes

$$W = K(1 + h/b_1)$$

in the terminology of this problem. Clearly, h depends upon r, rendering W dependent upon r. However, notice that as h varies between 3 m and zero, W varies between 1.03K and K. It is sufficiently accurate, therefore, to put $W=K=2.2 \times 10^{-5}$ cm/s.

Because the flow is steady, the discharge into the pond that will maintain a constant water level is

$$Q = \pi r_e^2 W \quad .$$

The problem, therefore, is to find r_e. Let $r=r_b$ be the radial distance to the point at which the water in the sand layer becomes unconfined. Then

$$Q(r) = -K(2\pi rb)\frac{dh}{dr} = Q_R - \pi(r^2-R^2)W \quad , \quad R \leq r \leq r_b \quad , \quad \text{(A)}$$

where Q_R is the discharge through the vertical wall of the pit. Also,

$$Q(r) = -K(2\pi rh)\frac{dh}{dr} = Q_b - \pi(r^2-r_b^2)W \quad , \quad r_b \leq r \leq r_e \quad \text{(B)}$$

where Q_b is the discharge at $r=r_b$. Integration of Eqs. A and B yields

$$2\pi Kb(d-b) = (Q_R + \pi R^2 W)\ell n\frac{r_b}{R} - \frac{\pi W}{2}(r_b^2-R^2) \quad \text{(C)}$$

and

$$\pi Kb^2 = (Q_b + \pi r_b^2 W)\ell n\frac{r_e}{r_b} - \frac{\pi W}{2}(r_e^2-r_b^2) \quad , \quad \text{(D)}$$

respectively.

The following relationships also apply:

$$Q_R = \pi(r_e^2-R^2)W$$

and

$$Q_b = \pi(r_e^2-r_b^2)W \quad ,$$

which are used to reduce Eqs. C and D to

$$2\pi Kb(d-b) = \pi r_e^2 W\ \ell n\frac{r_b}{r} - \frac{\pi W}{2}(r_b^2-R^2) \quad \text{(E)}$$

and

$$Kb^2 = r_e^2 W\ \ell n\ r_e/r_b - \frac{W}{2}(r_e^2-r_b^2) \quad . \quad \text{(F)}$$

At this point the problem has been reduced to the simultaneous solution of Eqs. E and F for the two unknowns r_b and r_e. This is accomplished by adding Eqs. E and F and rearranging to obtain

$$r_e^2(\ell n\, r_e/R - \frac{1}{2}) = \frac{K}{W}(2bd - b^2) - \frac{R^2}{2} \quad .$$

Putting the known values for each parameter in the right side of the above equation gives the result:

$$r_e^2(\ell n\, r_e/R - \frac{1}{2}) = \{ \frac{0.052}{2.2\text{x}10^{-5}} \}\{2(1.6)(3) - (1.6)^2\} - \frac{(5)^2}{2}$$

$$= 1.66\text{x}10^4 \quad .$$

By trial and error, the value of r_e which satisfies the above is 85 m and

$$Q = \pi(85)^2(2.2\text{x}10^{-7}) = 0.005 \text{ m}^3/\text{s}$$

is the required recharge into the pit.

4.7 HYDRAULICS OF TWO FLUIDS IN AQUIFERS

Ground water with a spatial variability of density due to differences in dissolved-solid content is an example of a nonhomogeneous fluid. Strictly speaking, essentially all ground waters are nonhomogeneous because of the variability of dissolved solids concentration. The effects of variable density and dissolved solids concentrations are usually neglected in hydraulic calculations because the variations are small. An important exception is the case of coastal and island aquifers in which the dissolved solids concentration can range from a few hundred milligrams per liter to that of sea water (≈30,000 mg/ℓ) with a corresponding range of density of 1-1.025 g/cm^3. Rigorous description of the flow of nonhomogeneous ground water is accomplished by solving, simultaneously, the equations for ground-water flow and the hydrodynamic dispersion equation (Bear, 1972). Even when the variable density is unimportant with respect to ground-water motion, hydrodynamic dispersion can be important with regard to the distribution of constituents dissolved in the water. A thorough treatment of hydrodynamic dispersion is beyond the scope of this basic text, however, and the following discussion is limited to those situations in which the nonhomogeneous fluid can be idealized as two distinct homogeneous fluids.

Two-Fluid Idealization of a Nonhomogeneous Fluid

An example of the vertical distribution of dissolved solids concentration in an island aquifer is shown in Fig. 4-25. The upper zone of small, nearly constant concentration is separated from the lower zone of large concentration by a transition zone. Provided that the vertical dimension of the transition zone is small compared to the aquifer thickness above the transition

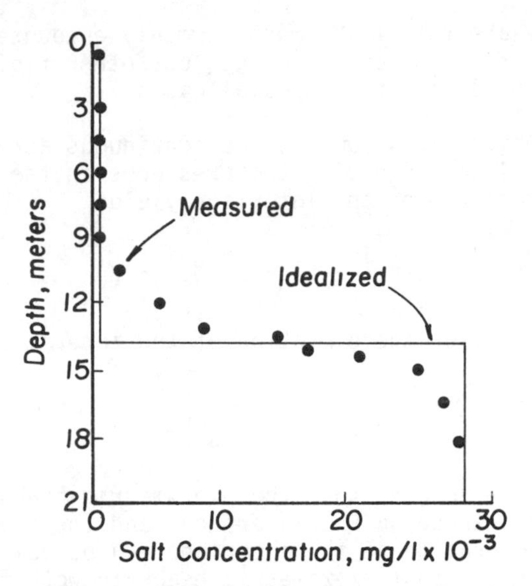

Figure 4-25. Vertical distribution of dissolved solids concentration in an island aquifer.

zone, the nonhomogeneous fluid can be treated as two immiscible fluids with different, but constant, densities separated by an interface. The elevation of the interface is taken as the elevation at which the average of the upper and lower zone concentrations occurs.

Idealization of the single nonhomogeneous fluid as two immiscible, homogeneous fluids permits the use of the piezometric head in each fluid as a force potential for that fluid (Fig. 4-26). It is not possible to define piezometric head as a single force potential applicable to both fluids. The two fluids in Fig. 4-26 were identified as being fresh water and salt water

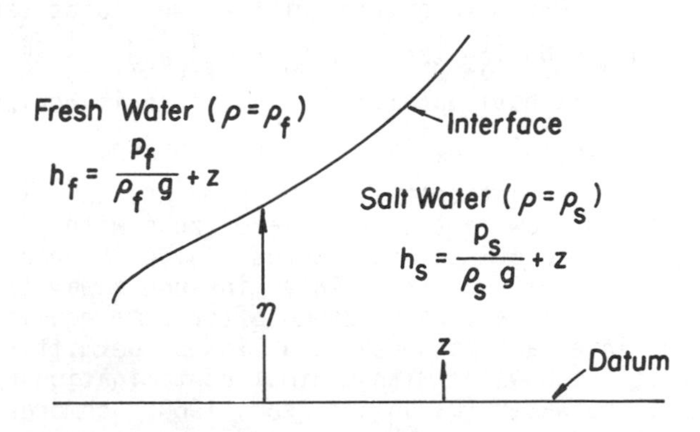

Figure 4-26. Two homogeneous fluids separated by an interface.

because they are the fluids most commonly encountered in the practice of ground-water hydrology, but other fluid systems may also be of interest in special cases.

The pressure is assumed to be continuous across the interface in Fig. 4-26. Equating the pressures in the two fluids at an elevation $z=\eta$ on the interface yields

$$\rho_f(h_f^{\,i} - \eta) = \rho_s(h_s^{\,i} - \eta) \quad , \qquad (4\text{-}85)$$

which is solved for the elevation of the interface η to obtain

$$\eta = \frac{\rho_s}{\Delta\rho} h_s^{\,i} - \frac{\rho_f}{\Delta\rho} h_f^{\,i} \quad . \qquad (4\text{-}86)$$

The superscript denotes that the piezometric heads under consideration are those on the interface, and the quantity $\Delta\rho$ is $\rho_s - \rho_f$. The elevation of the interface can be computed from Eq. 4-86 provided that piezometric heads in both fluids are known on the interface.

Let ℓ be a coordinate measured along the interface. Then,

$$\frac{\partial\eta}{\partial\ell} = \frac{\rho_s}{\Delta\rho}\frac{\partial h_s^{\,i}}{\partial\ell} - \frac{\rho_f}{\Delta\rho}\frac{\partial h_f^{\,i}}{\partial\ell} \quad , \qquad (4\text{-}87)$$

which becomes

$$\frac{\partial\eta}{\partial\ell} = \frac{\rho_f}{\Delta\rho}\frac{q_f^{\,i}}{K_f} - \frac{\rho_s}{\Delta\rho}\frac{q_s^{\,i}}{K_s} \qquad (4\text{-}88)$$

upon replacement of the derivatives of piezometric head with their equivalents from Darcy's Law. The velocities in Eq. 4-88 are the Darcy velocities of the two fluids tangent to the interface. Notice that, if both $q_f^{\,i}$ and $q_s^{\,i}$ are zero, the interface is horizontal. The interface is horizontal also for the case in which $q_f^{\,i}/\mu_f = q_s^{\,i}/\mu_s$. Of special practical importance is the fact that Eq. 4-88 predicts that it is possible to sustain flow in the fresh-water zone with no flow in the salt-water zone. In other words, Eq. 4-88 predicts that it is possible to extract fresh water in such a way that no flow occurs in the salt-water zone, after some equilibrium position of the interface has been established; permitting the exploitation of fresh water with minimum contamination by the underlying saline water (Dagan and Bear, 1968; Schmorak and Mercado, 1969; Chandler and McWhorter, 1975).

Under conditions of zero salt-water movement,

$$\frac{\partial \eta}{\partial \ell} = \frac{\rho_f}{\Delta\rho} \frac{q_f^{\,i}}{K_f} \quad . \tag{4-89}$$

Evidently, the elevation of the interface must increase in the direction of fresh-water flow when the underlying salt water is static. The left side of Eq. 4-89 is the sine of the angle between the interface and the horizontal plane. Thus, Eq. 4-89 is satisfied only if

$$|q_f^{\,i}| \leq \frac{\Delta\rho}{\rho_f} K_f \quad . \tag{4-90}$$

Should inequality 4-90 not be satisfied the salt water is not static. Inequality 4-90 is a very restrictive condition because $\Delta\rho \simeq 0.025$ g/cm^3 for a fresh-sea water system.

The Ghyben-Herzberg Equation

The interface between the fresh and salt water bodies constitutes a boundary to the two regions in which flow occurs. For the case of a static salt-water body, the interface represents a boundary to which the Darcy velocity must be tangent at all points. Certainly, the location and shape of the interface influences the distribution of piezometric head in the ground water. On the other hand, the distribution of piezometric head is required to establish the location of the interface. As in the case of flow bounded by a water table, the difficulty is circumvented by use of the Dupuit-Forchheimer approximations.

Equation 4-87 can be rewritten as

$$\Delta\eta = - \frac{\rho_f}{\Delta\rho} \Delta h_f^{\,i} \tag{4-91}$$

for the case of static salt water. Equation 4-91 relates the change of interface elevation to a change in the fresh-water piezometric head on the interface. According to the Dupuit-Forchheimer approximation, the piezometric head on a vertical line is constant. Therefore,

$$\Delta\eta \simeq - \frac{\rho_f}{\Delta\rho} \Delta h \tag{4-92}$$

approximately, where Δh is the change in piezometric surface elevation. In the case of a water-table aquifer, Δh represents the change of elevation of the water table. Provided that both z and h are measured from sea level,

$$z = \frac{\rho_f}{\Delta\rho} h \qquad (4\text{-}93)$$

in which z is the distance below sea level to the interface and h is the elevation of the piezometric surface or the water table above sea level. Equation 4-93 is the Ghyben-Herzberg equation and has been widely used for the analysis and description of flow in a fresh water zone overlying a static body of salt water. Note that $z=40\ h$ for $\Delta\rho=0.025$. Also note that the slope of the interface is forty times as great as the slope of the water table for $\Delta\rho=0.025$. Thus, the Ghyben-Herzberg equation, because it is based on the assumption of essentially horizontal flow, will yield accurate results only when the slope of the water table is, indeed, very small.

Fresh-Water Flow in a Coastal Aquifer

An unconfined coastal aquifer that outcrops in the sea is considered. A steady fresh water discharge at rate Q per unit length of outcrop from the aquifer to sea is assumed to occur (Fig. 4-27). The steady discharge is sustained by an inland

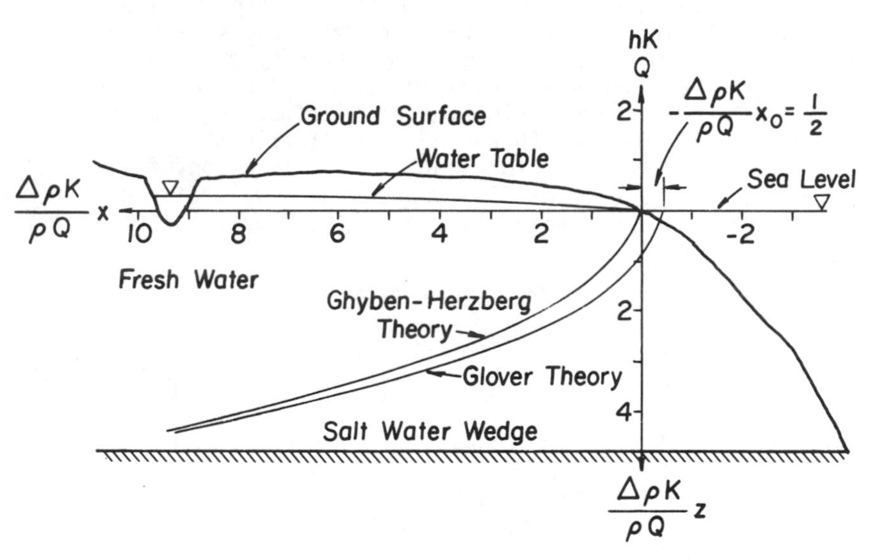

Figure 4-27. The water table and interface in a coastal aquifer.

source and, therefore, Q does not depend upon the coordinate distance x, measured positive inland from the coast. The discharge of fresh water is given by

$$Q = K(z+h)\,\frac{dh}{dx} \quad . \qquad (4\text{-}94)$$

The variable z is replaced by its equivalent from Eq. 4-93 and the result is integrated, subject to h=0 at x=0, to obtain,

$$h = \{ \frac{2Q\Delta\rho}{K(\rho+\Delta\rho)} x\}^{\frac{1}{2}} \simeq \{ \frac{2Q\Delta\rho}{K\rho} x\}^{\frac{1}{2}} \quad (4\text{-}95)$$

which is the equation for the water-table elevation relative to sea level. Using Eq. 4-93 again, provides

$$z = \frac{\rho}{\Delta\rho} \{ \frac{2Q\Delta\rho}{K(\rho+\Delta\rho)} x\}^{\frac{1}{2}} \simeq \{ \frac{2Q\rho}{\Delta\rho K} x\}^{\frac{1}{2}} \quad (4\text{-}96)$$

as the equation for the interface. The water table and interface predicted from Eqs. 4-95 and 4-96, respectively, are shown in dimensionless form in Fig. 4-27.

The above analysis predicts that the water table and the interface meet at the intersection of the water table with the sea. This physically unacceptable result is a consequence of using the Dupuit-Forchheimer approximation which neglects vertical components of flow. Glover (1959) provides a more rigorous (although still approximate) analysis for the flow in a coastal aquifer. Glover's equation for the water table is identical to Eq. 4-95, but the equation for the interface is

$$z^2 = \frac{2\rho Q}{\Delta\rho K} x + (\frac{\rho Q}{\Delta\rho K})^2 \quad . \quad (4\text{-}97)$$

A plot of Eq. 4-97 is also included in Fig. 4-27. A gap of length

$$x_o = - \frac{\rho Q}{2\Delta\rho K} \quad , \quad (4\text{-}98)$$

through which water escapes to the sea, is predicted by Glover's analysis. At large x, Eqs. 4-96 and 4-97 agree closely.

EXAMPLE 4-14

Fresh water is discharged from a coastal aquifer to the sea at a steady rate of 6.3×10^{-6} m^3/s per meter of coast line (idealized as a straight line). The source of the fresh water flow is inland as shown in Fig. 4-27. The unconfined aquifer rests on an approximately horizontal impervious substratum at a depth of 45 m below mean sea level. The hydraulic conductivity of the aquifer is 8.1×10^{-3} cm/s. Calculate the position of the toe of the salt-water wedge in the aquifer. Should ground-water development at inland locations reduce the sustained fresh-water flow to the sea by 95 percent, calculate the distance from the coast to which the salt water wedge will eventually intrude.

Solution:

The toe of the salt-water wedge is the intersection of the interface with the impervious base of the aquifer. The distance, L, from the coast to the toe of the wedge is calculated from Eq. 4-96 or Eq. 4-97 with z=45 m. From Eq. 4-97,

$$L = \frac{\Delta\rho K}{2\rho Q} \{z^2 - (\frac{\rho Q}{\Delta\rho K})^2\}$$

$$= \frac{(0.025)(8.1\text{x}10^{-5})}{(2)(1)(0.63\text{x}10^{-5})} [(45)^2 - \{ \frac{(1)(0.63\text{x}10^{-5})}{(0.025)(8.1\text{x}10^{-5})} \}^2]$$

$$= 324 \text{ m}$$

and from Eq. 4-96, L=325 m.

If ground-water development at inland locations reduces the fresh-water flow to the sea by 95 percent, then Q is reduced to $3.15\text{x}10^{-7}$ m^2/s. Repeating the above computation yields L=6510 m. This calculation demonstrates the phenomen of salt-water intrusion, and how it can be influenced by modifying the seaward gradient in the fresh-water zone. Sea-water intrusion, in response to inland development of ground water is actually a transient phenomenon, and the time required for the salt-water wedge to move from 300 m to 6.5 km from the coastline could be several years.

Interface Upconing Beneath Pumping Wells

Equation 4-92 predicts that the fresh-salt water interface rises in response to any depression in the piezometric surface or water table. Thus, there is a rise of the interface beneath a well pumping from the fresh-water zone because of the drawdown of the piezometric surface or water table in the vicinity of the well (Fig. 4-28). Furthermore, the Ghyben-Herzberg equation predicts that the rise of the interface is $\rho/\Delta\rho$ times the drawdown at any point. Although substantial vertical components of flow in the vicinity of a pumped well (Fig. 4-28) make inaccurate the predictions of the interface from the Ghyben-Herzberg equation, the conclusion that small drawdowns will result in large rises of the interface remains valid. Therefore, extraction of fresh water from a zone above an underlying salt-water body must be accomplished by creating very small drawdowns if salt water is to be prevented from upconing into the wells.

Laboratory and field observations and theoretical considerations have shown that, beneath a pumping well, there exists a critical elevation of the interface, above which the interface is not stable and salt water flows to the well (Dagan and Bear, 1968; Muskat, 1946; Schmorak and Mercado, 1969; Sahni, 1972).

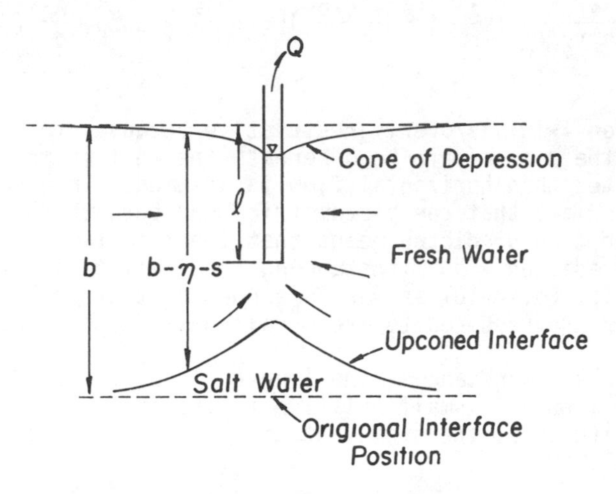

Figure 4-28. Upconing of the fresh-salt water interface beneath a pumping well.

Experimental and field observations indicate that the instability occurs even before inequality 4-90 is violated. The critical elevation is different for different thicknesses of the fresh-water zone and different well penetrations. As a practical matter, the interface is apparently stable for upconed heights less than or equal to about one-third of the distance between the bottom of the well and the original interface elevation (Dagan and Bear, 1968).

A first estimate of the maximum sustained well discharge that can be imposed without salt-water intrusion into the well can be made using the Ghyben-Herzberg equation (McWhorter, 1972). The steady discharge for the well in Fig. 4-28 is given by

$$Q = -2\pi K(b-s-\eta)r \frac{ds}{dr} \tag{4-99}$$

wherein b is the undisturbed fresh water thickness, s is the drawdown, and η is the interface elevation above the original level. Equation 4-99 is integrated, after putting $\eta=(\rho/\Delta\rho)s$ from Eq. 4-92, to obtain

$$Q = \frac{\pi K}{\ell n\ r_e/r_w} \{2bs_w - (1+ \Delta\rho/\rho)s_w^{\ 2}\} \quad . \tag{4-100}$$

The maximum safe discharge Q_m is obtained from Eq. 4-100 by substituting $s_w=(\Delta\rho/\rho)(b-\ell)/3$.

$$Q_m = \frac{\pi K b^2}{\ln r_e/r_w} \{ \frac{\Delta\rho}{\rho} (\frac{1 - \ell/b}{3})[2 - (1 + \Delta\rho/\rho) \frac{\Delta\rho}{\rho} (\frac{1 - \ell/b}{3})]\}. \tag{4-101}$$

Equation 4-101 is overly restrictive because, for a given drawdown, the piezometric head beneath the well is greater than predicted when horizontal flow is assumed. In view of Eq. 4-86, the fact that the piezometric head beneath the well is greater than predicted means that the interface rise is less than predicted. Notwithstanding the limitations of the above analysis, Eq. 4-101 serves to emphasize that small well discharges are required to avoid salt-water contamination.

In some instances, the length of the screened portion of the well bore is small relative to the depth of the well. In this situation, the steady-state interface elevation beneath the well is

$$\eta_w = \frac{Q\rho}{2\pi(b-\ell)\Delta\rho K} , \tag{4-102}$$

where ℓ is the distance between the top of the aquifer and the well screen for this case. Again using the condition that the interface should not be permitted to rise to more than one-third the distance between the bottom of the well and original interface elevation, the maximum safe, sustained discharge is

$$Q_m = \frac{2\pi}{3} (b-\ell)^2 \frac{\Delta\rho}{\rho} K \quad . \tag{4-103}$$

The vertical components of flow are accounted for, approximately, in the derivation of Eq. 4-103 (Dagan and Bear, 1968).

The discharges in Eqs. 4-101 and 4-103 are estimated sustained discharges that will eventually result in an upconing of the interface to a height $(b-\ell)/3$. A much larger discharge can be tolerated for short time periods because all of the fresh water occupying the volume between the original and critical interface positions must be displaced before the interface reaches the critical elevation. Intermittent operation of the well, causing the interface to rise and fall, also causes the transition zone between fresh and salt water to become larger, however, and the approximation of an abrupt interface becomes increasingly less satisfactory.

EXAMPLE 4-15

A well screen, 1 m in length, is located 16 m below the undisturbed water table in an aquifer in which K=17 m/d. An undisturbed fresh-sea water interface exists at a depth of 32 m below the water table. Estimate the maximum discharge that can be sustained from the well without causing the salt water to intrude into the well.

Solution:

The required discharge is computed from Eq. 4-103.

$$Q_m = \frac{2}{3} (32-16)^2 \frac{(0.025)}{1} (17) = 228 \text{ m}^3/\text{d}$$

$$Q_m = 2.64 \times 10^{-3} \text{ m}^3/\text{s} \quad .$$

The Island Fresh-Water Lens

Fresh ground water on small islands often occurs as a distinct body of water floating on underlying sea water. The fresh water accumulates as the result of recharge from precipitation. The theoretical developments of the foregoing sections predict that the thickness of the fresh water body is greatest near the center of the island and thins toward the coast where the fresh water is discharged into the sea. In cross-section, the fresh water accumulation has the shape of a lens (Fig. 4-29).

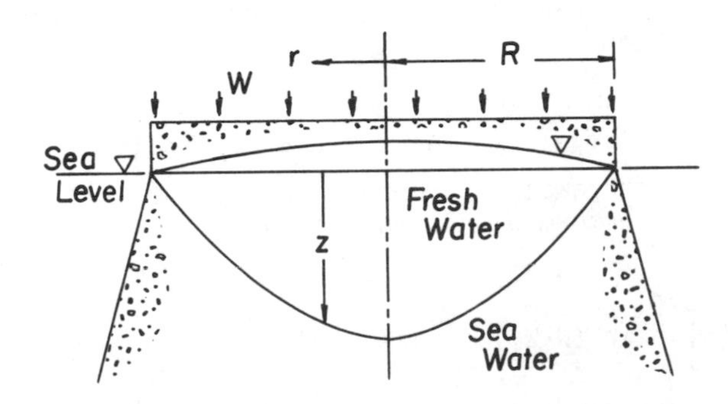

Figure 4-29. Fresh water lens on a circular island.

An equation relating the depth of the fresh-water lens to the hydraulic conductivity, the size of the island, and the recharge can be derived. The following development is for a circular island, but the approach applies equally well to a long, narrow island aquifer in which the flow is approximately one-dimensional, normal to the long axis of the island. Referring to Fig. 4-29, the discharge at a distance r from the center of the island is

$$Q(r) = -2\pi K(z+h)r \frac{dh}{dr} = \pi r^2 W \quad . \qquad (4\text{-}104)$$

Replacement of h by $(\Delta\rho/\rho)z$, and integration subject to z=0 at r=R, yields:

$$z = \left\{ \frac{W(R^2-r^2)}{2K\Delta\rho(1 + \frac{\Delta\rho}{\rho})} \right\}^{\frac{1}{2}} . \qquad (4\text{-}105)$$

Evidently, substantial fresh-water supplies are more likely to be found on large islands, with large recharge and small K.

EXAMPLE 4-16

On many islands, the major source of potable water is the fresh-water lens. Extraction systems often consist of large numbers of closely spaced shallow wells with very small discharges (on the order of 20 m^3/d). In some instances, shallow trenches are pumped in order to minimize the drawdown per unit of discharge. Consider an approximately circular lens of radius R on which the interface, at the center of the lens, is 13 m below sea level. The mean recharge is 30 cm per year. A well field is to be developed near the center of the lens. Plans call for the well field to cover 15 percent of the lens area. Calculate the withdrawal rate for the well field that will eventually result in a depth to the interface of 4 m at the center of the lens.

Solution:

The well field is idealized as a circle of radius R_i. Furthermore, withdrawal of ground water by the well field is simulated by a uniform withdrawal at rate W_i. In the well field

$$Q(r) = \pi r^2(W-W_i) = -2\pi K(z+h)r\frac{dh}{dr} \quad , \quad r \leq R_i \quad . \qquad (A)$$

Putting $h=(\Delta\rho/\rho)z$ into Eq. A and integrating subject to $z=z_o$ at $r=0$ and $z=z_i$ at $r=R_i$ gives

$$z_o^2 = \frac{\rho(W-W_i)R_i^2}{2K\Delta\rho(1 + \Delta\rho/\rho)} + z_i^2 \quad . \qquad (B)$$

The next step is to compute z_i^2. This is accomplished by analyzing the flow for $r \geq R_i$ where the discharge at any r is

$$Q(r) = (W-W_i)\pi R_i^2 + W\pi(r^2-R_i^2) = -2\pi K(z+h)r\frac{dh}{dr} \quad , \quad r \geq R_i \quad . \qquad (C)$$

Integration, subject to $z=z_i$ at $r=R_i$ and $z=0$ at $r=R$, gives

$$z_i^2 = \frac{\rho}{2K\Delta\rho(1 + \Delta\rho/\rho)} \{W(R^2-R_i^2) - 2W_iR_i^2 \ell n\, R/R_i\} \quad , \quad (D)$$

which is combined with Eq. B to yield

$$z_o^2 = \frac{\rho R^2}{2K\Delta\rho(1 + \Delta\rho/\rho)} [W-W_i(\frac{R_i}{R})^2\{1 + \ell n(R/R_i)^2\}] \quad , \quad (E)$$

after some arrangement.

Since the area, A, of the lens and the area, A_i, of the well field are proportional to their respective radii squared, Eq. E can be written

$$z_o^2 = \frac{\rho R^2}{2K\Delta\rho(1 + \Delta\rho/\rho)} [W-W_i(\frac{A_i}{A})\{1 + \ell n(\frac{A}{A_i})\}] \quad . \quad (F)$$

Before development of the well field, A_i=0. In the limit as A_i approaches zero, Eq. F reduces to

$$z_o^2 = \frac{\rho R^2 W}{2K\Delta\rho(1 + \Delta\rho/\rho)} \quad .$$

But, $z_o^2=(13)^2$ before development of the well field and W=0.3 m/yr. Therefore,

$$\frac{\rho R^2}{2K\Delta\rho(1 + \Delta\rho/\rho)} = \frac{(13)^2}{0.3} = 563 \text{ m-yr} \quad .$$

Thus, Eq. F becomes

$$z_o^2 = 563[W-W_i(\frac{A_i}{A})\{1 + \ell n(\frac{A_i}{A})\}] \quad ,$$

which can be solved for W_i:

$$W_i = \frac{W-(z_o^2/563)}{\frac{A_i}{A}\{1 + \ell n(\frac{A}{A_i})\}} \quad .$$

Putting A_i/A = 0.15, W = 0.3 m/yr, and z_o = 4 results in

$$W_i = \frac{0.3 - (16/563)}{0.15\{1 + \ell n(6.67)\}} = 0.62 \text{ m/yr} \quad .$$

Thus, the withdrawal rate is 6.25×10^5 m^3/yr per km^2. Approximately one-half of the annual withdrawal is contributed by recharge directly on the well field. The remaining portion is contributed by lateral inflow from lands outside of the well field. Notice that 31 percent of the total recharge on the lens is used. If the well field should be expanded to cover essentially all of the lens area, the maximum permissible withdrawal rate is reduced to

$$W_i = 0.3 - (16/563) = 0.27 \text{ m/yr} \quad ,$$

but the percent of recharge used is increased to 90 percent.

REFERENCES

Bear, J., 1972. Dynamics of Fluids In Porous Media. American Elsevier Publishing Co., Inc., New York. 764 p.

Chandler, R. A. and McWhorter, D. B., 1975. Interface Upconing Beneath a Pumping Well. Ground Water Vol. 13, No. 4, p. 354-359.

Churchill, R. V., 1960. Complex Variables and Applications. McGraw-Hill Book Company, Inc., New York. 297 p.

Dagan, G. and Bear, J., 1968. Solving the Problem of Local Interface Upconing in a Coastal Aquifer by the Method of Small Perturbations. Journal of Intern. Assoc. Hyd. Research, Vol. 6, No. 1, p. 15-44.

Ferris, J. G., Knowles, D. B., Brown, R. H., and Stallman, R. W., 1962. Theory of Aquifer Tests. U. S. Geol. Survey, Water Supply Paper 1536-E.

Glover, R. E., 1959. The Pattern of Fresh-Water Flow in a Coastal Aquifer. Journal of Geophysical Research, Vol. 64, No. 4, pp. 457-459.

Ground Water and Wells, 1972. Second printing, Johnson Division, Universal Oil Products Co., Saint Paul, Minnesota, 440 p.

Hantush, M. S., 1956. Analysis of Data from Pumping Tests in Leaky Aquifers. Trans. Am. Geophysical Union, Vol. 37, p. 702-714.

Hantush, M. S., 1962. On the Validity of the Dupuit-Forchheimer Well Discharge Formula. Journal of Geophysical Research, Vol. 67, No. 6, p. 2417-2420.

Hunt, B. W., 1970. Exact Flow Rates From Dupuit's Approximation. Proc. ASCE, Vol. 96, No. HY 3, p. 633-642.

McWhorter, D. B., 1972. Steady and Unsteady Flow of Fresh Water in Saline Aquifers. Water Management Techn. Report 20, Colorado State Univ., Fort Collins, Colorado. 49 p.

Mikels, F. C. and Klaer, F. H., Jr., 1956. Application of Groundwater Hydraulics to the Development of Water Supplies by Induced Infiltration. Intern. Assoc. Sci. Hydrology Symposium Darcy, Dijon, Publ. 41.

Muskat, M., 1946. The Flow of Homogeneous Fluids Through Porous Media. J. W. Edwards, Ann Arbor, Mich.

Polubarinova-Kochina, P. Ya., 1962. The Theory of Ground Water Movement. Translated by R. J. M. DeWiest, Princeton Univ. Press, Princeton, New Jersey. 613 p.

Sahni, B. M., 1972. Salt Water Coning Beneath Fresh Water Wells. Unpublished PhD Dissertation, Colorado State University, Fort Collins, Colorado.

Schmorak, S. and Mercado, A., 1969. Upconing of Fresh Water - Sea Water Interface Below Pumping Wells - Field Study. Water Resources Res., Vol. 5, No. 6, p. 1290-1311.

Shames, I. H., 1962. Mechanics of Fluids. McGraw-Hill Book Co., Inc., New York. 555 p.

Vallentine, H. R., 1959. Applied Hydrodynamics. Butterworth and Co., Limited, 272 p.

Walton, W. C., 1970. Groundwater Resource Evaluation. McGraw-Hill Book Co., Inc., New York. 664 p.

PROBLEMS AND STUDY QUESTIONS

1. Define equipotential contours and streamlines. Discuss their use in ground-water hydrology.

2. Prove that streamlines and equipotential lines intersect at right angles in nonhomogeneous, isotropic aquifers.

3. Identify several physical situations leading to mathematical boundary conditions of constant ψ constant h, and constant q.

4. Sketch the flow net, in the vertical plane, for confined, one-dimensional, uniform flow between parallel, fully penetrating channels, for h_L=30 m, h_o=10 m, L=50 m, b=4 m, and K=2 m/day. Label each equipotential and streamline with the correct numerical value. Calculate the pressure distribution along the top of the aquifer and the discharge per meter of length measured normal to the plane of your flow net. Answer: $Q = 3.2\ m^2/d$.

5. Show, mathematically, that the discharge per unit depth normal to the plane of flow in a streamtube is equal to the difference in the numerical value of the streamlines that form the boundaries of the streamtube.

6. Derive equations 4-24 and 4-25.

7. A fully penetrating well has been pumping from a large confined aquifer for a long time. The hydraulic conductivity of the aquifer is 0.023 cm/s. At a point 30 m from the well, the slope of the piezometric surface is 0.0216. If the aquifer thickness is 15 m, compute the well discharge in m^3/s. What is the drawdown between the points r=30 m and r=0.2 m? Answer: $0.014\ m^3/s$, 3.25 m.

8. Consider steady radial flow in a confined aquifer of thickness b. The hydraulic conductivity in the region between r_w and r_1 is K_1 and between r_1 and r_e is K. Derive an equation for the average permeability. Answer:

$$\bar{K} = \frac{KK_1\ \ell n(r_e/r_w)}{K_1\ \ell n(r_e/r_1) + K\ \ell n(r_1/r_w)} .$$

9. During the drilling of a well, drilling fluid infiltrates a confined aquifer in the region $r_w \leq r \leq 3$ m, and reduces the hydraulic conductivity in that region to 50% of the value in the remainder of the aquifer. If r_w=0.3 m and r_e=300 m,

estimate the steady-state specific capacity of the well with the damaged region relative to that which would be obtained with no damaged region. Answer: Specific capacity with damage is 75% of specific capacity without damage.

10. A city obtains water from a field of low capacity wells penetrating a confined sandstone aquifer. Estimate the increase in well discharge that could be obtained by increasing the hydraulic conductivity by a factor of 10 in the region $r_w \leq r \leq 15$ m. Assume r_w=0.3 m, r_e=300 m. Answer: the discharge could be doubled.

11. Derive a formula for average K of a confined layered aquifer in which radial flow is parallel to the layers. Answer: Eq. 3-34.

12. A well is pumped for a long time at a constant rate of 0.074 m^3/s from a confined aquifer. The difference in elevation of the piezometric surface in two observation wells located at r=6 m and r=46 m is 1.42 m. Calculate the transmissivity of the aquifer. Answer: T=0.017 m^2/s.

13. Two wells, spaced a distance of 75 m apart, pump at equal individual rates of 0.05 m^3/s from an aquifer with a transmissivity of 0.065 m^2/s. Assume that r_e is 1220 m. Calculate and plot the drawdown along the line joining the wells.

14. Consider two wells located a distance 2a apart in an infinite, confined aquifer. Assume that the two wells are pumped at equal rates, and that the origin of coordinates (x,y) is at the midpoint of the line joining the wells. The y-coordinate direction is normal to the line joining the wells (see Fig. 4-7). Show that $\partial h/\partial x = 0$ at all points on the line x=0, thus verifying that the line x=0 simulates an impermeable boundary. (Hint: see Example 4-5).

15. Derive an equation for the family of circles that represent streamlines for a pumped well located a perpendicular distance a from an infinitely long, fully penetrating stream with constant and horizontal water surface. (Hint: show that $\psi = -Q/2\pi b \{\tan^{-1}(y/a+x) + \tan^{-1}(y/a-x)\}$ by computing $\psi = \int \partial\Phi/\partial x \, dy$). Answer: $(y/a + c_i)^2 + (x/a)^2 = 1 + c_i^2$ where $c_i = -1/\{\tan(2\pi b\psi_i/Q)\}$.

16. Consider flow toward the horizontal drain shown in Fig. 4-10. Assume that the radius of the drain r_w is very small relative to both D and d. Using the first seven image drains located as shown in Fig. 4-9, derive an approximate formula for the drain discharge if the piezometric head at the ground surface is h_s and, at the drain, is h_d. Answer:

$$Q = \frac{-\ 2\pi K(h_s - h_d)}{\ell n\{r_w \dfrac{d(2D+d)}{D(d+D)(2d+D)}\}} \quad .$$

17. Show that the result in problem 16 reduces to the result in Eq. 4-36 as d becomes very large.

18. Suppose D=3 m, d=2 m, and r_w=0.15 m for the drain in problem 16. Draw a flow net and compute the quantity $Q/K(h_s-h_d)$ and compare with the results of problem 16. Answer: 1.88 from flow net and 1.85 from problem 16.

19. Streams never lie on an infinitely long straight line and the question arises as to what the length of a finite reach must be to yield results practically equal to those for an infinite reach. Calculate the length L of a reach of stream that contributes 95% of the discharge Q from a well located a distance a from the stream. (Hint: see Example 4-5). Answer: L = 25.4 a.

20. Draw several flow nets corresponding to different penetrations of the sheet piling in Fig. 4-16 and obtain a graphical relationship between depth of penetration (relative to the thickness of the aquifer) and Q/KH_t.

21. Show that $Q = n_s/n_\ell \sqrt{K_x K_y}\ H_t$ for seepage in an anisotropic aquifer.

22. Construct the flow net for the two-dimensional flow under the dam shown in the figure. Determine Q and the pressure distribution under the dam. Answer: Q = 1.8 m^3/d.

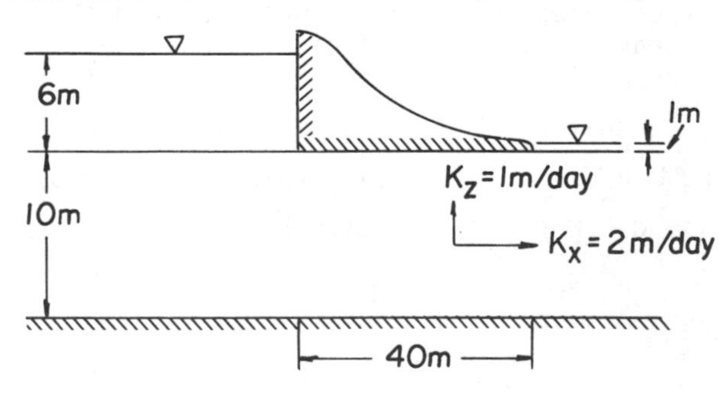

23. Resolve problem 22 with $K_z = K_x = 2$ m/d and compare the results with problem 22. Answer: $Q = 2\ m^2/d$.

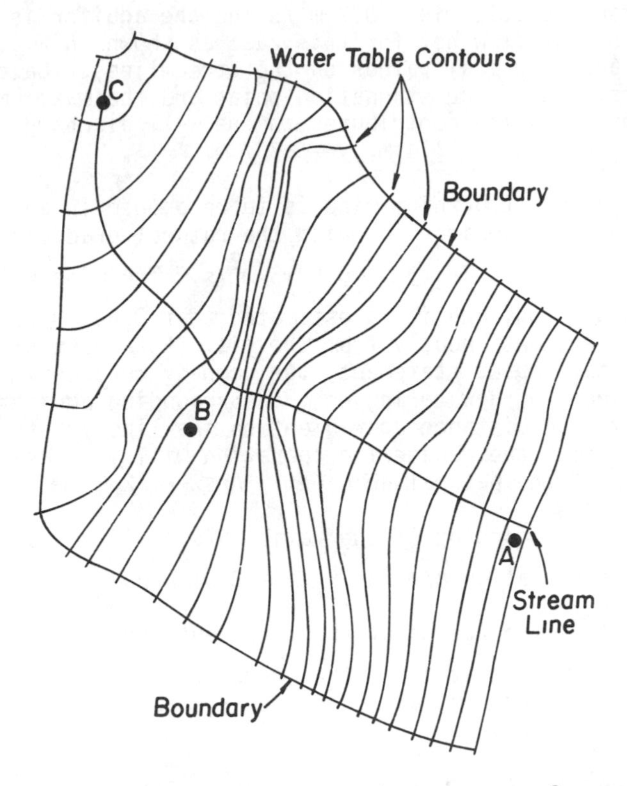

24. If the transmissivity at location A is 100 m^2/day, determine the transmissivity at locations B and C.

25. A city water supply is obtained from a fresh-water lens overlying salt-water. Because of the tendency for salt-water to cone up in response to drawdown, it is critical that the drawdown be minimized. One way to accomplish this is to pump many closely spaced wells at small discharges (distributed pumping). A total discharge of 3800 m^3 per day is required from an aquifer for which T=0.024 m^2/s and r_e = 1525 m. The maximum drawdown that can be tolerated is 45 cm. Estimate the required radius of the well field. The average spacing between wells on a grid pattern is 60 m. Calculate the discharge per well. Answer: r = 540 m.

26. A well is pumped at a constant discharge of 0.075 m^3/s for a long time in a uniform flow field in which the magnitude of the undisturbed gradient is 0.015. The aquifer transmissivity is 0.007 m^2/s and the aquifer is 10 m thick. Plot the flow net for this case as shown in Fig. 4-12. Place numerical values on the streamlines. Calculate the position of the stagnation point and the maximum width of aquifer which contributes to the well discharge.
Answer: $|x_s|$ = 114 m, y = ±357 m.

27. An open pit uranium mine is to be opened in an extensive sandstone aquifer in which the natural gradient is 0.008. The transmissivity is 0.0018 m^2/s. The mine will require dewatering, and it is estimated that 0.025 m^3/s will be pumped continuously from the pit. As a part of the environmental impact statement, the company must estimate the effects of dewatering on the surrounding aquifer. Calculate the distance upgradient on the line y=0 to the point at which the dewatering operation will increase $\partial h/\partial x$ by 10% of the natural gradient. Answer: 2.8 km.

28. Construct a plot of the ratio $(Q/s_w)_p/(Q/s_w)$ as a function of penetration ℓ/b for r_w/ℓ = 0.01 and r_w/ℓ = 0.005.

29. What is the radius of the catchment area for a drainage well pumping at a constant rate of 0.018 m^3/s from an aquifer subjected to a steady recharge rate of 0.082 cm/d?
Answer: 780 m.

30. A horizontal, impermeable layer exists 5 m below the surface of the soil in a humid region where the annual precipitation exceeds the annual evapotranspiration by 28 cm. A subsurface drainage system composed of equally spaced parallel drains is required that will maintain the maximum water-table elevation 1 m below the surface. If the drains are placed 2.1 m below and surface and $K=1.4\times10^{-4}$ cm/s, what should be the spacing between drains? Assume no surface runoff. Answer: 69 m.

31. A long narrow reservoir, 9.6 km in length, is formed in a canyon with sandstone walls with $K=2.1\times10^{-4}$ cm/s. Below the sandstone is a very very thick shale with $K=3.5\times10^{-6}$ cm/s. The contact between the sandstone and the shale is at the same elevation as the bottom of the reservoir. If the average reservoir depth is 50 m, estimate the steady-state leakage from the reservoir. Neglect the seepage directly out the bottom of the reservoir.
Answer: 22,500 m^3/d.

32. Is the water table a streamline in transient flow? In Problem 30? Explain!

33. A long straight fully penetrating canal flows parallel to a fully penetrating stream. Water seeps from the canal to the stream as indicated in Fig. 4-19, with h_o=4 m, h_L=3 m, L=15 m, and ϕ=0.40. A radio-active tracer was introduced into the canal and it appeared in the stream 48 hours later. Assume no dispersion and estimate the hydraulic conductivity. (Hint: put $dx_f/dt = Q/\phi h(x_f)$, where x_f is the distance to the tracer front, and use Eq. 4-74 to obtain $K = (4/3)\phi\, L^2(h_o^3 - h_i^3)/t(h_o^2 - h_i^2)^2$. Answer: 1.9 m/d.

34. A stream flows in the approximate center of an alluvial valley bounded by impermeable shale. The valley averages 1850 m in width and contains an alluvial aquifer in hydraulic connection with the stream. The hydraulic conductivity of the aquifer is 0.048 cm/s. The condition is one of steady state due to uniform recharge over the valley. Observation wells indicate that the water table at a distance of 100 m from the stream is 3 m above an impermeable shale and 1 m above the stream level. Compute the flow from the aquifer to the stream. Answer: 2.2 m^3/d per meter of stream.

35. Consider a flow situation similar to that depicted in Fig. 4-20. The recharge rate W in the interval x=±30 m is 0.03 cm/day due to deep percolation from irrigation. From x=±30 m to x=±120 m, evapotranspiration causes a net withdrawal of 0.015 cm/day. If K is 1×10^{-4} cm/s, L is 240 m, and h_o is 3 m, calculate and plot the water-table profile. Calculate the discharge between the ditches and the aquifer. Answer: Discharge from channels to aquifer is 9×10^{-3} m^2/d: one-half from each channel.

36. A lens of fresh water on an island can be idealized as a circle with an area of 550 ha. The depth of the interface below the sea level at the center of the lens is 18 m and the annual recharge rate is 40 cm per year. A plan for distributed pumping from the lens can be considered a reduction in recharge. What is the annual distributed pumping rate that will eventually result in new steady state with the interface 4 m below sea level? What is the average hydraulic conductivity of the aquifer?
Answer: Q=0.38 m/yr on 550 ha or 5730 m^3/d, K=1.4×10^{-3} m/s.

37. Consider the steady flow between two channels as shown in Fig. 4-20. Derive an equation for the height h of the water table if the water levels in the channels are h_L and h_R, $h_L > h_R$. Take the origin of the x-coordinate at the channel in which $h=h_L$.

Answer: $h^2 = h_L^2 - \frac{(h_L^2 - h_R^2)}{L} x + \frac{W}{K}(L-x)x$.

38. Derive formulas for the discharge rates to each channel and the x-coordinate of the ground-water divide for the conditions of problem 37.

Answer:
$$Q_L = \frac{K}{2L}(h_L^2 - h_R^2) - \frac{WL}{2}$$
$$Q_R = \frac{K}{2L}(h_L^2 - h_R^2) + \frac{WL}{2}$$
$$x_{divide} = \frac{L}{2} - \frac{K}{W}\left(\frac{h_L^2 - h_R^2}{2L}\right).$$

39. Derive an equation for the seepage rate from a long narrow pit under the circumstances indicated in Example 4-13 for a cylindrical pit.

Chapter V

UNSTEADY GROUND-WATER HYDRAULICS

Ground-water flows in which the piezometric head changes with time are called unsteady flows. Analyses of unsteady ground-water motion contained in this chapter are based on the differential equations developed in Chapter III. The developments for unconfined flow are, for the most part, limited to those cases in which the linearized Boussinesq equation applies. Because the linearized Boussinesq equation is formally identical to the differential equation for two-dimensional flow in confined aquifers, no mathematical distinction between the solutions for the two types of flow is required. The student should keep in mind, however, that the conditions under which the linearized Boussinesq equation was developed are significantly more restrictive than for confined flow.

The solutions presented in this chapter are but a few of the many that exist in the literature. Several additional solutions are presented in the appendix for easy reference.

5.1 RADIAL FLOW

Flow toward pumping wells and seepage to or from approximately circular reservoirs and pits are examples of flows which are essentially radial. Strictly speaking, radial flow in a plane is two-dimensional, but the use of the polar coordinate system takes advantage of the symmetry of radial flow and permits the differential equation to be written with one space variable; the radial coordinate, r.

Flow Toward A Fully Penetrating Well

The response to pumping at a constant rate from a single isolated well in an infinite, homogeneous, isotropic aquifer is given by the solution to

$$\alpha\left(\frac{\partial^2 s}{\partial r^2} + \frac{1}{r}\frac{\partial s}{\partial r} \right) = \frac{\partial s}{\partial t} \tag{5-1}$$

which is Eq. 3-60 written in terms of drawdown $s = h_0 - h$ for radial flow in the horizontal plane. The initial and boundary conditions are, respectively:

$$s(r,0) = 0$$
$$s(\infty,t) = 0$$
$$\lim_{r \to 0} r\frac{\partial s}{\partial r} = -\frac{Q}{2\pi T} \quad . \tag{5-2}$$

The parameter α is T/S for confined flow and T/S_{ya} for application in water-table aquifers and is sometimes called the *hydraulic diffusivity*. The first of conditions 5-2 is the initial condition and the second and third are boundary conditions. The third condition is an expression of the fact that the well is treated as a *line sink*. Also note that conditions 5-2 imply that the discharge abruptly changes from zero to Q at t=0.

The conditions under which Eq. 5-1 and following solutions (subject to 5-2) are derived are most nearly met in confined aquifers with a constant thickness. Application to unconfined flow is restricted to situations in which vertical components of flow are negligible and in which changes in aquifer storage by the mechanisms of water expansion and aquifer compression are negligible relative to gravity drainage of the pores as the water table falls in response to pumping. In practice, neither of these conditions are satisfied in the immediate vicinity of a well for a substantial period of time following an abrupt change of discharge. The response of unconfined aquifers in the vicinity of a pumped well is discussed in a subsequent section.

The solution to Eq. 5-1 subject to conditions 5-2 is obtained by introducing the *Boltzman variable*, u:

$$u = r^2/4\alpha t \quad . \qquad (5\text{-}3)$$

Use of the Boltzman variable permits the partial differential equation and the initial and boundary conditions to be expressed by

$$\frac{d^2 s}{du^2} + (1 + 1/u)\frac{ds}{du} = 0 \quad , \qquad (5\text{-}4)$$

and

$$s(\infty) = 0$$

$$\lim_{r \to 0} u\frac{ds}{du} = -\frac{Q}{4\pi T} \quad . \qquad (5\text{-}5)$$

A first integration of Eq. 5-4 yields

$$u\frac{ds}{du} = C_1 e^{-u} \quad , \qquad (5\text{-}6)$$

where C_1 is a constant of integration that is evaluated by use of the second of conditions 5-5. The result is

$$\frac{ds}{du} = -\frac{Q}{4\pi T}\frac{e^{-u}}{u} \quad . \qquad (5\text{-}7)$$

The desired solution is obtained by integration of Eq. 5-7 and use of the first of conditions 5-5:

$$s = \frac{Q}{4\pi T}\int_u^\infty \frac{e^{-x}}{x}\,dx \quad , \qquad (5\text{-}8)$$

where u is defined by Eq. 5-3 and x is a dummy variable of integration.

The integral in Eq. 5-8 is known as the *exponential integral*, values of which are tabulated in handbooks of mathematical tables. In ground-water literature (Theis, 1935) the exponential integral has become known as the *Theis well function* and is given the symbol W(u). Thus,

$$s = \frac{Q}{4\pi T} W(u) \quad . \qquad (5\text{-}9)$$

Values of the well function are presented in Table 5-1 for the range of values of u commonly encountered in ground-water work.

The well function can be expanded in an infinite series as follows:

$$W(u) = -0.5772 - \ell n\, u + u - \frac{u^2}{2\cdot 2!} + \frac{u^3}{3\cdot 3!} - \ldots . \qquad (5\text{-}10)$$

For $u \leq 0.01$, the well function is very closely approximated by the first two terms in the series and the equation for drawdown reduces to the simple closed form

$$s = \frac{Q}{4\pi T}\{\ell n\, 1/u - 0.5772\} \qquad u \leq 0.01 \quad . \qquad (5\text{-}11)$$

Actually, Eq. 5-11 is valid within 6.0 percent for $u \leq 0.10$. The reader should note that u is small for small r and/or large time. Therefore, Eq. 5-11 is accurate at large distances from the well only at large times.

A comparison of the Theis solution with drawdown data collected in a field test is shown in Fig. 5-1. Note that the measured drawdowns agree closely with the Theis solution, even at very small time.

EXAMPLE 5-1

Calculate the drawdown in a confined aquifer at r_1=0.3 m

TABLE 5.1. VALUES OF W(u) (After Wenzel, 1942)

N/u	$N\times10^{-15}$	$N\times10^{-14}$	$N\times10^{-13}$	$N\times10^{-12}$	$N\times10^{-11}$	$N\times10^{-10}$	$N\times10^{-9}$	$N\times10^{-8}$	$N\times10^{-7}$	$N\times10^{-6}$	$N\times10^{-5}$	$N\times10^{-4}$	$N\times10^{-3}$	$N\times10^{-2}$	$N\times10^{-1}$	N
1.0	33.9616	31.6590	29.3564	27.0538	24.7512	22.4486	20.1460	17.8435	15.5409	13.2383	10.9357	8.6332	6.3315	4.0379	1.8229	0.2194
1.5	33.5561	31.2535	28.9509	26.6483	24.3458	22.0432	19.7406	17.4380	15.1354	12.8328	10.5303	8.2278	5.9266	3.6374	1.4645	0.1000
2.0	33.2684	30.9658	28.6632	26.3607	24.0581	21.7555	19.4529	17.1503	14.8477	12.5451	10.2426	7.9402	5.6394	3.3547	1.2227	0.04890
2.5	33.0453	30.7427	28.4401	26.1375	23.8349	21.5323	19.2298	16.9272	14.6246	12.3220	10.0194	7.7172	5.4167	3.1365	1.0443	0.02491
3.0	32.8629	30.5604	28.2578	25.9552	23.6526	21.3500	19.0474	16.7449	14.4423	12.1397	9.8371	7.5348	5.2349	2.9591	0.9057	0.01305
3.5	32.7088	30.4062	28.1036	25.8010	23.4985	21.1959	18.8933	16.5907	14.2881	11.9855	9.6830	7.3807	5.0813	2.8099	0.7942	0.006970
4.0	32.5753	30.2727	27.9701	25.6675	23.3649	21.0623	18.7598	16.4572	14.1546	11.8520	9.5495	7.2472	4.9482	2.6813	0.7024	0.003779
4.5	32.4575	30.1549	27.8523	25.5497	23.2471	20.9446	18.6420	16.3394	14.0368	11.7342	9.4317	7.1295	4.8310	2.5684	0.6253	0.002073
5.0	32.3521	30.0495	27.7470	25.4444	23.1418	20.8392	18.5366	16.2340	13.9314	11.6280	9.3263	7.0242	4.7261	2.4679	0.5598	0.001148
5.5	32.2568	29.9542	27.6516	25.3491	23.0465	20.7439	18.4413	16.1387	13.8361	11.5330	9.2310	6.9289	4.6313	2.3775	0.5034	0.0006409
6.0	32.1698	29.8672	27.5646	25.2620	22.9595	20.6569	18.3543	16.0517	13.7491	11.4465	9.1440	6.8420	4.5448	2.2953	0.4544	0.0003601
6.5	32.0898	29.7872	27.4846	25.1820	22.8794	20.5768	18.2742	15.9717	13.6691	11.3665	9.0640	6.7620	4.4652	2.2201	0.4115	0.0002034
7.0	32.0156	29.7131	27.4105	25.1079	22.8053	20.5027	18.2001	15.8976	13.5950	11.2924	8.9899	6.6879	4.3916	2.1508	0.3738	0.0001155
7.5	31.9467	29.6441	27.3415	25.0389	22.7363	20.4337	18.1311	15.8280	13.5260	11.2234	8.9209	6.6190	4.3231	2.0867	0.3403	0.0000658
8.0	31.8821	29.5795	27.2769	24.9744	22.6718	20.3692	18.0666	15.7640	13.4614	11.1589	8.8563	6.5545	4.2591	2.0269	0.3106	0.0000376
8.5	31.8215	29.5189	27.2163	24.9137	22.6112	20.3086	18.0060	15.7034	13.4008	11.0982	8.7957	6.4939	4.1990	1.9711	0.2840	0.0000216
9.0	31.7643	29.4618	27.1592	24.8566	22.5540	20.2514	17.9488	15.6462	13.3437	11.0411	8.7386	6.4368	4.1423	1.9187	0.2602	0.0000124
9.5	31.7103	29.4077	27.1051	24.8025	22.4999	20.1973	17.8948	15.5922	13.2896	10.9870	8.6845	6.3828	4.0887	1.8695	0.2387	0.0000071

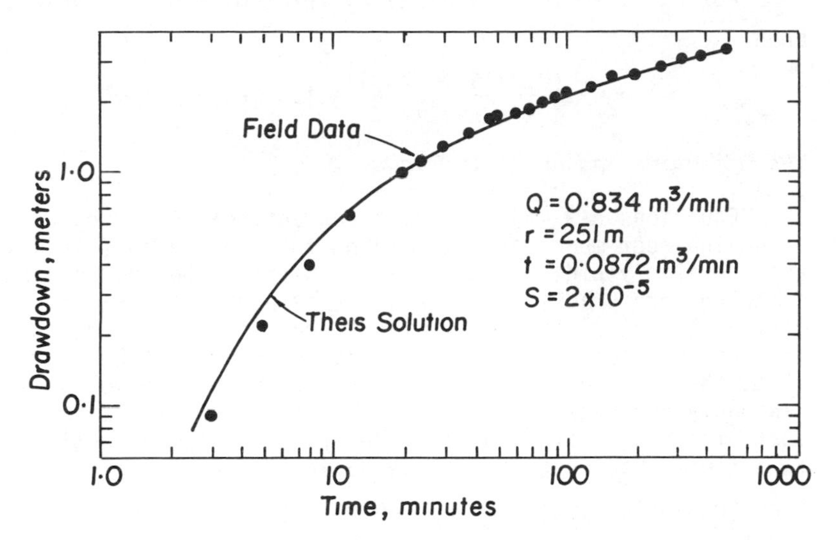

Figure 5-1. Measured and calculated drawdown-time curve for a confined aquifer (Adapted from Walton, 1962).

and r_2=25 m after 7 hours of pumping with a constant discharge of 0.0315 m^3/s. The aquifer properties are S=0.001 and T=0.0094 m^2/s.

Solution:

Numerical values for the variable u are computed from Eq. 5-3 with α = T/S = 9.4 m^2/s.

$$u_1 = \frac{(0.3)^2}{4(9.4)(7)(3600)} = 9.499 \times 10^{-8}$$

$$u_2 = \frac{(25)^2}{4(9.4)(7)(3600)} = 6.596 \times 10^{-4} \quad .$$

Because u_1 is less than 0.01, Eq. 5-11 can be used to compute the drawdown at r_1=0.3 m.

$$s_1 = \frac{0.0315}{4\pi(0.0094)} \{\ell n \left(\frac{1}{9.499 \times 10^{-8}} \right) - 0.5772\}$$

$$= 4.16 \text{ meters} \quad .$$

The above computation is verified by noting that W(9.499×10^{-8}) is 15.592 from Table 5-1. Thus, Eq. 5-9 also yields s_1=4.16 m.

For $u_2=6.596\times10^{-4}$, W=6.747 by interpolation in Table 5-1 and

$$s_2 = \frac{(0.0315)(6.747)}{4\pi(0.0094)} = 1.8 \text{ meters} \quad .$$

The Effective Radius Of Influence

Equations 5-9 and 5-11 find a great deal of practical use as an indirect means of determining the storage coefficient, the specific yield, and the transmissivity. Before proceeding to these very practical matters, it is useful to examine some of the characteristics of aquifer behavior as predicted by the present solution. First, it should be noted that Eq. 5-9 predicts that the cone of drawdown around the well develops instantaneously and extends to infinity. However, calculations show that, for practical purposes, the drawdown becomes negligible at a finite radius that will be designated by r_e. The practical radius of influence r_e can be regarded as the radius at which the drawdown is, say, 2 cm or any other arbitrary drawdown that can be regarded as negligible in the problem under consideration. Clearly r_e is a function of time because the cone of influence grows with increasing time. This phenomenon is demonstrated by the several profiles of piezometric head shown in Fig. 5-2.

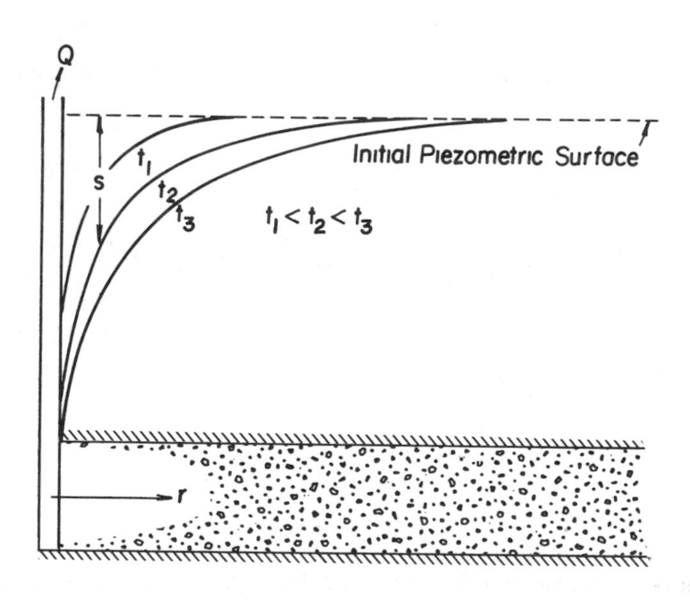

Figure 5-2. Drawdown profiles near a pumped well.

The dependence on time of the radius of influence is easily derived. The definition of r_e is that radius at which $s=s_e$, wherein s_e is some constant, arbitrarily small value. Thus, from Eq. 5-9, it follows that u must be constant and equal to

$$u = W^{-1} \left(\frac{4\pi T s_e}{Q} \right) = C \qquad (5\text{-}12)$$

where W^{-1} means the value of u for which W(u) is equal to the quantity in parentheses. For example, if the quantity in parentheses is equal to 1.823, then C is 0.1 from Table 5-1. From the definition of u, the radius of influence is found to be proportional to the square root of time.

$$r_e = (4\alpha C)^{\frac{1}{2}} t^{\frac{1}{2}} \quad . \qquad (5\text{-}13)$$

Equation 5-13 predicts that the radius of influence expands rapidly at small time and more slowly as the time since pumping began becomes large. Notice that, other things being equal, the radius of influence is larger in aquifers with small coefficients of storage as compared to those with large storage coefficients. For example, Eq. 5-13 predicts that the radius of influence in a confined aquifer with S=0.001 is 10 times greater than for a water table aquifer with S_{ya}=0.1, other factors equal. Consequently, even though the aquifer was assumed to be infinite in areal extent, Eq. 5-9 can be used in finite aquifers provided that the distance from the well to aquifer boundaries always exceeds r_e.

EXAMPLE 5-2

Compare the effective radii of influence at t=10 hrs for wells pumping at 0.0315 m^3/s in aquifers for which A) T=0.01 m^2/s, S_{ya}=0.1, B) T=0.1 m^2/s, S_{ya}=0.1, C) T=0.01 m^2/s, S=0.01. Let r_e be that radius at which $s=s_e$=0.05 m.

Solution:

$$\text{A)} \quad \frac{4\pi T s_e}{Q} = \frac{4\pi(0.01)(0.05)}{0.0315} = 0.1995 = W(u) \quad .$$

From Table 5-1, the value of u for which W(u)=0.1995 is 1.02 and from Eq. 5-13

$$r_e = \left\{ \frac{4(0.01)(1.02)(36000)}{0.1} \right\}^{\frac{1}{2}} = 121 \text{ meters} \quad .$$

A similar computation for case B yields r_e=109 m, and for case C, r_e=383 m. It is evident from this computation that the radius of influence is much more sensitive to the storage characteristics of the aquifer than to the transmissive capacity.

The Psuedo-Steady State

The boundary conditions imposed for the derivation of Eq. 5-9 do not provide for any recharge to the aquifer. It should not be surprising, therefore, that Eq. 5-9 predicts that the flow will remain in the transient state at all times. On the other hand, it was stated in Chapter IV that the flow toward an isolated well in a large aquifer can be accurately described by the steady radial flow equation in the vicinity of the well where a psuedo-steady state is developed. Now this statement can be justified, explicitly.

Provided that the well has been pumped at a constant rate for a sufficiently long time and that attention is restricted to a region near the well, then $u<0.01$ and the drawdown is given by Eq. 5-11. Suppose that the drawdowns are measured simultaneously in observation wells located at r_1 and r_2, where $r_2>r_1$. Then,

$$s_1 = \frac{Q}{4\pi T}\{\ell n\ 1/u_1 - 0.5772\} \qquad (5\text{-}14)$$

at r_1, and

$$s_2 = \frac{Q}{4\pi T}\{\ell n\ 1/u_2 - 0.5772\} \qquad (5\text{-}15)$$

at r_2. The drawdown at r_1 minus the drawdown at r_2 is h_2-h_1 so that

$$h_2-h_1 = \frac{Q}{4\pi T}\{\ell n\ (1/u_1) - \ell n\ (1/u_2)\}$$

or

$$h_2-h_1 = \frac{Q}{4\pi T}\ell n\ (r_2/r_1)^2 = \frac{Q}{2\pi T}\ell n\ (r_2/r_1)\ , \qquad (5\text{-}16)$$

which is Eq. 4-29 obtained by a steady-state analysis. It is important to note that the steady-state equation was derived here under the condition that u be less than 0.01 at r_2. Thus, the condition that

$$\frac{r_2^{\ 2}}{4\alpha t} < 0.01 \qquad (5\text{-}17)$$

can be used to determine when and where the steady-state equation can be applied. Actually, the above analysis does not imply that the flow is steady for $r<r_2$, but rather that the time rate of change of s in the region of $r<r_2$ is practically independent of r (i.e., the water table or piezometric surface is falling at nearly the same rate everywhere in the region $r<r_2$). In a true steady-state the water levels in the observation wells or piezometers do not change with time. The water levels are continuously declining in the psuedo-steady state and Eq. 5-16 is valid only if the water levels are observed simultaneously.

EXAMPLE 5-3

Calculate the limit of the psuedo-steady state region around the pumped well of Examples 2-5 and 4-2. Recall that $Q = 0.0312\ m^3/s$, $S_{ya} = 0.09$, $T = 0.0094\ m^2/s$, and t = 8 hrs and 22 minutes.

Solution:

From Eq. 5-17

$$r_2 = \left[\frac{(0.01)(4)(0.0094)(30120)}{0.09} \right]^{\frac{1}{2}} = 11 \text{ meters} \quad .$$

This computation indicates that the use of the steady-state equation for calculation of T in Example 4-2 is valid provided that observation wells less than 11 meters from the pumped well are used to measure water levels. Careful inspection of the data in Fig. 2-11 shows only a very small deviation from a linear relationship between h and log r out to $r \approx 20$ m . This observation indicates that the steady-state equation should be valid for $r<20$ m rather than $r<11$ m as just calculated. However, since the quantity ($\ln 1/u - 0.5772$) approximates W(u) to within 1.5 percent for values of u as great as 0.04, it is not surprising that h vs. log r appears linear for r as great as 20 meters. With $u = 0.04$, the limit for the psuedo-steady state region is 22 meters.

Response of An Unconfined Aquifer Near a Pumped Well

The drawdown, measured as a function of time, in two observation wells during a test of an unconfined aquifer near Fort Collins, Colorado is shown in Fig. 5-3. The measured relationship between drawdown and time at r = 20 m is very similar to that shown in Fig. 5-1 for a confined aquifer. The s vs. t data measured at r = 4.6 m bears few similarities to that at r = 20 m, however, and it cannot be expected that the data agree with the Theis solution.

When the well discharge is abruptly changed from zero to Q, the water level in the pumped well falls rapidly and there

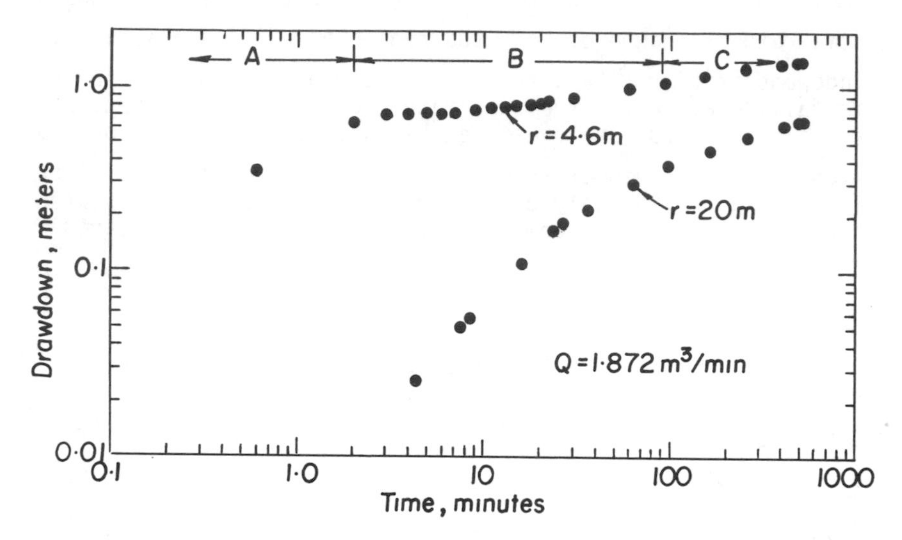

Figure 5-3. Measured drawdown - time curves at two radii during a test of an unconfined aquifer.

is a corresponding rapid decrease in the piezometric head in the aquifer near the pumped well. The water level in a completely penetrating observation well located near the pumped well reflects the average piezometric head over the vertical line on which it is located and, therefore, the drawdown increases rapidly as shown in Fig. 5-3 for r = 4.6 m during the time interval A. The water level in the observation well is not an indicator of the water-table elevation, however, the water-table level being substantially higher on the outside of the observation well than the level of water inside the observation well. The rate of vertical drainage is controlled by the difference in the water-table level and the piezometric head at points below the water, the average of which is reflected by the water level in the observation well.

Because the water table does not fall significantly in the first few minutes following the start of the pump, the volume of water discharged by the pump cannot be accounted for by pore drainage (i.e., by the apparent specific yield) and must be the result of aquifer compaction and water expansion. Therefore, the drawdown-time relationship for small r during the time period A in Fig. 5-3 is more nearly characteristic of that for a confined aquifer with storage coefficient S than of an unconfined aquifer with apparent specific yield S_{ya}.

During the time interval B in Fig. 5-3, vertical drainage in response to the difference in the piezometric head on the water table and the reduced piezometric head at points below the water table becomes important relative to the rate of change of storage characterized by water expansion and aquifer compaction. Thus, the rate at which drawdown increases slows substantially. The situation is not unlike that of vertical leakage into a confined aquifer. Finally, during time interval C in Fig. 5-3, vertical components of the gradient of piezometric head become small and the drawdown-time relationship becomes that which is characterized by the apparent specific yield and the Theis solution with $\alpha = T/S_{ya}$ should apply. Even though vertical components of flow are sufficiently small during time period C to make the horizontal flow approximation valid, the small vertical flow may still substantially reduce the apparent specific yield from the value that would be obtained after a very long pumping period.

All of the above described phenomena can be expected to be less pronounced at large r because the piezometric head is not reduced so greatly nor so rapidly as at small r. Indeed, the data for r = 20 m in Fig. 5-3 do not show the above described behavior.

The above description of the drawdown response to a pumped well in unconfined aquifers is essentially that of Walton (1960). The phenomena described have been collectively called *delayed yield from storage* (Boulton, 1954). Approximate analytical solutions that are capable of simulating drawdown-time curves for unconfined aquifers have been developed (Boulton, 1954; Boulton and Pontin, 1971; Neuman, 1972; Streltsova, 1972). Streltsova (1972) writes a mass conservation equation using the procedures discussed in Chapter II that accounts for changes in storage by water expansion and aquifer compaction and desaturation or drainage of the pores. Streltsova's differential equation is

$$T\left(\frac{\partial^2 s}{\partial r^2} + \frac{1}{r}\frac{\partial s}{\partial r} \right) = S\frac{\partial s}{\partial t} + S_{ya}\frac{\partial s^o}{\partial t} \qquad (5\text{-}18)$$

where T is an average transmissivity, s is the drawdown as indicated by the difference in the initial, static water-table level and the water level in a fully penetrating observation well that reflects the average piezometric head over the saturated thickness. The variable s^o is the drawdown of the water table. The approximate average value of the Darcy velocity in the vertical direction is, from Darcy's Law,

$$q_z = -K_z \frac{(s-s^o)}{b_a} \qquad (5\text{-}19)$$

where b_a is the vertical distance between the water table and the point at which the average drawdown s occurs. The value of b_a can be taken as 1/3 of the initial saturated thickness b as a first approximation. Also, consideration of the velocity of the declining water table requires

$$q_z = -S_{ya} \frac{\partial s^o}{\partial t} \quad . \tag{5-20}$$

Combining Eqs. 5-18 through 5-20 yields

$$\frac{\partial^2 s}{\partial r^2} + \frac{1}{r}\frac{\partial s}{\partial r} = \frac{S}{T}\frac{\partial s}{\partial t} + \frac{s - s^o}{B^2} \tag{5-21}$$

where

$$B^2 = \frac{Kbb_a}{K_z} \quad . \tag{5-22}$$

The student should note the similarity between Eq. 5-21 and Eq. 3-81 derived for a leaky aquifer in Chapter III. It must be understood, however, that s^o is a function of both time and radial distance in Eq. 5-21, while h_o in Eq. 3-81 is a constant. Streltsova (1972) gives the solution to Eq. 5-21 as

$$s = \frac{Q}{4\pi T}\int_0^\infty \frac{2J_o(x, r/B)}{} \left[1 - \frac{1}{x^2+1} \exp\{-\frac{r^2}{4B^2}\left(\frac{1}{u}\right)\frac{x^2}{x^2+1}\}\right]dx \tag{5-23}$$

for the case in which S = 0 and subject to the initial and boundary conditions given by Eqs. 5-2. The function J_o is the Bessel function of the first kind and order zero, u is the same as defined in Eq. 5-3, and x is a dummy variable of integration. The same result was obtained by Boulton (1963). Because S was set equal to zero in the derivation of Eq. 5-23, the drawdown time relationship for small time is not simulated by the result. This is evident on the comparison of the curves in Fig. 5-4 with the measured drawdown at r=4.6 m as shown in Fig. 5-3. It is clear from Fig. 5-4 that the drawdown-time relationship predicted by Eq. 5-23 reduces to the Theis solution for sufficiently small u. The family of curves shown in Fig. 5-4 does not cover the entire range of conditions encountered in practice. Table 5-2 contains data from which the reader can prepare curves similar to those shown in Fig. 5-4 for a wider range of r/B values.

Radial Flow In a Leaky Aquifer

The differential equation for flow in a confined aquifer with vertical accretion was derived in Chapter III and is given

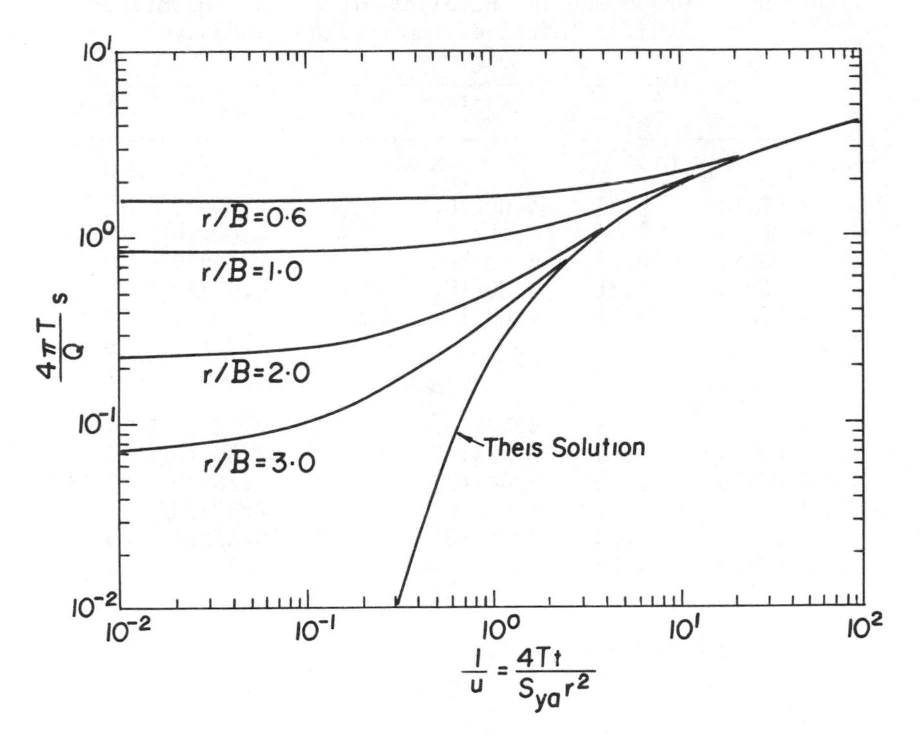

Figure 5-4. Graphical presentation of Eq. 5-23.

in polar coordinates and in terms of drawdown by

$$\frac{\partial^2 s}{\partial r^2} + \frac{1}{r}\frac{\partial s}{\partial r} - \frac{s}{\mathcal{B}^2} = \frac{S}{T}\frac{\partial s}{\partial t} \qquad (5\text{-}24)$$

where

$$\mathcal{B} = \frac{Tb_a}{K_a} \quad . \qquad (5\text{-}25)$$

The solution of Eq. 5-24 subject to conditions given by Eq. 5-2 is (Hantush and Jacob, 1955):

$$\frac{4\pi T}{Q} s = 2K_o(r/\mathcal{B}) - 2\int_0^\infty [x/(x^2+1)]J_o(x,r/\mathcal{B})\exp[-\frac{r^2}{4\mathcal{B}^2}(\frac{1}{u})(x^2+1)]dx. \qquad (5\text{-}26)$$

The right side of Eq. 5-26 is given the symbol $W(u,r/\mathcal{B})$ and is called the well function for an infinite leaky aquifer with no change in storage in the aquitard. Values of $W(u,r/\mathcal{B})$ are given in Table 5-3.

Table 5-2. Drawdown-Time Relationship For an Unconfined Aquifer (Adapted from Boulton, 1963).

r/B	u^{-1}	$\frac{4\pi T}{Q}s$
r/B=0.01	4.00×10^{2}	9.45
	4.00×10^{3}	9.54
	4.00×10^{4}	10.23
	4.00×10^{5}	12.31
	4.00×10^{6}	14.61
r/B=0.1	4.00×10^{0}	4.86
	4.00×10^{1}	4.95
	4.00×10^{2}	5.64
	4.00×10^{3}	7.72
	4.00×10^{4}	10.01
r/B=0.2	4.00×10^{-1}	3.51
	4.00×10^{0}	3.54
	2.00×10^{1}	3.69
	4.00×10^{1}	3.85
	1.50×10^{2}	4.55
	4.00×10^{2}	5.42
r/B=0.4	1.00×10^{-1}	2.23
	1.00×10^{0}	2.26
	5.00×10^{0}	2.40
	1.00×10^{1}	2.55
	3.75×10^{1}	3.20
	1.00×10^{2}	4.05
r/B=0.6	4.44×10^{-1}	1.586
	2.22×10^{0}	1.707
	4.44×10^{0}	1.844
	1.67×10^{1}	2.448
	4.44×10^{1}	3.255
r/B=0.8	2.50×10^{-2}	1.133
	2.50×10^{-1}	1.158
	1.25×10^{0}	1.264
	2.50×10^{0}	1.387
	9.37×10^{0}	1.938
	2.50×10^{1}	2.704
r/B=1.0	4.00×10^{-2}	0.844
	4.00×10^{-1}	0.901
	4.00×10^{0}	1.356
	4.00×10^{1}	3.140
r/B=1.5	7.11×10^{-2}	0.444
	3.55×10^{-1}	0.509
	7.11×10^{-1}	0.587
	2.67×10^{0}	0.963
	7.11×10^{0}	1.569
r/B=2.0	4.00×10^{-2}	0.239
	2.00×10^{-1}	0.283
	4.00×10^{-1}	0.337
	1.50×10^{0}	0.614
	4.00×10^{0}	1.111
r/B=2.5	2.56×10^{-2}	0.1321
	1.28×10^{-1}	0.1617
	2.56×10^{-1}	0.1988
	9.60×10^{-1}	0.399
	2.56×10^{0}	0.7977
r/B=3.0	1.78×10^{-2}	0.0743
	8.89×10^{-2}	0.0939
	1.78×10^{-1}	0.1189
	6.67×10^{-1}	0.2618
	1.78×10^{0}	0.5771

At large time, the integral in Eq. 5-26 approaches zero and

$$s = \frac{Q}{2\pi T} K_0(r/B) \quad . \tag{5-27}$$

This corresponds to the case in which the replenishment to the aquifer via leakage through the aquitard equals the pumping rate and the flow is steady. Equation 5-27 is the result obtained in Chapter IV when steady flow was assumed at the outset.

Drawdown With Variable Pumping Rates

The Theis solution predicts the drawdown in response to a step change in pumping rate from zero to Q at t=0; the discharge

Table 5-3. Values of $W(u,r/B)$ For a Leaky Confined Aquifer (Adapted from Hantush, 1956).

	W(u,r/B)							
u \ r/B=	0.01	0.05	0.2	0.5	1.0	1.5	2.0	2.5
5×10^{-6}	9.441							
1×10^{-5}	9.418							
5×10^{-5}	8.883							
1×10^{-4}	8.398	6.228						
5×10^{-4}	6.975	6.082						
1×10^{-3}	6.307	5.796	3.505					
5×10^{-3}	4.721	4.608	3.457					
1×10^{-2}	4.036	3.980	3.288	1.849				
5×10^{-2}	2.468	2.458	2.311	1.493	0.841			
1×10^{-1}	1.823	1.818	1.753	1.442	0.819	0.427	0.228	
5×10^{-1}	0.560	0.559	0.553	0.521	0.421	0.301	0.194	0.117
1×10^{0}	0.219	0.219	0.218	0.210	0.186	0.151	0.114	0.080
5×10^{0}	0.001	0.001	0.001	0.001	0.001	0.001	0.001	0.001

remaining constant thereafter. The Theis solution is not restricted to a step change at t=0, however. For example,

$$s_i = \frac{\Delta Q_i}{4\pi T} W\{ \frac{r^2}{4\alpha(t-t_i)} \} \quad , \quad t \geq t_i \tag{5-28}$$

is the drawdown in response to a step change ΔQ_i in discharge at $t=t_i$. Furthermore, since Eq. 5-1 is linear, the drawdown given by Eq. 5-28 can be added to the drawdown that would have existed, had the change ΔQ_i not occurred, to obtain the total drawdown. In other words, the incremental response to a change in pumping rate is independent of the previous history of discharge. The drawdown at any time $t \geq t_n$ in response to n step changes in pumping rate is, therefore,

$$s = \frac{1}{4\pi T} \sum_{i=1}^{n} \Delta Q_i \, W\{ \frac{r^2}{4\alpha(t-t_i)} \} \quad , \quad t \geq t_n . \tag{5-29}$$

Application of Eq. 5-29 to a case in which the pumping rate is changed from zero to $Q=Q_1$ at $t_1=0$ and then from Q_1 to Q_2 at t_2 is illustrated in Fig. 5-5. The indicated drawdown is computed from

$$s = \frac{Q_1}{4\pi T} W(\frac{r^2}{4\alpha t}) \quad , \quad t \leq t_2 \tag{5-30}$$

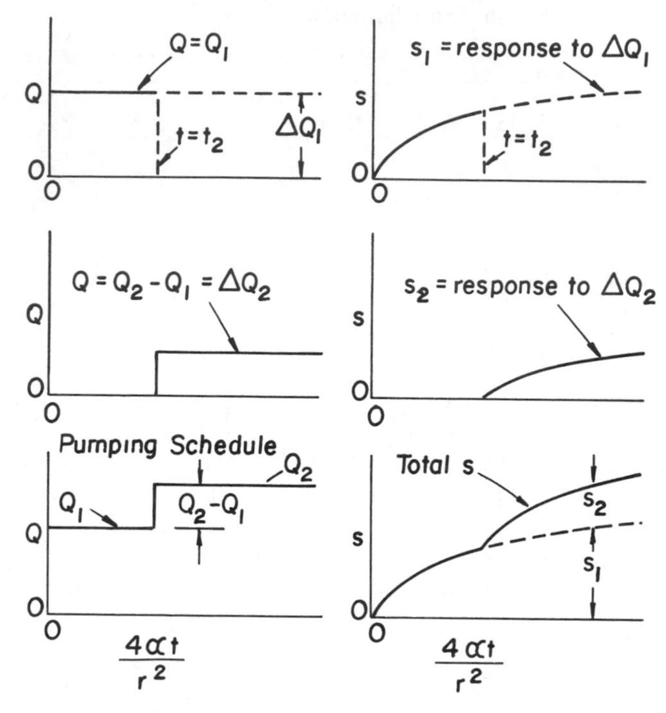

Figure 5-5. Superposition of drawdowns to obtain drawdown after a step change in discharge.

and

$$s = \frac{Q_1}{4\pi T} W\left(\frac{r^2}{4\alpha t} \right) + \frac{Q_2 - Q_1}{4\pi T} W\left\{ \frac{r^2}{4\alpha(t-t_2)} \right\} \quad , \quad t \geq t_2 \quad . \tag{5-31}$$

The above computation is equivalent to calculating the drawdown due to pumping at rate Q_1 beginning at $t=t_1=0$ from one well and adding to it the drawdown due to the pumping at rate Q_2-Q_1 beginning at $t=t_2$ from a second, imaginary well at the same location as the real well. The computation is equally valid for a reduction in discharge rate at $t=t_2$.

An important use of Eq. 5-29 is to compute the response to intermittent operation of wells. The special case given by Eq. 5-31 can be used to calculate the recovery of water levels in observation wells following the shut-down, at time t_2, of a pumped well. Following the shut-down of the well, $Q_2=0$ and Eq. 5-31 becomes

$$s = \frac{Q_1}{4\pi T} [W(\frac{r^2}{4\alpha t}) - W\{ \frac{r^2}{4\alpha(t-t_2)} \}] \quad , \quad t \geq t_2 \quad . \tag{5-32}$$

In this case, t_2 is the pumping interval. Provided $r^2/4\alpha(t-t_2)$ is less than 0.01, Eq. 5-32 can be written in the particularly simple form:

$$s = \frac{Q_1}{4\pi T} \ln \frac{t}{t-t_2} \quad . \tag{5-33}$$

Equation 5-33 implies the drawdown is practically the same (i. e., a horizontal piezometric surface) in the region for which $r^2/4\alpha(t-t_2) \leq 0.01$ during the recovery period. The region in which the piezometric surface is practically horizontal enlarges with increasing time after shutdown until, eventually, the initial static condition is reached if the aquifer is, indeed, very large.

Equation 5-29 can be used to compute the drawdown in response to a continuously changing pumping rate by approximating the function Q(t) by a series of step changes of discharge. It is possible to write a mathematically exact equation for the drawdown in response to a continuous function Q(t), however. For example, since

$$\Delta Q_i = \left. \frac{\Delta Q}{\Delta t} \right|_{t=t_i} \Delta t \quad , \tag{5-34}$$

Eq. 5-29 becomes

$$s = \frac{1}{4\pi T} \sum_{i=1}^{n} \left. \frac{\Delta Q}{\Delta t} \right|_{t=t_i} W\{ \frac{r^2}{4\alpha(t-t_i)} \} \Delta t \quad . \tag{5-35}$$

The interpretation of Eq. 5-35 is the cumulative response at time t to a sequence of rate changes that occurred at times t_i. In the limit as the rate changes continuously,

$$s = \frac{1}{4\pi T} \int_0^t \frac{dQ}{d\tau} W\{ \frac{r^2}{4\alpha(t-\tau)} \} d\tau \tag{5-36}$$

where τ is an integration variable that represents the time at which the change dQ occurred and t is the time at which the response is desired. The function $W\{r^2/4\alpha(t-\tau)\}$ is the response at time t to a unit step change in discharge that occurred at time τ

Integration of Eq. 5-36 by parts yields an equivalent expression

$$s = \frac{1}{4\pi T} \int_0^t \frac{Q(\tau)}{t-\tau} \exp\{-r^2/4\alpha(t-\tau)\}d\tau \quad . \qquad (5\text{-}37)$$

The quantity $(1/4\pi T)\exp\{-r^2/4\alpha(t-\tau)\}\{1/(t-\tau)\}$ has the significance of being the aquifer response at radial distance r and at time t to an instantaneous withdrawal at r=0 and at time τ of a unit volume of water. This quantity is often called the instantaneous sink solution. On the other hand, Eq. 5-37 is the response at radius r and at time t to a continuous withdrawal at rate $Q(\tau)$ and is called the continuous-sink solution (Morel-Seytoux and Daly, 1975).

Ferris and Knowles (1963) adapted the instantaneous sink solution to the problem of determining the transmissivity of confined aquifers by measuring the response to the injection of a slug of water of volume V at t=0. In this case a buildup of piezometric head is obtained upon injection, followed by a dissipation of the buildup. The difference between the elevation of the piezometric surface before injection of the slug and the elevation at any time after injection is

$$s = -\frac{V}{4\pi Tt} \exp(-r^2/4\alpha t) \quad . \qquad (5\text{-}38)$$

Provided r is sufficiently small, the argument of the exponential rapidly approaches zero in a confined aquifer with a small storage coefficient. Thus,

$$s \simeq -\frac{V}{4\pi Tt} \quad , \qquad (5\text{-}39)$$

which suggests that measured values of s can be plotted against 1/t and the value of transmissivity computed from the slope of the resulting straight line. Analysis of slug tests has been improved by obtaining a solution that does not assume an infinitesimal diameter of the line on which the slug is injected (Cooper et al., 1967; Papadopulos et al., 1973).

5.2 AQUIFER TEST APPLICATIONS

A practical use of the material discussed in the previous section is made in the determination of the hydrogeologic parameters of aquifers. Aquifer properties are obtained from an aquifer test by determining the values of storage coefficient (apparent specific yield), hydraulic conductivity, and transmissivity which cause the drawdowns predicted from theoretical solutions to most nearly agree with drawdowns measured in one or more observation wells. The hydrogeologic parameters may be used for designing well fields, estimating the effects of pumping on surface water, assessing ground-water supply, estimating inflow into mines, and a host of other purposes. Aquifer tests require a substantial investment in time and finances and, to some degree, the type of test, duration, equipment, and number of observations are dictated by non-scientific considerations. This is all the more reason to plan the test carefully so that a maximum of information is derived from the investment.

Preliminary Considerations In Aquifer Testing

Because the aquifer properties are determined by matching measured drawdowns with those predicted by theoretical equations, it is important that the aquifer geometry, boundary conditions, and initial conditions at the test site match those assumed in the theoretical equations as closely as possible. These factors should be considered in the selection of the test site if the site is not dictated by other factors. Geologic information assists in the location of aquifer boundaries and the assessment of stratification in the aquifer, and the degree of homogeniety and isotropy that can be expected.

The construction and location of observation wells is an important consideration. Observation wells are usually cased with perforated pipe. The diameter of observation wells is usually dictated by the method to be used for measuring the water level, but are normally from 5-15 cm in diameter. In unconfined aquifers it is desirable that the observation well casing completely penetrate the saturated thickness and that the casing be perforated over the entire portion that extends below the water table. This helps to insure that the water level in the observation well properly indicates the average piezometric head over the vertical section; an important consideration if Eq. 5-23 is to be used to analyze the data. In confined aquifers, care should be taken that the observation well itself does not provide a conduit by which water from the aquifer can flow into overlying strata. This may require sealing the annular space between the observation-well casing and wall of the hole in which it is placed. Over the perforated section, water must be able to transfer from the observation well to the aquifer with negligible resistance. It is desirable to surge and pump the water in the observation well a few times to establish good transfer between the well and the aquifer.

The location and number of observation wells depends upon the requirements of the test, the finances available, and the particular conditions at the test site. It is highly desirable to have two or more observation wells. If conditions limit the number of observation wells to, say, 2 or 3 they should be located at different points on a single radial line measured from the well to be pumped. If several observation wells are to be used, they should be located on two radial lines making an approximate 90 degree angle with one another. This procedure will help to ascertain any deviations from symmetry that the cone of depression may exhibit.

The appropriate spacing of observation wells is dictated by the degree to which the pumped well penetrates the thickness of the aquifer, whether the aquifer is confined or unconfined, and the anticipated duration of the test. Vertical components of flow, caused by partial penetration of the aquifer by the pumped well, are usually negligible at distances exceeding about 1.5 times the aquifer thickness. Correction of observed drawdowns for the effects of vertical flow is not required, therefore, for observation wells located more than 1.5 b from the partially penetrating pumped well. Should it be necessary to locate observation wells nearer than 1.5 b to a partially penetrating well, the observation well should be perforated in the same depth interval as the pumped well.

Because the effective radius of influence expands much more rapidly in confined aquifers than in unconfined aquifers, the maximum distance between the observation wells and the pumped well can be larger for tests of confined aquifers. Samples of the aquifer material should be collected (during the installation of the first observation well and/or the pumped well) and a log of the strata prepared. Estimates of the hydraulic conductivity and the transmissivity can be made from the grain-size distributions of the aquifer material. Rough estimates of the storage coefficient or the apparent specific yield can also be made. These estimates, coupled with the anticipated pumping rate and duration of the test, permit one to estimate the drawdown that can be expected at various distances from the pumped well and to site the observation wells so that significant and accurately measurable drawdowns are probable during the test.

Provision must be made to measure and control the pumping rate. Measurement can be accomplished by one of several available methods including a velocity gage, orifice plate, water meters, and direct measurement of the volume pumped over a measured time interval. As the drawdown increases in the pumped well, the total pumping head increases causing the discharge to become less if it is not controlled. Control of the pumping rate by a valve requires that the discharge rate during the test be somewhat less than the long term pumping capacity with the valve fully open. Thus, the test must be started with the valve partially closed so that it can be opened subsequently to offset the increase in pumping head caused by increased drawdown.

It should be recognized by the student that the discharge Q that appears in all of the foregoing equations is the discharge from or to the aquifer and is not necessarily the pumping rate. For example, one can pump at a constant rate from a well, but a portion of the pumped discharge is derived directly by lowering the water level in the well and the remainder is derived from the aquifer. The relative contribution from well-bore storage and inflow from the aquifer changes with time, even though the pumping rate is constant. Changes in well-bore storage are usually negligible but may have to be considered if the well diameter exceeds a meter or so, especially if the pumping rate is small.

Provision must be made for disposing of the pumped water in a manner that will not effect the test results. It is best to convey the pumped water via a pipe or lined channel to a distance that exceeds the maximum effective radius of influence that is anticipated during the test. This procedure guards against the possibility of recharge of the pumped water back to the aquifer in the vicinity of the test. Appropriate disposal of the water is an important consideration because the duration of tests is sometimes 72 hours or more and quite large volumes of water may be involved, depending upon the pumping rate.

Water levels in the observation wells must be measured, of course. An accurate method is to use a chalked steel tape. The procedure is to coat the first meter or so with chalk and lower the chalked end into the well until a portion of the chalked interval extends below the water level. The tape is lowered until an integer division mark on the tape is at a convenient reference elevation (usually the top of the observation well casing). The tape is removed and the tape division that corresponds to the top of the wetted portion is subtracted from the tape division at the reference level to obtain the depth to water. This method becomes inconvenient at large depths (about 50 m) and electronic sounders may be more useful.

Water levels in the observation wells and in the pumped well should be monitored for several days prior to the test when possible. Any pre-test trends of the water levels, determined during this period, can be extrapolated into the test period and observed drawdowns corrected accordingly. Also, it is desirable to determine the elevations (relative to an arbitrary datum) of each of the reference or measuring points on the observation wells. The depth to water data can then be referenced to the common datum and any pre-test gradient of piezometric head determined.

The depth to water in the observation wells is recorded periodically during the test. Drawdown first becomes apparent in the observation wells nearest the pumped well, of course. In these wells, the depth to water should be recorded at 1 or 2 minute intervals for the first several minutes of the test and gradually increased to 10 minute intervals after about one hour. One-half hour recordings are sufficient after 2 hours and hourly recordings are adequate between 5 and 12 hours of pumping time. Beyond 12 hours, the time interval between recording can be gradually increased to several hours. A plot of drawdown vs. time on logarithmic paper in the field can be used to determine appropriate times for subsequent depth measurements. A similar procedure is used for wells more remote from the pumped well. In confined aquifers, substantial drawdown can become apparent, even at small time, in the observation wells most remote from the pumped well.

The duration of the test depends upon the use and accuracy required of the data. Usually, the best estimates of the aquifer propreties can be obtained from tests that are conducted for more than 24 hours. This is particularly true for unconfined aquifers because of the influence of vertical drainage on the apparent specific yield. Sometimes aquifer tests are conducted for 72 hours or more. Also, it is good practice to continue measuring the water levels in the observation wells (particularly those nearest the pumped well) after pumping has ceased. The water-level recovery data can be analyzed to provide still another determination of the aquifer properties.

Analysis of Aquifer Test Data Using the Theis Solution

At the completion of the test, the drawdown vs. time data at each observation point can be analyzed in a number of ways. The Theis method makes use of the following procedure.

1) Prepare a plot, on transparent log-log paper, of drawdown vs. r^2/t, where r is the distance between the observation point and the pumped well.

2) Prepare, from Table 5-1, a plot on log-log paper of W(u) vs. u. The length of each log cycle of the paper used in this plot must be the same as used for the data plot. The W(u) vs. u plot is often called the type curve.

3) Superimpose the data plot on the type curve, keeping the drawdown axis parallel to the W(u) axis. Adjust until most of the data points fall on the type curve. Corresponding coordinate axes must be kept parallel during the adjustment.

4) Select any arbitrary point (not necessarily on the curves) and record the W(u) and u coordinates and the corresponding s and r^2/t coordinates of the point.

5) The transmissivity is computed from

$$T = \frac{QW(u)}{4\pi s} \tag{5-40}$$

and the storage coefficient (apparent specific yield) from

$$S = \frac{4Ttu}{r^2} \tag{5-41}$$

using the coordinates determined in step 4.

Equations 5-40 and 5-41 follow from a rearrangement of Eqs. 5-9 and 5-3, respectively.

The procedure described above is nothing more than a graphical method of determining values of S and T which will cause Eq. 5-9 to match the measured data. Data from all wells can be plotted on one graph. Often the aquifer properties, as determined at each observation well will not agree precisely for several reasons including variability of aquifer properties, delayed yield, vertical components of flow, and experimental errors. It should be noted that the aquifer test method yields values of S and T that represent averages in some aquifer volume. Thus, small scale heterogenieties are masked or averaged.

Aquifer test data can also be analyzed using the Jacob method which is based on and subject to the same restrictions as Eq. 5-11. The measured drawdown is plotted on the coordinate axis of semi-log paper vs. time on the log axis. Provided the test was conducted for a sufficiently long period (i.e., $u<0.01$), the data should approximate a straight line at large times. The slope of the line is related to the transmissivity as shown below. Equation 5-11 can be written in terms of common logarithms as

$$s = \frac{2.303Q}{4\pi T}\{\log t + \log(\frac{2.246\alpha}{r^2})\} \tag{5-42}$$

which shows that the slope of s vs. log t is the coefficient $2.303Q/4\pi T$. Thus,

$$T = \frac{2.303Q}{4\pi\Delta s} \tag{5-43}$$

where Δs is the drawdown per log cycle. Note that Eq. 5-43 is identical to that obtained in Example 4-2.

The storage coefficient can also be obtained by the Jacob method. The procedure is to extrapolate the straight line on the data plot to the time t_o at which $s=0$. From Eq. 5-42 with

s=0,

$$S = \frac{2.246Tt_o}{r^2} \quad . \qquad (5\text{-}44)$$

Prior to the determination of T and S, one cannot determine the time range for which $u<0.01$. Thus, there is no assurance that the data used to establish the straight line in Jacobs' method actually satisfy the requirement that u be less than 0.01. It is recommended practice, therefore, to compute the maximum value of u for the data used to determine the first estimate of the straight line. Values of T and S determined from first estimate of the straight line are used to compute u. Should it be discovered that data points for which $u>0.01$ were used in the determination of the first straight line, a second line is determined that does not consider those data points and the procedure is repeated.

When data from 2 or more observation wells are available, a variation of the Theis curve fitting procedure can be used. The procedure is to select a particular time at which the drawdowns in all the observation wells are known. The drawdown is plotted versus r^2/t on log-log paper, where t is the selected time and r is the distance to each observation point. The Theis type curve is superimposed on the data plot as discussed previously; the coordinates W(u), u, s, and r^2/t of an arbitrary match point are recorded; and, Eqs. 5-40 and 5-41 are used to calculate the aquifer parameters. This method is called the drawdown-distance method.

EXAMPLE 5-4

The following data were collected from an observation well during a test of an unconfined aquifer near Fort Collins, Colorado. Estimate the transmissivity and apparent specific yield.

r = 20 meters , Q = 1.872 m^3/min

s meters	0.025	0.050	0.055	0.110	0.170	0.180	0.220
r^2/t - m^2/min	88.9	53.3	47.1	25.0	16.7	15.1	11.1
	0.300	0.370	0.450	0.530	0.620	0.640	0.650
	6.25	4.12	2.47	1.55	0.98	0.82	0.78

Solution:

The first step is to plot the given data on logarithmic paper as shown in Fig. 5-6. Note that the data could have been plotted in the form s vs. t/r^2 as well; the difference being that drawdown would increase toward the right instead of toward the left. It is observed that slow vertical drainage, although possibly occurring, is not sufficient to cause a noticeable reduction in the rate at which the drawdown increases (i.e., there is no apparent flattening of the curve at small times). Therefore, the Theis method can be used to obtain the required parameters. Following the procedures outlined previously, the data plot is superimposed on the W(u) vs. u plot as shown in Fig. 5-6. The coordinates of the match point are:

$$W(u) = 1.0 \qquad u = 0.1$$
$$s = 0.183 \qquad r^2/t = 6.2 \quad .$$

From Eqs. 5-40 and 5-41

$$T = \frac{(1.872)(1.0)}{4\pi(0.183)} = 0.814 \text{ m}^2/\text{min} = 0.0136 \text{ m}^2/\text{s} \quad ,$$

and

$$S_{ya} = \frac{(4)(0.814)(0.1)}{6.2} = 0.053 \quad .$$

Note that u is greater than 0.01 for the entire test period and the use of the Jacob method could produce errors in T and S. It would be instructive for the student to use the Jacob method and compare results with Example 5-4.

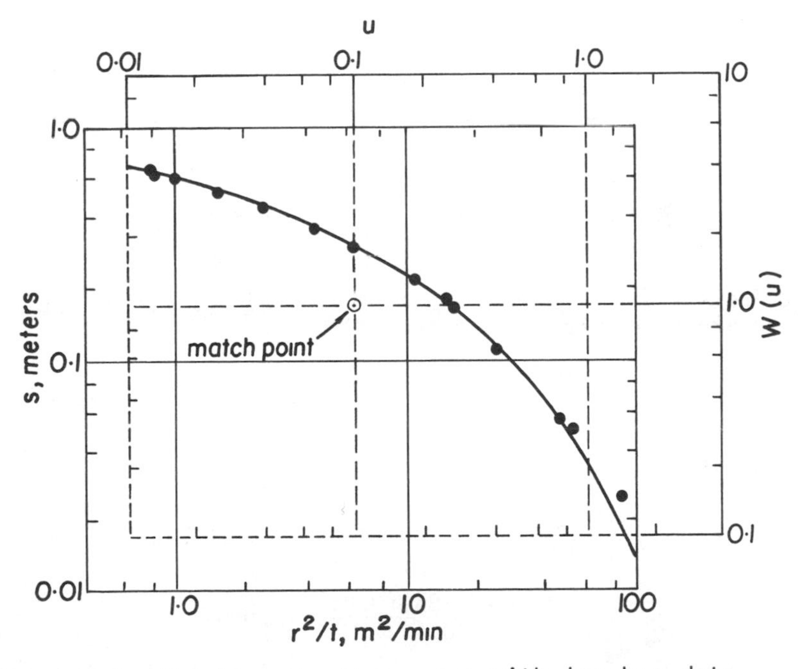

Figure 5-6. Matching the type curve with drawdown data.

EXAMPLE 5-5

The following data were collected during the same aquifer test discussed in Example 5-4. Use the drawdown-distance method to determine the transmissivity and apparent specific yield.

$t = 502$ minutes , $Q = 1.872\ m^3/min$

s-meters	1.350	0.940	0.650	0.410	0.200	0.140
r-meters	4.6	10.4	20	34.5	65.8	91.8

Solution:

A plot of drawdown versus r^2/t is shown in Fig. 5-7.

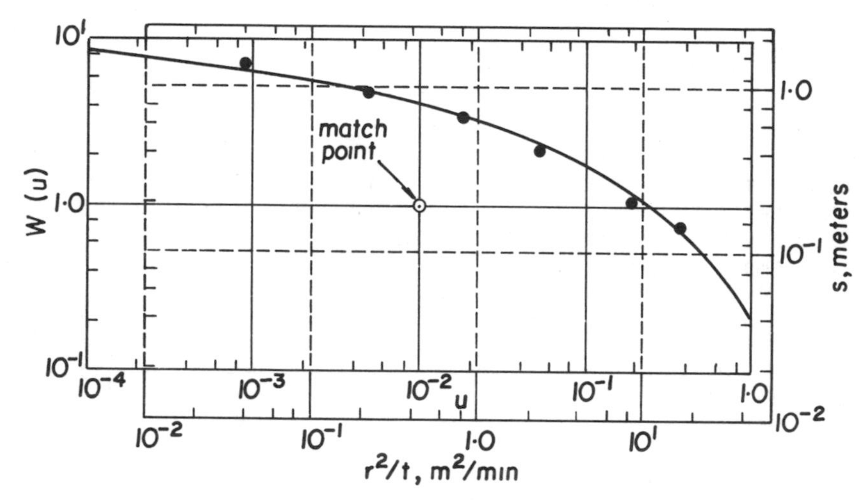

Figure 5-7. Matching the type curve with distance-drawdown data.

The Theis type curve has been superimposed and the coordinates of an arbitrary match point are

$$W(u) = 1.0 \qquad u = 0.01$$
$$s = 0.19 \qquad r^2/t = 0.46 \quad .$$

Using Eqs. 5-40 and 5-41,

$$T = \frac{(1.872)(1.0)}{4\pi(0.19)} = 0.784\ m^2/min = 0.0131\ m^2/s$$

$$S_{ya} = \frac{(4)(0.784)(0.01)}{0.46} = 0.068 \quad .$$

Because all drawdowns used in the drawdown-distance method were determined near the end of the test period and because observed drawdowns at several points in the aquifer were used, the apparent specific yield determined by the distance-drawdown method is a more reliable value than determined in Example 5-4. Note that the transmissivity values agree closely.

EXAMPLE 5-6

The following data were collected during a test of a confined aquifer by the U. S. Geological Survey. Determine the transmissivity and the storage coefficient.

$r = 61$ m , $Q = 1.894$ m^3/min

Time, min	1	2	3	4	5	6	8
s, m	0.200	0.300	0.370	0.415	0.450	0.485	0.530
	10	12	14	18	24	30	40
	0.570	0.600	0.635	0.670	0.720	0.760	0.810
	50	60	80	100	120	150	180
	0.850	0.875	0.925	0.965	1.000	1.045	1.070
	210	240					
	1.100	1.120					

Solution:

The given data are plotted on semi-logarithmic paper as shown in Fig. 5-8. A slight curvature is evident in the data plot at small time and these points are ignored in the determination of the straight line. The slope of the straight line is $\Delta s = 0.4$ m. Therefore, from Eq. 5-43

$$T = \frac{(2.303)(1.894)}{4\pi(0.4)} = 0.868 \text{ m}^2/\text{min} \quad .$$

The straight line is extrapolated to s=0 and t_0 determined to be 0.4 minutes. From Eq. 5-44

$$S = \frac{(2.246)(0.868)(0.4)}{(61)^2} = 2.0 \times 10^{-4} \quad .$$

The time corresponding to u=0.01 is

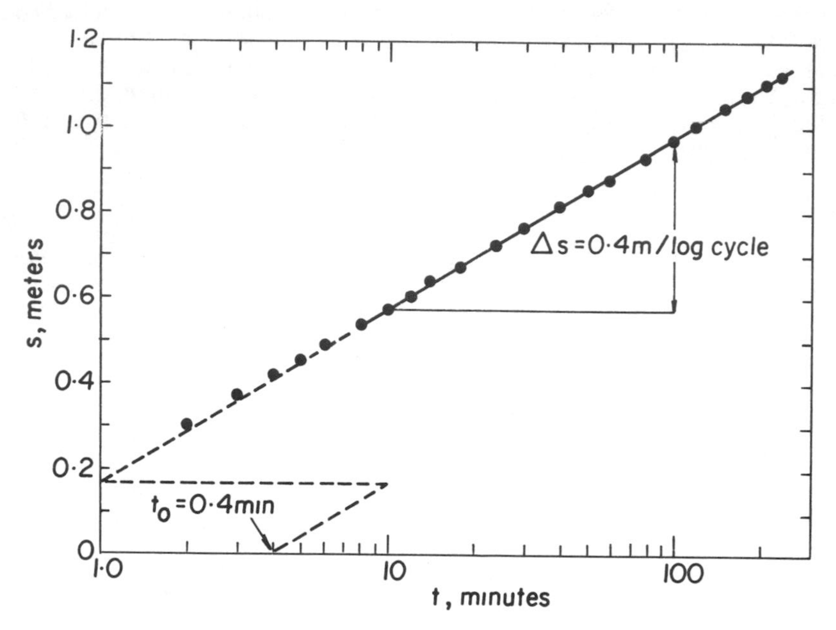

Figure 5-8. Example of the Jacob method for determining aquifer properties.

$$t = \frac{r^2 S}{4Tu} = \frac{(61)^2(2 \times 10^{-4})}{4(0.868)(0.01)} = 21 \text{ minutes} \quad .$$

Thus, data points for t<21 minutes should not be included in the determination of the straight line. As mentioned previously, however, the deviation of the logarithmic approximation from the Theis well function is only about 6.0 percent for u=0.10. This explains the fact that data points for t<21 min in Fig. 5-8 apparently fall on the straight line.

Analysis of Data Influenced By Delayed Water-Table Response

The procedure for determining the aquifer properties from drawdown-time data affected by delayed yield is, again, a curve matching procedure. A type curve for various values of r/B is prepared on logarithmic paper from the values of dimensionless drawdown and u given in Table 5-2. The result is a family of curves as shown in Fig. 5-4. The type curve is superimposed upon the drawdown-time data plot and adjusted to provide the best match between one of the theoretical curves and the data. Again, corresponding coordinate axis must be kept parallel. The procedure and computations are illustrated in Example 5-7.

EXAMPLE 5-7

During the aquifer test of Examples 5-4 and 5-5, the following data were collected at an observation well 4.6 m from the pumped well. Estimate the transmissivity, apparent specific yield, and the vertical hydraulic conductivity of the aquifer.

$r = 4.6$ m, Initial Saturated Thickness = $b = 11$ m,
$Q = 1.872\ m^3/min$

s - meters	0.340	0.635	0.705	0.715	0.705	0.715
t - minutes	0.5	2	3	4	5	6
	0.720	0.745	0.765	0.770	0.800	0.805
	7	9	11	13	15	18
	0.820	0.845	0.870	0.885	0.995	1.060
	20	22	30	32	61	94
	1.140	1.230	1.315	1.340	1.355	
	156	253	400	482	509	

Solution:

The drawdown-time data plot on logarithmic paper (Fig. 5-9) is quite flat indicating a rather small value of r/B.

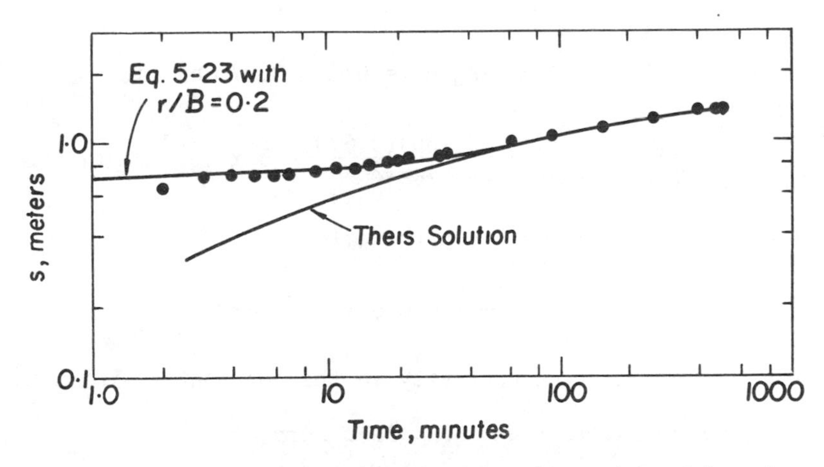

Figure 5-9. Drawdown-time relationship affected by delayed water-table response.

Therefore, the type curve prepared included curves for $r/B=0.2$, $r/B=0.4$, and $r/B=0.6$. Superposition of the data plot and the type curve gave a match point with the following coordinates

$$u^{-1} = 2.7 \qquad t = 1.0 \text{ minutes}$$

$$\frac{4\pi Ts}{Q} = 4.9 \qquad s = 1.0 \text{ meter}$$

$$r/B = 0.2 \quad .$$

Since $4\pi Ts/Q = 4.9$ and $s = 1.0$, then

$$T = \frac{(4.9)Q}{4\pi s} = \frac{(4.9)(1.872)}{4\pi(1)} = 0.730 \text{ m}^2/\text{min} \quad ,$$

and from the definition of u and the above coordinates

$$S_{ya} = \frac{4Ttu}{r^2} = \frac{(4)(0.730)(1)}{(4.6)^2(2.7)} = 0.051 \quad .$$

The degree of agreement between the theoretical result (i.e., Eq. 5-23) and the measured data with $T = 0.730$ m^2/min and $S_{ya} = 0.051$ is shown in Fig. 5-9. The student should also note the rather close agreement between the aquifer properties determined in this example with those determined in Examples 5-4 and 5-5.

Since $r/B = 0.2$, then

$$B = \frac{r}{0.2} = \frac{4.6}{0.2} = 23 = \left(\frac{Tb_a}{K_z}\right)^{1/2}$$

from Eq. 5-22. Estimating $b_a = b/3 = 3.67$, an estimate for K_z is

$$K_z = \frac{Tb_a}{(23)^2} = \frac{(0.730)(3.67)}{(23)^2} = 5 \times 10^{-3} \text{ m/min} \quad .$$

The above value for K_z compares with

$$K = \frac{T}{b} = \frac{0.73}{11} = 6.6 \times 10^{-2} \text{ m/min}$$

for the hydraulic conductivity in the horizontal direction.

Analysis of Recovery and Slug Test Data

After pumping has ceased in an aquifer test, the water level in observation wells begins to rise. This is known as recovery. The drawdown during the recovery period is given by

Eq. 5-32 or by Eq. 5-33 when u is sufficiently small. An important application of the recovery method is to make an estimate of the transmissivity by measuring the recovery in the pumped well, itself, when conditions do not permit the construction of observation wells. More precise data can be obtained during the recovery period than during the pumping period because the water in the well is not disturbed by the pump. The procedure for analysis of recovery data when $u<0.01$ is to plot the drawdown on the coordinate scale of semi-logarithmic paper and the corresponding value of $t/(t-t_2)$ on the log scale. Recall that t is the time since pumping began and t_2 is the duration of the pumping period. From Eq. 5-33 the slope Δs per log cycle and the transmissivity are related by

$$T = \frac{2.303Q}{4\pi\Delta s} \quad . \tag{5-45}$$

Slug tests involve injection of a known volume of water into the aquifer over a time interval sufficiently short to be considered as an instantaneous injection. Slug tests are sometimes used to estimate the transmissivity of aquifers with low hydraulic conductivity and when a full scale aquifer test is not justified (Papadopulos et al., 1973). The transmissivity obtained from a slug test is that of a rather limited aquifer volume in the vicinity of the injection well. The response (i.e. the residual buildup of piezometric head) is often observed in the injection well, itself. When u is sufficiently small, the transmissivity can be computed from the slope of a straight line passed through the observed residual buildups plotted on coordinate paper against the corresponding values of $1/t$ as indicated by Eq. 5-39. It should be clear that the theory applies equally well for a slug removal as might be accomplished with a bail, for example.

EXAMPLE 5-8

Calculate the transmissivity from the following recovery test data.

$Q = 1.790\ m^3/min$, Pumping interval = 443 minutes , $r = 4.6$ meters

s - meters	1.640	1.595	1.535	1.490	1.445	1.400
t = minutes	443.5	444	444.5	445	445.5	446

	1.305	1.235	1.200	1.060	0.930	0.845
	447	447.5	448.5	451	455	459

0.755	0.700	0.590	0.521	0.451	0.384
464	469	479	489	499	514

Solution:

A plot of drawdown versus the quantity $t/(t-t_2)$ is shown in Fig. 5-10. The slope of the line is 0.58 m per log cycle which yields a transmissivity of (Eq. 5-45)

$$T = \frac{2.303Q}{4\pi\Delta s} = \frac{(2.303)(1.790)}{4\pi(0.58)} = 0.566 \text{ m}^2/\text{min} \quad .$$

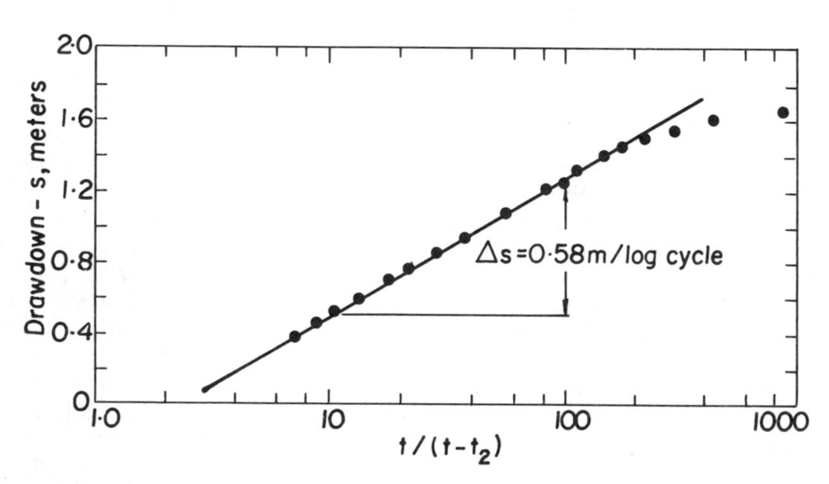

Figure 5-10. Water levels in a recovery test.

EXAMPLE 5-9

Determine the transmissivity from the following data (adapted from Ferris and Knowles, 1963).

$$V = 0.148 \text{ m}^3$$

$-s$, cm	7.9	7.6	6.1	5.2	4.9	4.6
$1/t$, min^{-1}	0.800	0.750	0.667	0.521	0.461	0.435

4.3	3.7	3.4	3.0	2.8	2.4
0.413	0.361	0.333	0.300	0.265	0.231

2.1	1.8	1.5	1.2	0.9
0.212	0.183	0.146	0.117	0.077

Solution:

Fig. 5-11 shows the above data plotted on coordinate paper. More weight was accorded the data at large time than at small time in the determination of the line. Note the line must pass through the origin, a point that corresponds to infinite time after the injection. Selecting arbitrarily the point on the line with coordinates $-s$ = 6.3 cm and 1/t = 0.6 and using Eq. 5-39 gives

$$T = \frac{V}{4\pi t(-s)} = \frac{(0.148)(0.6)}{4\pi(6.3/100)} = 0.11 \; m^2/min \quad .$$

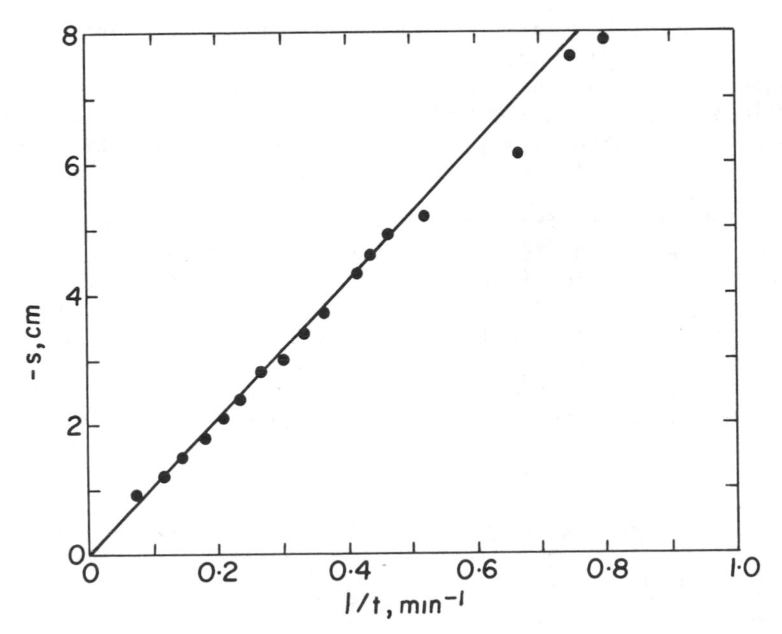

Figure 5-11. Response to a slug injection.

5.3 PUMPING NEAR HYDRO-GEOLOGIC BOUNDARIES

The principal of superposition of solutions to approximate the effects of aquifer boundaries is equally valid for unsteady flow as for steady flow. In fact the derivation of the solutions for variable pumping rates in the previous section made use of superposition in the time domain. The mathematical expression for the unsteady flow caused by a well pumping near a constant head or impermeable boundary can be obtained using the image well method, just as for the case of steady flow.

Pumping Near a Stream or Impermeable Boundary

The drawdown at any distance r from a pumped well in a semi-finite aquifer bounded by a long straight stream on which the drawdown is zero is given by (refer to Figs. 4-7 and 4-8),

$$s = \frac{Q}{4\pi T}[W(\frac{r^2}{4\alpha t}) - W(\frac{r_i^2}{4\alpha t})] \quad . \qquad (5\text{-}46)$$

The second term in Eq. 5-46 represents the buildup due to an image recharge well positioned as shown in Fig. 4-8. As before r_i is the distance from the image well to the point at which the drawdown is desired. Clearly, the drawdown is zero on a line that perpendicularly bisects the line joining the wells because $r=r_i$ on such a line and the two values of the well functions in Eq. 5-46 are identical. Thus, the locus of points for which $r=r_i$ simulates the stream on which the drawdown is zero.

A well pumping from an aquifer that is intersected by a stream derives a portion of its discharge from aquifer storage and part from induced flow from the stream. At small time, practically all of the pumped water is derived from storage because the effective radius of influence is less than the distance to the stream. At large pumping times, the drawdown caused by pumping induces an increasingly larger inflow from the stream as the radius of influence expands. The drawdown after a steady state has been established is given by Eq. 4-35 in Chapter IV.

The discharge from the stream to the aquifer induced by the pumped well can be derived by the procedures of Example 4-4 and is given by

$$Q_s = T \int_{-\infty}^{\infty} \frac{\partial s}{\partial x}\bigg|_{x=0} dy \quad , \qquad (5\text{-}47)$$

in which $\partial s/\partial x$ is computed from Eq. 5-46. As before, the line x=0 represents the effective line on which recharge occurs. The result is (Glover and Balmer, 1954),

$$Q_s = Q\{1 - \text{erf}(\frac{a}{\sqrt{4\alpha t}})\} \quad , \qquad (5\text{-}48)$$

where a is the perpendicular distance from the well to the stream and erf(x) stands for the error function defined by

$$\text{erf}(x) = \frac{2}{\sqrt{\pi}} \int_0^x \exp(-y^2)dy \quad . \qquad (5\text{-}49)$$

Values of the erf(x) are presented in Table 5-4. Notice that as the argument of the error function becomes small (i.e., at large t), the function approaches zero and Eq. 5-48 predicts that the discharge from the stream to the aquifer approaches the well discharge, Q.

Simulation of a well pumping a distance a from an infinitely long, linear, impermeable boundary is accomplished by placing an image well at x=-a, y=0. The drawdown at a point a distance r from the real well and distance r_i from the image well is given by

$$s = \frac{Q}{4\pi T} [W(\frac{r^2}{4\alpha t}) + W(\frac{r_i^2}{4\alpha t})] . \qquad (5\text{-}50)$$

The student should verify that $\partial s/\partial x\big|_{x=0}$ is zero on the line x=0 (i.e., on the line that perpendicularly bisects the line joining the wells and represents the impermeable boundary) as required to simulate the impermeable boundary.

Table 5-4. Values of the Error Function

x	erf x
0.00	0.000
0.20	0.223
0.40	0.428
0.60	0.604
0.80	0.742
0.90	0.797
1.00	0.843
1.10	0.880
1.20	0.910
1.30	0.934
1.40	0.952
1.50	0.966
1.60	0.976
1.70	0.984
1.80	0.989
1.90	0.993
2.00	0.995
∞	1.000

If the pumping rate changes with time the corresponding discharge from the stream to the aquifer is computed from

$$Q_s = \int_0^t \frac{dQ(\tau)}{d\tau} \operatorname{erfc}(\frac{a}{\sqrt{4\alpha(t-\tau)}}) \, d\tau \qquad (5\text{-}51)$$

where erfc(x) stands for the complimentary-error function and is equal to 1-erf(x). Equation 5-51 is derived by the procedures used in the discussion of variable pumping rates in the previous section. The more practical form of Eq. 5-51 is a finite difference form for the integration because of the difficulty in obtaining closed form expressions for the indicated integration.

EXAMPLE 5-10

A well, located 200 m from a stream, discharges at $Q=2\ m^3/min$ for one week, at which time the pump is shut down.

The aquifer transmissivity is 1.0 m^2/min and the apparent specific yield is 0.10. Compute the discharge from the stream to the aquifer after 96 hr of pumping and after the pump has been shut down for 36 hours.

Solution:

From the information provided

$$\frac{a}{\sqrt{4\alpha t}} = \frac{200}{\{\frac{(4)(1)(96)(60)}{0.1}\}^{\frac{1}{2}}} = 0.417 \quad ,$$

and from interpolation in Table 5-4, erf(0.417) = 0.44. The discharge from the stream to the aquifer after 96 hr of pumping is

$$Q_s = Q(1-0.44) = (2)(0.56) = 1.12 \text{ m}^3/\text{min} \quad .$$

After the pump has been shut down, the water level does not recover immediately, of course, and inflow from the stream continues. In fact, provided the well is not pumped again, all of the water withdrawn from the aquifer storage will eventually be replenished from the stream. The rate of inflow from the stream to the aquifer after pumping has ceased is calculated by using the procedures developed during the discussion of variable pumping rates. The well is assumed to continue discharging at $Q=2.0$ m^3/min indefinitely beyond the real pumping period of one week and conceptually a second well, located in the same position as the real well, is assumed to begin recharging at $Q=-2.0$ m^3/min at $t=t_2=1$ week=10080 min. The real flow rate from the stream is the sum of the flow rates induced by the two wells.

$$Q_s = Q\{1 - \text{erf}\,\frac{a}{\sqrt{4\alpha t}}\} - Q\{1 - \text{erf}\,\frac{a}{\sqrt{4\alpha(t-t_2)}}\}$$

$$= Q\{\text{erf}\,\frac{a}{\sqrt{4\alpha(t-t_2)}} - \text{erf}\,\frac{a}{\sqrt{4\alpha t}}\}$$

$\alpha=10$ m^2/min, $t=1$ wk + 36 hr=12240 min, $t_2=10080$ min

$$\frac{a}{\sqrt{4\alpha(t-t_2)}} = 0.680 \quad , \quad \frac{a}{\sqrt{4\alpha t}} = 0.286$$

$$Q_s = (2.0)\{\text{erf}(0.680) - \text{erf}(0.286)\}$$
$$= (2.0)\{0.665 - 0.315\} = 0.70 \text{ m}^3/\text{min} \quad .$$

Aquifer Tests in the Vicinity of Hydro-Geologic Boundaries

Drawdowns measured in observation wells during an aquifer test are influenced by streams or impervious boundaries that intersect the aquifer in the vicinity of the test. The drawdown in an observation well for which $r \ll r_i$ will increase for small time, just as if the boundary were not present. At large time, the drawdown will increase more rapidly than for an infinite aquifer if the boundary is an impervious one. The drawdown will increase less rapidly if the boundary is a constant head type. An example of the drawdown as a function of time in the presence of both impervious and constant head boundaries is shown in Fig. 5-12.

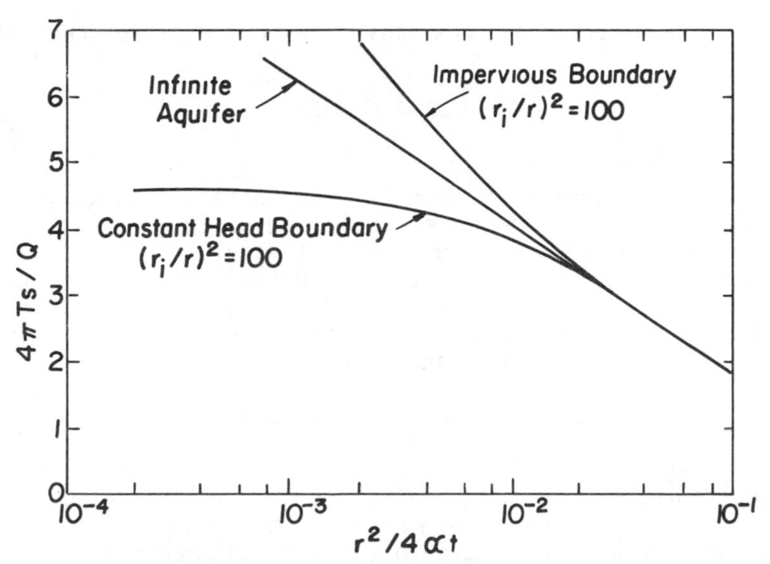

Figure 5-12. Drawdown as a function of time in the presence of both impervious and constant head boundaries.

A constant head boundary causes the drawdown to stabilize at the observation well, and the long term drawdown approaches

$$s = \frac{Q}{4\pi T} \ln\left(\frac{r_i}{r}\right)^2 = \frac{Q}{2\pi T} \ln \frac{r_i}{r} \quad , \qquad (5\text{-}52)$$

where r is the distance from the pumped well to the observation point and r_i is the distance from the image well to the observation point. At large time, the logarithmic approximations for the well functions in Eq. 5-50 can be used and the drawdown in an observation well in the vicinity of an impervious boundary becomes

$$s = \frac{Q}{4\pi T}\left[\ell n\{(\frac{4\alpha t}{r^2})(\frac{4\alpha t}{r_i^2})\} - 2(0.5772)\right] \quad . \tag{5-53}$$

or

$$s = \frac{Q}{2\pi T}\ell n\ t + \text{constant} \quad . \tag{5-54}$$

It is evident that, at large time, the slope of a plot of drawdown versus time on semi-log paper is exactly twice the slope that is observed at small time when the drawdown-time relationship is that of an infinite aquifer (i.e., Eq. 5-11).

5.4 ONE-DIMENSIONAL FLOW

One-dimensional, unsteady flows are described by solutions to

$$\alpha \frac{\partial^2 s}{\partial x^2} = \frac{\partial s}{\partial t} \quad , \tag{5-55}$$

subject to appropriate boundary and initial conditions. Equation 5-55 is the one-dimensional form of Eq. 3-62. Examples of approximately one-dimensional flows of interest to hydrologists include the interchange of water between a stream and the aquifer in response to a change in stage, ground water return flow to a stream in response to recharge to the area adjacent to the stream, and recharge from canals.

Flow Toward a Plane on Which the Piezometric Head is Prescribed

Figure 5-13 shows an idealization of flow from an aquifer that forms the banks of a stream or reservoir. A sudden drop of the stage in the stream or reservoir produces the flow shown in Fig. 5-13. As before, it is assumed that the drawdown s is everywhere small compared to b so that Eq. 5-55 applies. The initial and boundary conditions corresponding to a step change of stage at t=0 are, respectively,

$$\begin{aligned} s(x,0) &= 0 \\ s(\infty,t) &= 0 \\ s(0,t) &= s_o \end{aligned} \tag{5-56}$$

where s_o is the change of stage.

The solution is obtained in a manner quite similar to that used to derive Eq. 5-8. A new variable

$$y = \frac{x}{\sqrt{4\alpha t}} \tag{5-57}$$

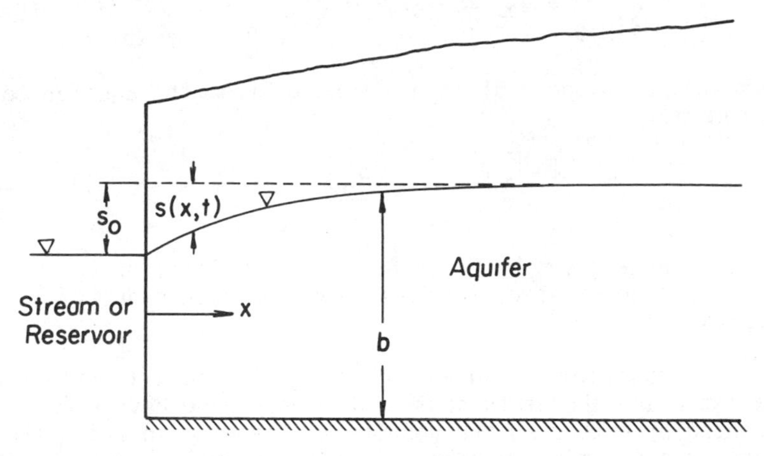

Figure 5-13. Flow from bank storage.

is introduced which permits Eq. 5-55 to be written as the ordinary differential equation

$$\frac{d^2s}{dy^2} + 2y\,\frac{ds}{dy} = 0 \qquad (5\text{-}58)$$

subject to the conditions that $s = s_o$, $y = 0$, and $s = 0$, $y = \infty$. Integration of Eq. 5-58 yields

$$s = s_o\{1 - \mathrm{erf}(\frac{x}{\sqrt{4\alpha t}})\} = s_o\,\mathrm{erfc}(\frac{x}{\sqrt{4\alpha t}}) \quad . \qquad (5\text{-}59)$$

Equation 5-59 was derived for an aquifer that extends to infinity but it also provides a good approximation for finite aquifers in the vicinity of the stream until the effective extent of influence x_e reaches the aquifer boundary. The quantity x_e is defined as the distance to the moving point at which s is some arbitrarily small value.

The discharge from the aquifer to the channel, per unit of channel length, as a result of the reduction of stage is computed from

$$Q = -T\,\frac{\partial s}{\partial x}\bigg|_{x=0} \qquad (5\text{-}60)$$

in which the derivative is computed from

$$\left.\frac{\partial s}{\partial x}\right|_{x=0} = -s_0 \frac{d}{dy} (\text{erf } y) \left.\frac{\partial y}{\partial x}\right|_{x=0} = - \frac{s_0}{\sqrt{\pi \alpha t}} \quad . \qquad (5\text{-}61)$$

From Eqs. 5-60 and 5-61, the discharge from the aquifer to the stream is

$$Q = \frac{s_0 T}{\sqrt{\pi \alpha t}} \quad . \qquad (5\text{-}62)$$

The discharge given by Eq. 5-62 must be multiplied by two if the reduction of stage induces flow from both sides of the channel.

The condition of an abrupt change in stage is rather unrealistic and the usefulness of the foregoing developments is enhanced by extending the results to the case in which the stage is a function of time. This is accomplished using the same concepts used to derive Eqs. 5-29 and 5-36. In other words, the quantity $\text{erfc}(x/\sqrt{4\alpha t})$ is regarded as the response of the aquifer to a unit step change of stage that occurred at $t=0$ (Hall and Moench, 1972). The drawdown in the aquifer in response to a continuous change of stage is

$$s = \int_0^t \frac{ds_0}{d\tau} \text{erfc}\left(\frac{x}{\sqrt{4\alpha(t-\tau)}} \right) d\tau \qquad (5\text{-}63)$$

and the discharge from the aquifer to the channel is (Cooper and Rorabaugh, 1963),

$$Q = \frac{T}{\sqrt{\pi \alpha}} \int_0^t \frac{ds_0}{d\tau} (t-\tau)^{-\frac{1}{2}} d\tau \quad . \qquad (5\text{-}64)$$

In practical applications, the change of stage as a function of time, $s_0(t)$, is often too complicated to permit the above indicated integrations to be carried out analytically. The corresponding finite difference forms

$$s = \sum_{i=1}^{n} \Delta s_{oi} \; \text{erfc}\left(\frac{x}{\sqrt{4\alpha(t-t_i)}} \right) \quad , \quad t > t_n \qquad (5\text{-}65)$$

and

$$Q = \frac{T}{\sqrt{\pi \alpha}} \sum_{i=1}^{n} \Delta s_{oi} (t-t_i)^{-\frac{1}{2}} \quad , \quad t > t_n \qquad (5\text{-}66)$$

can be used in such cases. The change of stage can be positive or negative, of course, in all of the above developments.

The volume of water that has been interchanged between the aquifer and one side of the channel per unit length of channel at any time t is

$$V = \int_0^t Q \, dt \quad . \tag{5-67}$$

In some instances the variation of s_0 with time may be periodic and expressible as a sine or cosine function. An example is the case of a confined aquifer outcropping beneath the surface of the sea. Tides induce a sinusoidal variation in piezometric head on the face of the aquifer exposed to the sea. The sinusoidal boundary condition induces a periodic response in the aquifer given by (Ferris, 1963)

$$s = s_a \exp\left(-x\sqrt{\frac{\pi S}{t_0 T}}\right) \sin\left(\frac{2\pi t}{t_0} - x\sqrt{\frac{\pi S}{t_0 T}}\right) \quad . \tag{5-68}$$

In Eq. 5-68, s_a is the amplitude of the sinusoidal variation of stage at x=0, s is the increase or decrease of piezometric head in the aquifer relative to the mean value, and t_0 is the period of the sinusoidal boundary condition. Equation 5-68 represents a strongly damped wave. Ferris (1963) used the above result as a basis for determining the transmissivity of the alluvial aquifer adjacent to the Platte River near Lincoln, Nebraska. Carr and Van Der Kamp (1969) determined aquifer properties from tidal induced fluctuations of piezometric head.

EXAMPLE 5-11

An unlined canal is located in a large alluvial valley. Before diversion of water into the canal, the water table in the adjacent alluvial aquifer is at the same elevation as the floor of the canal. Water is diverted into the canal and the stage-time relationship is approximated by

$$-s_0 = 0.0167\ t \quad , \quad 0 \le t \le 120 \text{ minutes}$$

$$-s_0 = 2 \text{ meters} \quad , \quad t \ge 120 \text{ minutes} \quad .$$

The hydraulic properties of the aquifer are $T=1.0\ m^2/min$ and $S_{ya}=0.1$. Calculate the discharge from the canal to the aquifer as a function of time.

Solution:

The rise of stage in the canal causes a water table buildup, $-s_0=y$. From Eq. 5-64,

$$Q = \frac{-2T(0.0167)}{\sqrt{\pi\alpha}} \int_0^t (t-\tau)^{-\frac{1}{2}} d\tau \quad , \quad t \leq 120$$

and

$$Q = \frac{-2T(0.0167)}{\sqrt{\pi\alpha}} \int_0^{120} (t-\tau)^{-\frac{1}{2}} d\tau \quad , \quad t \geq 120 \quad .$$

wherein both sides of the channel have been considered. Also, $\alpha=10\ m^2/min$, and

$$Q = -0.0119\ t^{\frac{1}{2}} \quad , \quad t \leq 120$$

$$Q = 0.0119\ \{(t-120)^{\frac{1}{2}} - t^{\frac{1}{2}}\} \quad , \quad t \geq 120 \quad .$$

The discharge into the aquifer from the canal is shown graphically in Fig. 5-14.

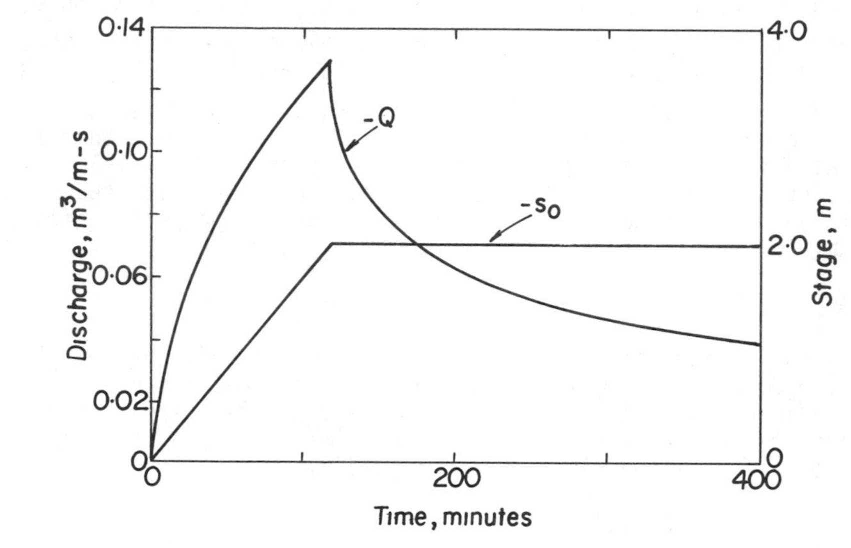

Figure 5-14. Discharge from a canal into bank storage in response to an increase of stage in the canal.

Flow Toward a Plane at Which the Discharge is Prescribed

In the previous section, the drawdown (and therefore the piezometric head) was prescribed on the banks of the channel, and the drawdown and discharge in the aquifer were calculated. Suppose, however, that the discharge to or from the aquifer at the channel bank is known, rather than the drawdown. An example of such an occurrence is pumping at a constant rate from a narrow tranch where inflow to the trench from the aquifer provides the water to be pumped. Another application is the analysis of recharge from a narrow channel located above the water table. This application will be demonstrated in Example 5-12.

The initial and boundary conditions are

$$s(x,0) = 0$$
$$s(\infty,t) = 0 \tag{5-69}$$
$$\partial s/\partial x = Q/2T \quad \text{at} \quad x=0 \quad .$$

The required solution is obtained by noting that Eq. 5-55 can be differentiated with respect to x and Darcy's Law introduced to obtain

$$\alpha \frac{\partial^2 q}{\partial x^2} = \frac{\partial q}{\partial t} \quad , \tag{5-70}$$

where q is the Darcy velocity. The initial and boundary conditions become

$$q(x,0) = 0$$
$$q(\infty,t) = 0 \tag{5-71}$$
$$q(0,t) = Q/2b \quad .$$

Note that Eqs. 5-70 and 5-71 are identical in form to Eqs. 5-55 and 5-56 for which a solution has already been written. Therefore,

$$q = \frac{Q}{2b} \operatorname{erfc}\left(\frac{x}{\sqrt{4\alpha t}} \right) \quad . \tag{5-72}$$

The drawdown is obtained by replacing q in Eq. 5-72 with $K\, \partial s/\partial x$ and integrating to yield (Ferris et al., 1962)

$$s = \frac{Qx}{2T} \left\{ \frac{\exp(-x^2/4\alpha t)}{\sqrt{\pi}(x/\sqrt{4\alpha t}\,)} + \operatorname{erf}\left(\frac{x}{\sqrt{4\alpha t}} \right) - 1 \right\} \quad . \tag{5-73}$$

The drawdown on the plane x=0 is

$$s_0 = \frac{Q\sqrt{\pi\alpha t}}{\pi T} \quad . \tag{5-74}$$

Equation 5-73 is called the *plane-source* solution. If required, it can be modified in a now familiar way to calculate the drawdown in response to a discharge rate that varies in time.

EXAMPLE 5-12

An ephemeral stream intersects, at right angles, an alluvium-filled channel as shown in Fig. 5-15. Prior to the spring flows, the water table in the alluvium is 5 m below the floor of the stream. During the spring flow, the ephemeral stream is

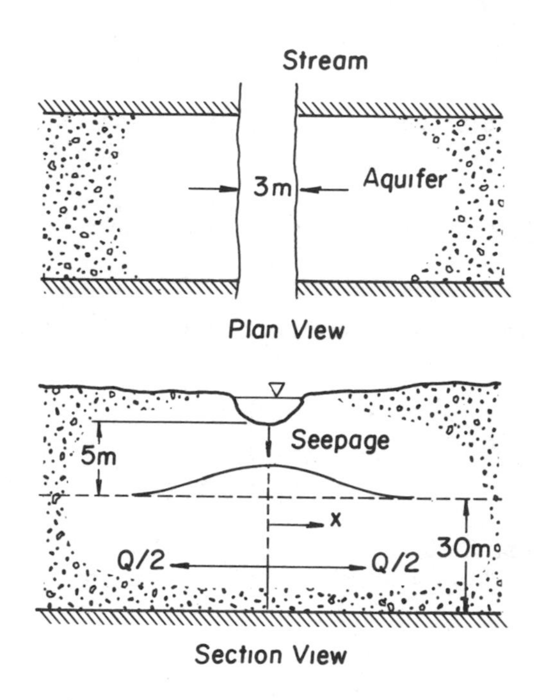

Figure 5-15. Definition sketch for Example 5-12.

3 m wide and very shallow. The saturated thickness of the alluvium is 30 m, K=10 m/day, and S_{ya}=0.2. Estimate the volume of recharge per unit of stream length between the time flow begins and the time at which the water table is 1 m below the floor of the stream.

Solution:

The situation is idealized as shown in Fig. 5-15. The vertical seepage from the stream is assumed to constitute a prescribed discharge across a vertical plane (x=0) located below the stream channel. The seepage rate Q is estimated by use of developments in Example 3-6 in which it was shown that

$$q_z = -K(1 + z/L) \quad .$$

For the application in the present example, z/L is small relative to unity because the depth of water, z, in the stream is small compared to the depth L to the water table. The seepage discharge per unit of stream length is, therefore,

$$Q = q_z \text{Area} = -K(\text{width of stream})$$
$$= -(10)(3) = -30 \text{ m}^3/\text{m-day} \quad .$$

The time at which the water table has built up beneath the stream to within 1 m of the surface follows from Eq. 5-74 with s_o=-4 m. Thus,

$$t = \left(\frac{\pi T s_o}{Q} \right)^2 \frac{1}{\pi\alpha} = \left[\frac{\pi(10)(30)(-4)}{-30} \right]^2 \frac{1}{\pi(10)(30)/0.2}$$
$$= 3.35 \text{ days} \quad .$$

The volume of recharge per unit of stream length is

$$V = Qt = (30)(3.35) = 100 \text{ m}^3/\text{m} \quad .$$

5.5 ONE-DIMENSIONAL FLOW WITH DISTRIBUTED RECHARGE

Vertical percolation of water to the water table as a result of precipitation or irrigation is an important source of recharge in many aquifers. Unlike the recharge in Example 5-12, the recharge under consideration in this section is distributed over large areas. The corresponding case for steady flow was treated in Chapter IV.

Flow Between Parallel Drains

Consider parallel channels in which the water level is maintained a distance b above the floor of the aquifer. For time less than t_i, the water-table elevation is the same as the water level in the channels. It is assumed that the water-table level is increased instantaneously at time t_i to a uniform height z_i above the constant water level in the channels (Fig. 5-16). The cause of the abrupt increase in water-table elevation is the instantaneous, uniform addition of a volume of water to the aquifer. In terms of volume per unit area (i.e., depth) the recharge is I_i and the corresponding instantaneous buildup of the water table is $z_i = I_i/S_{ya}$. Following the recharge event at $t=t_i$, ground water flows toward the channels and the water-table elevation decreases.

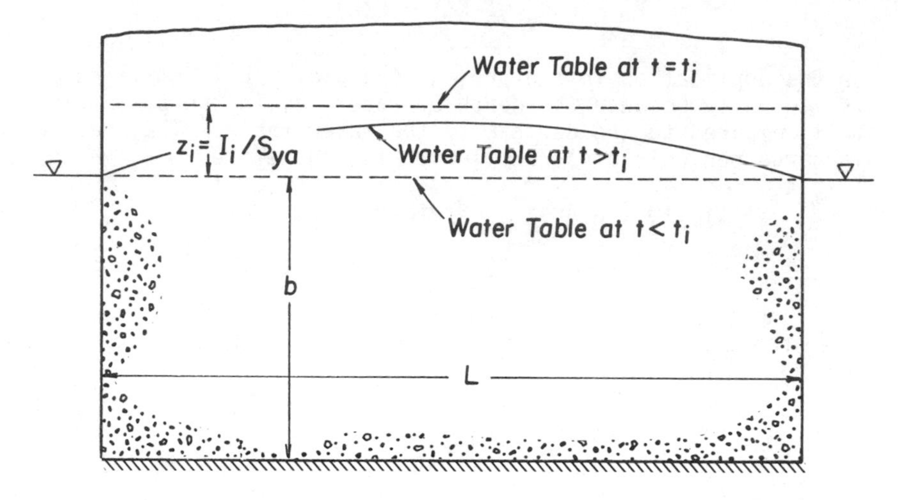

Figure 5-16. Flow toward parallel channels in response to an instantaneous recharge increment.

Relative to the water level in the drains, the water table is z(x,t) and the linearized Boussinesq equation can be written

$$\alpha \frac{\partial^2 z}{\partial x^2} = \frac{\partial z}{\partial t} \quad , \tag{5-75}$$

which is restricted to those cases in which z<<b. The boundary conditions are z(0,t)=z(L,t)=0, where L is the distance between the channels. The solution (Glover, 1953) obtained by separation of variables, is the Fourier series

$$z = \frac{4I_i}{\pi S_{ya}} \sum_{n=1,3,5,\ldots}^{\infty} (1/n) \exp\{ \frac{-\alpha n^2 \pi^2}{L^2} (t-t_i)\} \sin \frac{n\pi x}{L} \quad . \tag{5-76}$$

The water-table elevation midway between the channels (x=L/2) is

$$z_c = \frac{I_i}{S_{ya}} G\{ \frac{\alpha(t-t_i)}{L^2} \} \quad , \tag{5-77}$$

where

$$G\{\frac{\alpha(t-t_i)}{L^2}\} = \frac{4}{\pi}\sum_{n=1,3,5,...}^{\infty} (1/n)\ \exp\{\frac{\alpha n^2\pi^2}{L^2}(t-t_i)\}\ \sin\frac{n\pi}{2}\ . \qquad (5\text{-}78)$$

Values of the G function, for $\alpha(t-t_i)/L^2 \leq 0.05$, are tabulated in Table 5-5. The infinite series in Eq. 5-78 converges rapidly and G can be calculated from

$$G = \frac{4}{\pi}\exp\{\frac{-\alpha\pi^2}{L^2}(t-t_i)\} \qquad (5\text{-}79)$$

with less than one percent error for $\alpha(t-t_i)/L^2 > 0.05$.

The discharge to either channel from one side is

$$Q = \frac{4TI_i}{S_{ya}L}\sum_{n=1,3,5,...}^{\infty} \exp\{\frac{-\alpha n^2\pi^2}{L^2}(t-t_i)\} \qquad (5\text{-}80)$$

which is derived by calculating

$$Q = T\left.\frac{\partial z}{\partial x}\right|_{x=0}\ . \qquad (5\text{-}81)$$

Provided $\alpha(t-t_i)/L^2$ is greater than about 0.06, the discharge is closely approximated by the first term in the series of Eq. 5-80, and combination with the equation for z_c yields the approximate discharge formula:

$$Q = \frac{\pi T z_c}{L}\ . \qquad (5\text{-}82)$$

The total discharge from the aquifer between the channels is twice the value computed from Eqs. 5-80 or 5-82.

It is sometimes desirable to know the fraction P of the volume recharged that remains in the aquifer at any time after the recharge event. The fraction P is known as the *part remaining* and is defined by

$$P = \frac{S_{ya}\int_0^L z\ dx}{I_i L}\ . \qquad (5\text{-}83)$$

The integral in Eq. 5-83 is evaluated using Eq. 5-76 with the result

$$P = \frac{8}{\pi^2} \sum_{n=1,3,5,...}^{\infty} (1/n)^2 \exp\{-\frac{n^2\pi^2\alpha}{L^2}(t-t_i)\} \quad .(5\text{-}84)$$

As in the previous cases, P is satisfactorily approximated by the first term in the series if $\alpha(t-t_i)/L^2$ is greater than 0.05. Table 5-6 presents values of P for small arguments. The water-table elevation z at time t that results from any arbitrary sequence of instantaneous recharge events of magnitude I_i that occurred at times t_i is obtained by summing the individual responses given by Eq. 5-76 (Maasland, 1959; McWhorter, 1977).

Table 5-5. Values of the Function G for $\alpha(t-t_i)/L^2 \leq 0.05$.

$\alpha(t-t_i)/L^2$	G
0.000	1.000
0.010	0.999
0.015	0.992
0.020	0.975
0.025	0.949
0.030	0.918
0.035	0.882
0.040	0.846
0.045	0.809
0.050	0.772

Table 5-6. Values of the Part Remaining for $\alpha(t-t_i)/L^2 \leq 0.05$.

$\alpha(t-t_i)/L^2$	G
0.000	1.000
0.001	0.929
0.002	0.899
0.003	0.876
0.004	0.857
0.005	0.840
0.006	0.825
0.007	0.811
0.008	0.798
0.009	0.786
0.010	0.774
0.020	0.681
0.030	0.609
0.040	0.549
0.050	0.496

EXAMPLE 5-13

An aquifer has formed on an impervious stratum located 12 m below the soil surface. Artificial, parallel, relief drains are to be installed at a depth of 2 m to keep the water table from rising to a height that would endanger crop productivity. Recharge increments from irrigation average 2 cm and are spaced three weeks apart. The hydraulic conductivity of the aquifer is $2.1\text{x}10^{-3}$ cm/s and S_{ya}=0.12. Calculate a drain spacing that will cause 70 percent of each increment of recharge volume to be drained before the next increment occurs.

Solution:

The spacing is to be determined for which the part remaining is 0.30, three weeks following the recharge event. Inspection of Table 5-6 indicates that, for P=0.3, $\alpha(t-t_i)/L^2$ is greater than 0.05 and the first term in the series of Eq. 5-84 can be used. Therefore,

$$P = (8/\pi^2)\ \exp\{\frac{-\pi^2\alpha(t-t_i)}{L^2}\} \quad ,$$

or

$$L = [\frac{-\pi^2\alpha(t-t_i)}{\ell n(P\pi^2/8)}]^{\frac{1}{2}} \quad .$$

From the information provided:

$$T = (12-2)(2.1\text{x}10^{-5})(86400) = 18.14\ \text{m}^2/\text{day}$$

$$\alpha = 18.14/0.12 = 151.2\ \text{m}^2/\text{day}$$

$$t-t_i = 3\ \text{weeks} = 21\ \text{days}$$

and

$$L = 178\ \text{meters} \quad .$$

EXAMPLE 5-14

The irrigation season in the problem of Example 5-13 extends over a 12 week period resulting in 5 recharge increments of 2 cm each. Presuming that the water-table elevation, relative to the drain elevation, is everywhere zero before the first irrigation of the season, compute the water-table elevation z_c midway between drains immediately following the last irrigation of the season and just prior to the first irrigation in the second season. Assume no recharge occurs in the winter months between irrigation seasons and L=100 m.

Solution:

The principle of superposition of solutions is applicable. In other words, the water-table height z_c at any time t, resulting from all the recharge events that have occurred to time t, is the sum of the z_{ci} that result from each individual increment of recharge. It is convenient to mark time from the first recharge event. Therefore, the time at which the water-table height is desired is t=12 weeks or 84 days for the first part of the problem. Recall that the t_i in Eqs. 5-77 and 5-78 are the times at which the recharge events occurred. The computation procedure is illustrated in Table 5-7, with α/L^2 = 0.01512 from Example 5-13. The values G are calculated from Eq. 5-79, except for the last one in the column. The values in the last column follow from Eq. 5-77 with each I_i equal to 2 cm and S_{ya}=0.12. The water-table elevation midway between drains is 17.6 cm above the drains immediately following the last recharge event of the season.

The computation of the water-table height just prior to the first irrigation of the second season can be computed as above with t=365. However, it is evident from Table 5-7 that the effect on the water table of each event lasts only on the order of 50 days. The effect of all the recharge events will have dissipated by the beginning of the second season, therefore, and the water-table elevation z will again be zero for practical purposes. The student should recognize that, had

Table 5-7. Calculation of the Water Table Height In Response To a Sequence of Instantaneous Recharge Events

Recharge Event No.	t_i days	$t-t_i$ days	$\alpha(t-t_i)/L^2$	G	z_{ci} cm
1	0	84	1.270	0.000	0.00
2	21	63	0.953	0.000	0.00
3	42	42	0.635	0.002	0.03
4	63	21	0.318	0.055	0.93
5	84	0	0.000	1.000	16.67
					17.63

the effect of the first season's recharge not been completely dissipated before the second season began, the residual water-table height would have to be accounted for in the calculation of z_c in subsequent seasons.

Flow in Response to Continuously Varying Recharge

The principal of superposition can be used to extend the developments of the previous section to apply when the recharge is continuous at rate W(t). The procedure is the same as that used to develop Eq. 5-36. In the case at hand, the instantaneous increment of recharge I_i is replaced by $d\tau(dI/d\tau)$ which is the increment of recharge dI that occurred at time τ over the interval $d\tau$. The quantity $dI/d\tau$ is the recharge rate $W(\tau)$, however, and the water-table elevation midway between the channels is

$$z_c = \frac{1}{S_{ya}} \int_0^t W(\tau)\, G\{ \frac{\alpha(t-\tau)}{L^2} \} d\tau \quad . \qquad (5\text{-}85)$$

The corresponding result for the discharge to the channels is

$$Q = \frac{4T}{S_{ya} L} \int_0^t W(\tau) \sum_{n=1,3,5,...}^{\infty} \exp\{ \frac{-\alpha n^2 \pi^2 (t-\tau)}{L^2} \} d\tau \quad . \qquad (5\text{-}86)$$

It is permissible to interchange the order of integration and summation.

EXAMPLE 5-15

A river is located near the center of a valley that contains an alluvial aquifer. The edges of the valley form impervious boundaries of the aquifer and are practically parallel to the river. The distance from the river to the edges of the valley is 3 km. The aquifer properties are $T=0.14\ m^2/s$ and $S_{ya}=0.14$. Irrigation is practiced on the land between the river and the edges of the valley. Recharge from irrigation and canal seepage averages 0.003 m/day for 100 days of the growing season, and zero thereafter. Calculate the discharge per unit of stream length from the aquifer to the stream that results from the recharge during one season.

Solution:

The vertical plane located midway between the channels of Fig. 5-16 is a ground-water divide through which no flow occurs. Thus, the plane simulates an impervious boundary and the flow on either side of the river is simulated by the flow between $x=0$ and $x=L/2$ in Fig. 5-16. The discharge from the aquifer to the river from one side is given by Eq. 5-86 in which $W(\tau)$ is a constant equal to 0.003 m/day for $0 \leq t \leq 100$, and zero thereafter. L in this problem is the valley width equal to 6 km. For $0 \leq t \leq 100$, the discharge from both sides of the river is

$$Q = \frac{8TW}{S_{ya} L} \sum_{n=1,3,5,...}^{\infty} \int_0^t \exp\{ \frac{-\alpha n^2 \pi^2 (t-\tau)}{L^2} \} d\tau$$

$$= WL\{ \frac{8}{\pi^2} \sum_{n=1,3,5,...}^{\infty} (1/n)^2$$

$$- \frac{8}{\pi^2} \sum_{n=1,3,5,\ldots}^{\infty} (1/n)^2 \exp\left(\frac{-\alpha n^2 \pi^2 t}{L^2} \right)\}$$

$$= WL\{1-P\}$$

where P is given by Eq. 5-84.

For $t \geq 100$ days, the integration yields

$$Q = WL[P\{ \frac{\alpha(t-100)}{L^2} \} - P(\frac{\alpha t}{L^2})] \quad .$$

The discharge is easily computed with W=0.003 m/day, L=6000 m, and α=1.0 m^2/s = 86400 m^2/day. The computations are shown in Table 5-8. Time is marked from the beginning of the recharge. The discharge values in Table 5-8 are in addition to any discharges that may have existed independent of the recharge under consideration.

Table 5-8. Computation of Irrigation Return Flow

t days	$\alpha t/L^2$	$\alpha(t-100)/L^2$	$P(\alpha t/L^2)$	$P\{\alpha(t-100)/L^2\}$	Q m^3/m-day
0	0.000	-	1.000	-	0.0
10	0.024	-	0.650	-	6.3
20	0.048	-	0.506	-	8.9
40	0.096	-	0.314	-	12.4
60	0.144	-	0.196	-	14.5
80	0.192	-	0.122	-	15.8
100	0.240	0.000	0.076	1.000	16.6
110	0.264	0.024	0.060	0.650	10.6
120	0.288	0.048	0.047	0.506	8.3
140	0.336	0.096	0.029	0.314	5.1
160	0.384	0.144	0.018	0.196	3.2
180	0.432	0.192	0.011	0.122	2.0
200	0.480	0.240	0.007	0.076	1.2

REFERENCES

Boulton, N. S., 1954. Unsteady Radial Flow to a Pumped Well Allowing for Delayed Yield From Storage. Inter. Assoc. Sci. Hydrology, Rome.

Boulton, N. S., 1963. Analysis of Data From Nonequilibrium Pumping Tests Allowing for Delayed Yield From Storage. Proc. Inst. Civil Engrs. (London), V. 26, p. 469-482.

Boulton, N. S. and Pontin, J. M. A., 1971. An Extended Theory of Delayed Yield From Storage Applied to Pumping Tests in Unconfined Anisotropic Aquifers. J. Hydrology, V. 14, N. 1.

Carr, P. A. and Van Der Kamp, G. S., 1969. Determining Aquifer Characteristics by the Tidal Method. Water Res. Res., V. 5, N. 5, p. 1023-1031.

Cooper, H. H. Jr. and Rorabaugh, M. I., 1963. Ground-Water Movements and Bank Storage Due to Flood Stages in Surface Streams. U. S. Geol. Survey, Water Supply Paper 1536-J.

Cooper, H. H. Jr., Bredehoeft, J. D. and Papadopulos, I. S., 1967. Response of a Finite Diameter Well to an Instantaneous Charge of Water. Water Res. Res., V. 3, N. 1, p. 263-269.

Ferris, J. G., 1963. Cyclic Water-Level Fluctuations as a Basis for Determining Aquifer Transmissibility. In: Methods of Determining Permeability, Transmissibility, and Drawdown, U. S. Geol. Survey, Water Supply Paper 1536-J.

Ferris, J. G. and Knowles, D. B., 1963. The Slug-Injection Test for Estimating the Coefficient of Transmissibility of an Aquifer. In: Methods of Determining Permeability, Transmissibility and Drawdown, U. S. Geol. Survey, Water-Supply Paper 1536-J.

Glover, R. E., 1953. Formulas for Movement of Ground Water, Oahe Unit Missour River Basin Project. U. S. Bureau of Rec. Memo. N, 657, Sec. D, p. 35-46.

Glover, R. E. and Balmer, G. G., 1954. River Depletion Resulting From Pumping a Well Near a River. Trans. Am. Geophys. Union, V. 35, p. 468-470.

Hall, F. R. and Moench, A. F., 1972. Application of the Convolution Equation to Stream-Aquifer Relationships. Water Res. Res., V. 8, N. 2, p. 487-493.

Hantush, M. S. and Jacob, C. E., 1955. Non-Steady Radial Flow in an Infinite Leaky Aquifer. Trans. Am. Geophys. Union, V. 36, N. 1, p. 95-100.

Hantush, M. S., 1956. Analysis of Data From Pumping Tests in Leaky Aquifers. Trans. Am. Geophys. Union, V. 37, N. 6.

Maasland, M., 1959. Water Table Fluctuations Induced by Intermittent Recharge. J. Geophys. Res., V. 64, N. 5, p. 549-559.

McWhorter, D. B., 1977. Drain Spacing Based on Dynamic Equilibrium. ASCE, J. Irr. and Dr. Div., V. 103, N. IR2, Proc. Paper 13022, June.

Morel-Seytoux, H. J. and Daly, C. J., 1975. A Discrete Kernel Generator for Stream-Aquifer Studies. Water Res. Res., V. 11, N. 2, p. 253-260.

Neuman, S. P., 1972. Theory of Flow in Unconfined Aquifers Considering Delayed Response of the Water Table. Water Res. Res., V. 8, N. 4, p. 1031-1045.

Papadopulos, I. S., Bredehoeft, J. D., and Cooper, H. H. Jr., 1973. On the Analysis of Slug Test Data. Water Res. Res., V. 9, N. 4, p. 1087-1089.

Streltsova, T. D., 1972. Unsteady Radial Flow in an Unconfined Aquifer. Water Res. Res., V. 8, N. 4, p. 1059-1066.

Theis, C. V., 1935. The Relation Between the Lowering of the Piezometric Surface and the Rate and Duration of Discharge of a Well Using Groundwater Storage. Trans. Am. Geophys. Union, V. 16, p. 519-524.

Walton, W. C., 1960. Application and Limitation of Methods Used to Analyze Pumping Test Data. Water Well J., Feb.-Mar.

Walton, W. C., 1962. Selected Analytical Methods for Well and Aquifer Evaluation. Illinois State Water Survey Bull. 49.

Wenzel, L. K., 1942. Methods for Determining Permeability of Water-Bearing Material, With Special Reference to Discharging-Well Methods. U. S. Geol. Survey, Water-Supply Paper 887, 192 p.

PROBLEMS AND STUDY QUESTIONS

1. The transmissivity and storage coefficient for a confined aquifer are 1.78 m^2/min and 2.1×10^{-4}, respectively. A fully penetrating well is pumped at a constant discharge of 3.1 m^3/min. Calculate and plot the drawdown as a function of time at r=30 m up to a time of 600 min. For t=600 min., plot a drawdown-distance curve.

2. Repeat the calculations of problem 1 with S_{ya}=0.021 and compare with the results of the previous calculation. Provide a physical explanation for the differences observed.

3. Repeat the calculations of problem 1 with T=10 m^2/min and compare with the results from problem 1. Provide a physical explanation for the differences observed.

4. Two wells, A and B, are located 200 m apart and pump from an aquifer for which T = 1.1 m^2/min and S_{ya}=0.11. The wells begin pumping simultaneously at rates Q_A=2 m^3/min and Q_B=3 m^3/min. Compute the drawdown in an observation well located a distance of 50 m from well A and 206 m from well B after 7 days of continuous pumping from both wells. Answer: 1.04 m.

5. Let x be a coordinate measured along a straight line between two wells, A and B, that began pumping simultaneously at rates Q_A and Q_B. The origin of the coordinates is midway between the two wells and well A is located at x=a, y=0 and well B at x=-a, y=0. At some coordinate $(x_d,0)$ the quantity $\partial s/\partial x = 0$, and x_d represents the location of the ground-water divide on the line joining the two wells. Show that

$$\frac{Q_A}{Q_B} = \left(\frac{1 - x_d/a}{1 + x_d/a} \right) \exp\{-\left(\frac{a^2}{\alpha t} \right) \frac{x_d}{a} \} \quad .$$

Hint:

$$\left.\frac{\partial s}{\partial x}\right|_{y=0} = \frac{ds}{du} \left.\frac{\partial u}{\partial x}\right|_{y=0} \quad .$$

6. Prepare a graphical representation of Q_A/Q_B vs. x_d/a for several values of $a^2/\alpha t$ and discuss the results.

7. The transmissivity and storage coefficient for a confined aquifer are 0.1 m^2/min and 2.7×10^{-4}, respectively. A well pumped at 0.967 m^3/min is located a perpendicular distance of 500 m from an impermeable boundary. Estimate the time interval from the start of pumping over which the aquifer responds as if it were infinite for practical purposes. Answer: 56 min, assuming a drawdown of s_e=0.01 m is negligible.

8. What is the drawdown at the point of intersection of the barrier and a line normal to the barrier and passing through the well at a pumping time of 56 min for the conditions of problem 7? Answer: s=0.02 m.

9. Repeat problem 7 with S_{ya}=0.08. Answer: 11.6 days.

10. Replace the impermeable boundary of problem 7 with an infinitely long constant head boundary. What is the rate of aquifer recharge from the boundary at a pumping time of 56 min? Answer: 1.4% of the well discharge.

11. A fully penetrating observation well is located 14 m from a fully penetrating well discharging at a constant rate of 2.0 m^3/min from an unconfined aquifer. The aquifer is composed of small scale horizontal strata. The average hydraulic conductivity in directions parallel to the horizontal bedding is 0.05 m/min and 3.1×10^{-3} m/min across the bedding planes (i.e., in a vertical direction). The saturated thickness of the aquifer is 30 m and S_{ya}=0.061. Calculate the drawdown in the observation well at 8 min, 40 min, 80 min, 800 min, and 8000 min after the start of pumping. Answer: 0.38 m, 0.39 m, 0.41 m, 0.57 m, 0.82 m.

12. Discuss the effects of delayed water-table response in relation to aquifer thickness, other factors remaining equal.

13. Write a volume balance equation for a unit area of water table displaced vertically by Δs^o and derive Eq. 5-20 therefrom.

14. The differential equation 5-21 contains two dependent variables, s and s^o. Combine Eqs. 5-19 and 5-20 to form a linear, first-order, nonhomogeneous differential equation:

$$\frac{\partial s^o}{\partial t} + \alpha s^o = \alpha s \quad , \quad \alpha \equiv \frac{K_z}{S_{ya} b_a} \quad .$$

Solve this equation subject to $s^{o}=s=0$ at t=0 by standard methods to obtain

$$s-s^{o} = \int_{0}^{t} \frac{\partial s}{\partial t} \exp\{-\alpha(t-\tau)\}d\tau \quad ,$$

and write Eq. 5-21 in terms of the single dependent variable s.

15. The following data were collected during a test of a coarse textured, alluvial, unconfined aquifer near Crook, Colorado. Determine the transmissivity, hydraulic conductivity and apparent specific yield. The pumped well is partially penetrating.

$b = 19$ m $\quad Q = 3.78$ m^3/min

r = 6.2 m		r = 30.5 m	
Time min	Drawdown m	Time min	Drawdown m
1	0.003	5	0.037
2	0.006	9	0.052
4	0.012	14	0.064
7	0.030	19	0.076
10	0.043	24	0.082
15	0.061	29	0.085
20	0.079	39	0.107
25	0.088	49	0.116
30	0.104	59	0.128
40	0.128	69	0.143
45	0.146	94	0.165
50	0.158	152	0.192
55	0.171	222	0.229
60	0.180	378	0.293
70	0.195	588	0.348
92	0.226	1457	0.420
150	0.284	1816	0.491
221	0.332	2026	0.521
374	0.412	2973	0.537
593	0.473	3315	0.540
1460	0.598		
1813	0.610		
2029	0.625		
2976	0.656		
3316	0.692		

Do these data suggest a delayed water-table response? Provide a discussion and explanation of your findings.

16. The following recovery data were collected after pumping ceased in the aquifer test of the previous problem. Total pumping time was 3317 min. Analyze the recovery data.

r = 6.2 m

Time since pumping ceased min	Drawdown m
0	0.692
1	0.482
9	0.442
17	0.348
42	0.317
59	0.280
958	0.271

r = 30.5 m

Time since pumping ceased min	Drawdown m
0	0.540
4	0.503
12	0.491
20	0.491
39	0.445
57	0.406
960	0.384
2566	0.351
3875	0.281

17. The following data were collected during a test of an unconfined aquifer. The pumped well fully penetrated the aquifer and the observation wells were constructed so that the observed drawdowns are average values over the entire saturated thickness. Plot the data (i.e., s vs. r^2/t) for all three observation wells on a single graph and obtain the aquifer properties using the Theis type curve.

$$Q = 0.568 \text{ m}^3/\text{min}$$

	Drawdown		
Time min	r = 30.5 m	r = 61 m	r = 122 m
19.5	0.030	0.010	-
794	0.150	0.075	0.025
1605	0.205	0.105	0.045
3045	0.240	0.145	0.075
4485	0.260	0.180	0.090
5955	0.275	0.180	0.100
7155	0.300	0.205	0.120

18. The following data were collected during a test of a leaky aquifer. The pumped well fully penetrates the aquifer. Plot the data (s vs. r^2/t) from all 4 observation wells on one graph and determine T, S, and B by using a type curve prepared from Table 5-3.

$$Q = 1.284 \text{ m}^3/\text{hr}$$

Time hrs	Drawdown - meters r=30.5m	r=61m	r=122m	r=240m
1.0	1.145	0.745	0.395	0.135
6.0	1.600	1.190	0.790	0.440
43.5	2.195	1.770	1.340	0.960
340.0	2.485	2.010	1.600	1.160

19. A well discharges from a confined aquifer at a constant rate of 2.1 m^3/min for a period of 2 days. The aquifer properties are T=0.6 m^2/min and S=$4x10^{-3}$. Compute and display graphically the drawdown vs. distance for 2 hrs, 1 day, and 2 days after pumping ceased.

20. Suppose the well in problem 19 is pumped at Q=2.1 m^3/min for 2 days and then the discharge is decreased to 1.0 m^3/min. Calculate and plot the drawdown vs. time curve at r=20 m.

21. A well is to be pumped for a period t_p and then pumping ceases for a period t_o followed by a second pumping period of length t_p and another rest period of length t_o. This cycle is to be repeated for n cycles, each cycle consisting of a pumping period t_p and a rest period t_o. Assume that the logarithmic approximation for the Theis well function applies and derive

$$s = \frac{Q}{4\pi T} \ln \left\{ \frac{n!}{\prod_{i=0}^{n-1} (n-p-i)} \right\}$$

for the drawdown at the end of the rest period in the nth cycle. The parameter $p = t_p/(t_o+t_p)$. Hint: To find the effect of the first cycle add the drawdown due to pumping a time = $n(t_o+t_p)$ and subtract the buildup due to recharge for a time $n(t_o+t_p)-t_p$. The effect of the second cycle is computed for a pumping time $(n-1)(t_o+t_p)$ and a recharge time $(n-1)(t_o+t_p)-t_p$ and so on.

22. An irrigation well is to be pumped at Q=4 m^3/min for 3 days, followed by a rest period of 7 days. This cycle is to be repeated for 7 cycles during the irrigation season. The effective well radius is 0.3 m and the aquifer properties are 1.2 m^2/min and S_{ya}=0.09. Compute the drawdown in the

well at the end of the seventh rest period and the end of the eighth pumping period.
Answer: 0.23 m, 0.23+3.76 = 3.99 m.

23. Integrate Eq. 5-36 by parts to derive Eq. 5-37.

24. A well discharges at a constant rate of 4 m^3/min from an aquifer intersected by a stream. The aquifer properties are $T=0.5\ m^2/min$ and $S_{ya}=0.10$. The distance from the well to the stream (measured normal to the stream) is 500 m. Calculate the discharge from the stream to the aquifer at 2 days, 7 days, 21 days, and 42 days after pumping began.
Answer: $\simeq$ 0, 0.46 m^3/min, 1.45 m^3/min, 2.08 m^3/min.

25. Use the results of problem 24 to prepare a graph of Q_s vs. t. Graphically or numerically integrate to obtain the total volume of water taken from the stream after 42 days of pumping.

26. The pumping in problem 24 ceases after 42 days of pumping. Compute Q_s for 7 days, 21 days, and 42 days after the pumping has ceased. Answer: 1.74 m^3/min, 1.38 m^3/min, 0.52 m^3/min.

27. A well is pumped at a constant rate near a stream for a time t_1 and then pumping ceases. Show that

$$\int_0^\infty Q_s\, dt = Qt_1 \quad .$$

28. A well is located relative to impervious and constant head boundaries as shown in the figure. Estimate the drawdown in the well after 1 week of continuous pumping at a constant rate of 2.8 m^3/min for $T=0.4\ m^2/min$, $S_{ya}=0.11$ and $r_w=0.30$ m. Answer: 8.01 m.

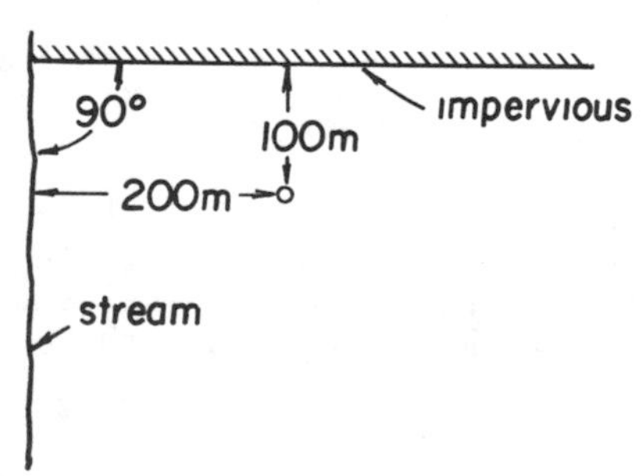

29. The water table in the aquifer that forms the banks of a long, narrow reservoir is at the same elevation as the water level in the reservoir. Inflow to the reservoir causes the stage to increase by 2 m in a period of 100 days, and thereafter the stage is constant. Assume one-dimensional flow and that the gradual rise of stage can be replaced by an abrupt increase of 2 m that occurs 50 days after inflow to the reservoir begins. If the length of shoreline is 20 km, compute the total volume of water that has been discharged into bank storage 100 days after inflow begins and 180 days after inflow begins. The aquifer properties are T=720 m^2/day and S_{ya}=0.07.
Answer: 2.3×10^6, 3.65×10^6 m^3.

30. Rework problem 29 assuming that the reservoir stage increases linearly with time for $0 \leq t \leq 100$ days and remains constant thereafter. Compare the result with problem 29.
Answer: 2.14×10^6 m^3, 3.63×10^6 m^3.

31. Show by direct substitution that Eq. 5-68 satisfies Eq. 5-55.

32. A horizontal water table exists an average depth of 10 m below a canal. When water is diverted into the canal, leakage occurs at an average rate of 1.7 m^3/min per km of canal length. Estimate the time at which the water table directly beneath the canal will have risen to 2 m above the original level. The properties of the aquifer are T=0.2 m^2/min and S_{ya}=0.05. Answer: 30 days.

33. Water is diverted from the canal in problem 32 at t=30 days, and recharge ceases. Compute and show graphically the rise and decay of the ground-water mound directly beneath the canal.

34. A flood wave in a channel bounded on both sides by an alluvial aquifer is approximated by the square wave: s_0=0, $t \leq 0$; s_0=2 m, $0 \leq t \leq 2$ days; s_0=0, $t \geq 2$ days, where s_0 is the stage measured relative to the pre-flood stage. The properties of the aquifer are T=0.3 m^2/min and S_{ya}=0.15. Calculate and display graphically the discharge into and from the aquifer, assuming that the water table in the aquifer is at the same elevation as the water surface in the stream prior to the flood.

35. Compute and plot the water level (relative to the pre-flood level) in an observation well located 15 m from the stream for the conditions in problem 34.

36. A long straight drain is installed 1 m below an initially horizontal water table. The transmissivity of the aquifer is 0.2 m^2/min and S_{ya}=0.04. Assuming that the water pressure in the drain is atmospheric calculate the drain discharge per meter of length 1 day, 2 days, and 20 days after the drain is opened.
Answer: 3.8 m^3/d-m, 2.7 m^3/d-m, 0.86 m^3/d-m.

37. Estimate the perpendicular distance on either side of the drain in problem 36 to a point where the drawdown of the water table is 0.1 m after 20 days of drainage.
Answer: 1.4 km.

38. Two parallel ditches are installed to lower the water table in the strip of aquifer between them. The initial saturated thickness is 10 m, S_{ya}=0.1, and K=0.001 m/min. The water level in the drain ditches is maintained at a level 3 m below the initial level. The drain ditches are 250 m apart. Estimate the drainage time required to lower the water table 2 m on the line midway between and parallel to the ditches. Answer: 60 days.

39. Compute and plot the water-table level z_c, midway between parallel drains, as a function of time following an instantaneous recharge of I=0.02 m if L=200, T=0.02 m^2/min, and S_{ya}=0.10.

40. Repeat problem 39 with two recharge events spaced 5 days apart.

41. Consider n instantaneous recharge events between parallel drains, each of magnitude I and spaced by equal drainout periods of lengths t_d. Use a procedure similar to that used in problem 21 for cyclic pumping to derive

$$z_c = \frac{4I}{\pi S_{ya}} \left\{ \frac{1 - \exp(-\frac{n\pi^2\alpha}{L^2} t_d)}{\exp(\frac{\pi^2\alpha}{L^2} t_d) - 1} \right\}$$

for the water-table level at the end of the nth drainout period for those cases in which $\alpha t_d/L^2 > 0.05$. Hint: Refer to Example 5-14 and superimpose the z_{ci} for each recharge event to obtain

$$z_c = \frac{4I}{\pi S_{ya}} \{\exp[-\frac{n\pi^2\alpha}{L^2} t_d] + \exp[-\frac{(n-1)\pi^2\alpha}{L^2} t_d] + \ldots$$

$$+ \exp[-\frac{\pi^2\alpha}{L^2} t_d]\}$$

which is a geometric progression with a common ratio of

$$\exp(-\frac{\pi^2\alpha}{L^2} t_d) \quad .$$

42. Use the results of problem 41 to compute the water-table elevation midway between the drains of Example 5-14 immedately prior to the last irrigation of the season. Also compute the water-table elevation immediately following the last irrigation. Answer: 0.96 cm, 17.6 cm.

43. A stream flows near the center of an alluvial valley bounded by impervious strata. The valley is 3 km wide and contains an alluvial aquifer for which $T=0.6\ m^2/min$ and $S_{ya}=0.15$. Recharge from a rain averages 0.06 m over the valley. Compute the volume of recharged water that has returned to the stream per km 0 days, 10 days, 30 days, and 100 days following the rain.
Answer: $0\ m^3$, $3.3x10^4\ m^3$, $5.6x10^4\ m^3$, $1.0x10^5\ m^3$.

44. Compute the return flow discharge per km to the stream in problem 43, 10 days, 30 days, and 100 days following the rain. Answer: $1625\ m^3/d\text{-}km$, $940\ m^3/d\text{-}km$, $490\ m^3/d\text{-}km$.

45. Because the recharge in problem 43 is assumed to be instantaneous, the water-table elevation relative to the stream level is increased instantaneously by 0.4 m. Mathematically, this is similar to an instantaneous drop in the stream stage of 0.4 m. Use Eq. 5-62 (which was derived for inflow from an infinite aquifer to one side of a stream whose stage was decreased by s_0 at t=0) to compute the return flow discharge for the conditions of problem 43 for t=10 days, t=30 days, t=100 days. Discuss any difference observed.
Answer: $1625\ m^3/d\text{-}km$, $940\ m^3/d\text{-}km$, $514\ m^3/d\text{-}km$.

46. From the results of problems 44 and 45, draw conclusions relative to the use of Eq. 5-62 to approximate the discharge given by Eq. 5-80 for small $\alpha t/L^2$.

Chapter VI

FINITE-DIFFERENCE METHODS

Numerical methods offer another viable procedure for evaluating flows in ground water systems. Only the finite-difference technique is presented here. The numerical solution procedure by finite differences is fairly simple and straightforward but the application of the model to a given physical system can be complex and requires considerable judgment and skill in setting up the problem and in interpretation of results. Finite-difference methods should not be viewed as a replacement for the previously presented analytic methods, but rather as a tool for the evaluation of complex groundwater flows. The selection of any technique (analytical, numerical or physical models) depends upon the users preference, experience, ability, the complexity of the particular problem to be solved and, among other things, the availability of computers, mathematical tables and data. For example, one would not normally use a numerical method to evaluate the performance of a single well in a homogeneous, isotropic, confined aquifer (except perhaps to confirm the adequacy of a particular numerical method). On the other hand, numerical methods are used advantageously to evaluate the performance of multiple wells in a stream-aquifer system of nonhomogeneous anisotropic materials.

No attempt is made here to develop or present all of the numerical methods that are available and are presently in use. Instead, a simple and basic development of the difference form of the parabolic equation is presented and solved by both explicit and implicit techniques. The alternating direction implicit and the Crank-Nicholson method is also presented, as these methods are commonly being used by many investigators. The basic objective of this chapter is to develop a model for the practical analysis of ground-water flow problems.

The finite-element method is another powerful procedure for solving ground-water problems, but is beyond the scope of this text. Students interested in the finite-element method are referred to the text book written by Segerlind (1976).

6.1 ONE-DIMENSIONAL FLOW - CONFINED AQUIFER

The one-dimensional confined aquifer case is presented first and subsequently generalized for the two-dimensional unconfined aquifer case. Three-dimensional cases, although very important, are not considered. The difference equations are developed by mass balance considerations rather than by using Taylor's series expansion of the derivatives (von Rosenberg, 1969). Although both methods lead to identical results, the mass balance approach emphasizes the application of the Darcy and continuity equations.

Consider the non-steady, one-dimensional flow in a confined aquifer shown in Fig. 6-1a. Since the flowlines are parallel and invarient with time, it is convenient to analyze a unit width of the confined aquifer ($\Delta x=1$ units). The region of flow is overlaid by a grid system shown in Fig. 6-1b. For each grid,

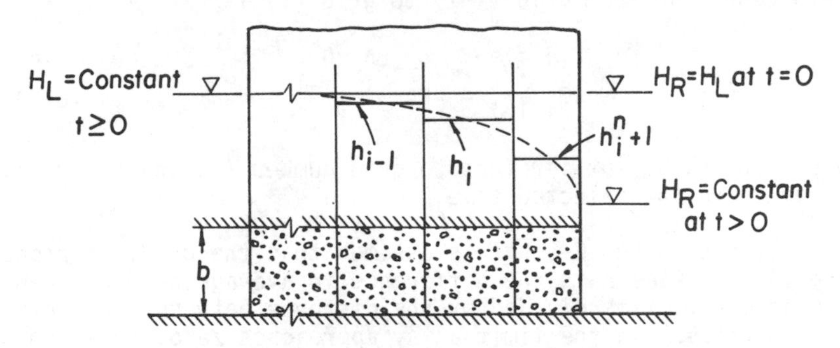

Figure 6-1a. Section view of grid overlay in a confined aquifer.

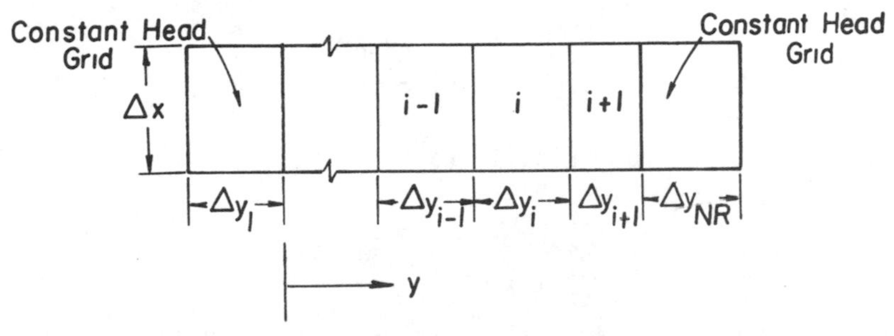

Figure 6-1b. Plan view of grid overlay.

values of hydraulic conductivity K_i, grid dimension Δy_i, aquifer thickness b_i, storage coefficient S_i, and initial values of head h_i need to be specified. Mass balance for grid (i) requires that inflow ($Q_{i-1 \to i}$) from grid (i-1) to grid (i) minus the outflow ($Q_{i \to i+1}$) from grid (i) to grid (i+1), must equal the rate of change in storage ($\Delta V_i/\Delta t$) in grid (i).

For discussion purposes, consider a homogeneous and isotropic aquifer of constant thickness in which constant values of K, S, b, and Δy are known. That is:

$$K_1 = K_2 = \ldots = K_\ell = K$$
$$S_1 = S_2 = \ldots = S_\ell = S \qquad (6\text{-}1)$$
$$b_1 = b_2 = \ldots = b_\ell = b$$

and $\Delta y_1 = \Delta y_2 = \ldots = \Delta y_\ell = \Delta y$

where subscript ℓ refers to the total number of grids. For this case flow from grid (i-1) to grid (i) is

$$Q_{i-1\to i} = -KA\,\frac{h_i^n - h_{i-1}^n}{\Delta y} \qquad (6\text{-}2)$$

where the subscript i refers to grid number (i) and superscript n refers to the selected time.

To obtain Eq. 6-2, it is assumed that the gradient producing flow at time n is the difference in average head between two adjacent grids divided by the distance between the center of the grids. In the limit as Δy approaches zero, this approximation becomes exact.

The area A is the product of Δx and b, and since Δx was set equal to unity and b is a constant, Eq. 6-2 becomes:

$$Q_{i-1\to i} = -T\,\frac{h_i^n - h_{i-1}^n}{\Delta y} \quad . \qquad (6\text{-}3)$$

Likewise, flow from grid (i) to (i+1) is:

$$Q_{i\to i+1} = -T\,\frac{h_{i+1}^n - h_i^n}{\Delta y} \quad . \qquad (6\text{-}4)$$

The rate of change in storage of water in grid (i) for the time interval Δt is:

$$\frac{\Delta V_i}{\Delta t} = S\Delta y \left[\frac{h_i^{t-\Delta t} - h_i^t}{\Delta t} \right] \qquad (6\text{-}5)$$

Continuity for grid (i) is:

$$Q_{i-1\to i} - Q_{i\to i+1} = \frac{\Delta V_i}{\Delta t} \quad , \qquad (6\text{-}6)$$

which upon substitution of Eq. 6-3, 6-4, and 6-5 results in:

$$\left(-T\,\frac{h_i^n - h_{i-1}^n}{\Delta y}\right) - \left(-T\,\frac{h_{i+1}^n - h_i^n}{\Delta y}\right) = S\Delta y\left(\frac{h_i^{t+\Delta t} - h_i^t}{\Delta t}\right) \quad . \tag{6-7}$$

This can be rearranged to yield

$$h_{i+1}^n - 2h_i^n + h_{i-1}^n = \frac{S}{T}\,\frac{(\Delta y)^2}{\Delta t}\,(h_i^{t+\Delta t} - h_i^t) \quad . \tag{6-8}$$

In Eq. 6-8, if n=t (the present or known value) th n the explicit or forward difference equation is obtained. On the other hand, if n=t+Δt (the future values to be calculated) then the implicit or backward difference equation results.

Forward Difference Equation - Explicit Solution

The forward difference equation is obtained by setting n=t and rearranging Eq. 6-8 to obtain

$$h_i^{t+\Delta t} = \frac{T\Delta t}{S(\Delta y)^2}\,(h_{i+1}^t + h_{i-1}^t) + h_i^t\left[1 - \frac{2T\Delta t}{S(\Delta y)^2}\right] . \tag{6-9}$$

The space derivatives are centered at the beginning of the time step and the only unknown is $h_i^{t+\Delta t}$ which results from the time derivative. Thus, Eq. 6-9 can be solved explicitly at each grid for the head at the new time level (t+Δt). Since the solution is dependent only upon known values of heads in adjacent grids at the beginning of the time period t, the computation for $h_i^{t+\Delta t}$ in any grid can be made in any order without regard to values of $h_i^{t+\Delta t}$ for any other grid.

The solution represented by Eq. 6-9 is only an approximation to the exact solution, of course. The degree to which Eq. 6-9 approximates the exact solution is not necessarily uniform in time and space and depends upon the selection of Δy and Δt. It is possible, for example, to select Δy and Δt such that the difference between the approximate and exact solution grows as t increases. The approximate solution is said to be *unstable* in this case. A stable solution is insured, however, in a one-dimensional homogeneous case if

$$\frac{T\Delta t}{S(\Delta y)^2} < \frac{1}{2} \quad . \tag{6-10}$$

Consequently, the time increment cannot be selected independently of the space increment. It is emphasized that the above inequality insures stability only and does not insure an accurate approximation to the exact solution.

EXAMPLE 6-1

As an example of the use of the explicit method, consider the problem in Fig. 6-1. For this case, let Δy=3 m, b=1.5 m, $H_L=h_1$=6.1 m for t>0, $H_R=h_5$=1.5 m for t>0, number of grids = 5, K=0.5 m/d, and S=0.02. The initial condition is $h_2^o=h_3^o=h_4^o$=6.1 m.

Solution:

To satisfy the stability requirement of Eq. 6-10, the maximum time step Δt is computed as

$$\Delta t < \frac{1}{2}\,\frac{S(\Delta y)^2}{T} = \frac{1}{2}\,\frac{(0.02)(3)^2}{0.75} = 0.12 \text{ days} \quad .$$

Therefore, the time increment is selected as 0.1 days.

For the first time step only grid (4) is affected and Eq. 6-9 becomes:

$$h_4^{t+\Delta t} = \frac{T\Delta t}{S(\Delta y)^2}\,(h_5^o+h_3^o) + h_4^o[1-2\,\frac{T\Delta t}{S(\Delta y)^2}\,]$$

or

$$h_4^{0.1} = \frac{0.75(0.1)}{(0.02)(3)^2}\,(1.5+6.1) + 6.1[1 - \frac{2(0.75)(0.1)}{(0.02)(3)^2}\,]$$

$$h_4^{0.1} = 3.2 + 1.0 = 4.2 \text{ m}$$

and

$$h_3^{0.1} = h_2^{0.1} = 6.1 \text{ m} \quad .$$

For the second time step t+Δt=0.2 days, one obtains for grid (4)

$$h_4^{0.2} = \frac{T\Delta t}{S(\Delta y)^2}\,(h_5^{0.1}+h_3^{0.1}) + h_4^{0.1}[1 - \frac{2T\Delta t}{S(\Delta y)^2}\,]$$

$$= 0.42(1.5+6.1) + 4.2(1-0.83) = 3.9 \text{ m} \quad .$$

And, for grid (3)

$$h_3^{0.2} = \frac{T\Delta t}{S(\Delta y)^2}[h_4^{0.1} + H_2^{0.1}] + h_3^{0.1}[1 - \frac{2T\Delta t}{S(\Delta y)^2}]$$

$$= 0.42(4.2+6.1) + 6.1(1-0.83) = 5.3 \text{ m} \quad .$$

Grid (2) is not affected until the third time step.

The above process is repeated until the head for each grid is calculated at the desired time. Figure 6-2 shows the calculated value for the head (h_4^t) in grid (4) as a function of

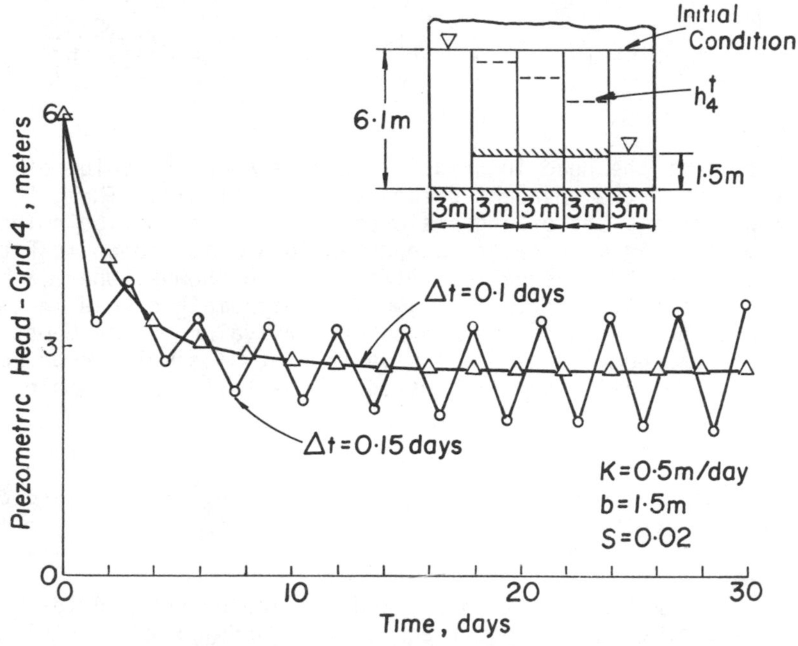

Figure 6-2. Piezometric head in grid (4) calculated by the forward difference equation.

time. To illustrate the stability problem, a set of calculations was made in which Δt was selected to be 0.15 days so that the expression for stability results in

$$\frac{T\Delta t}{S(\Delta y)^2} = 0.63 \nleq \frac{1}{2} \quad .$$

The calculated head in grid (4) as a function of time is shown in Fig. 6-2. The computed values fluctuate with each time step for Δt=0.15 giving completely erroneous results. Also, note that the amplitude of the fluctuation increases with increasing time.

Backward Difference Equation - Implicit Solution

The backward difference equation (implicit equation) is obtained by setting n=t+Δt in Eq. 6-8 and is

$$h_{i+1}^{t+\Delta t} - 2h_i^{t+\Delta t} + h_{i-1}^{t+\Delta t} = \frac{S(\Delta y)^2}{T\Delta t}(h_i^{t+\Delta t} - h_i^t) \quad .(6\text{-}11)$$

In contrast to the explicit solution technique, the space derivatives are centered at time t+Δt. Rearranging Eq. 6-11 so that all of the known values are on the right hand side of the equal sign results in

$$h_{i+1}^{t+\Delta t} - (2 + \frac{S(\Delta y)^2}{T\Delta t})h_i^{t+\Delta t} + h_{i-1}^{t+\Delta t} = -\frac{S(\Delta y)^2}{T\Delta t} h_i^t \quad . \qquad (6\text{-}12)$$

Note that the head in grid (i) depends upon the value of head at t+Δt in the adjacent grids, (i+1) and (i-1). Thus, Eq. 6-12 represents a set of algebraic equations that must be solved simultaneously. For the one-dimensional case shown in Fig. 6-1, there are (NR-2) equations with (NR-2) unknowns, where NR is the number of grids. The values of piezometric head in the first and last grid are known boundary values. The Gauss elimination scheme is used to solve the set of simultaneous equations (Eq. 6-12) as illustrated in the following example.

EXAMPLE 6-2

Solve the problem of Example 6-1 using the backward difference equation.

Solution:

Note that grids (1) and (5) are boundary grids and values of head are specified as 6.1 m and 1.5 m respectively for t>0. Equation 6-12 need only be written for the three interior grids. For the first time step, Δt=0.1 days, the following equations for grids (2), (3), and (4) are obtained:

Grid (2) $6.1 - 4.4\, h_2^{0.1} + h_3^{0.1} = -2.4(6.1)$

Grid (3) $h_2^{0.1} - 4.4\, h_3^{0.1} + h_4^{0.1} = -2.4(6.1)$

Grid (4) $h_3^{0.1} - 4.4\, h_4^{0.1} + 1.5 = -2.4(6.1)$.

Rearranging the above equations so that all known values are placed on the right hand side and adding to the equation for grid (2) zero times $h_4^{0.1}$ and to grid (4) zero times $h_2^{0.1}$, the following equations are obtained:

$$\text{Grid (2)} \quad -4.4\, h_2^{0.1} + h_3^{0.1} + 0\, h_4^{0.1} = -20.7$$

$$\text{Grid (3)} \quad h_2^{0.1} - 4.4\, h_3^{0.1} + h_4^{0.1} = -14.6$$

$$\text{Grid (4)} \quad 0\, h_2^{0.1} + h_3^{0.1} - 4.4\, h_4^{0.1} = -16.1 \quad .$$

In matrix form this set of equations may be written as:

$$\begin{bmatrix} -4.4 & 1 & 0 \\ 1 & -4.4 & 1 \\ 0 & 1 & -4.4 \end{bmatrix} \begin{bmatrix} h_2^{0.1} \\ h_3^{0.1} \\ h_4^{0.1} \end{bmatrix} = \begin{bmatrix} -20.7 \\ -14.6 \\ -16.1 \end{bmatrix} \quad .$$

Note that this is a tridiagonal coefficient matrix with non-zero values on the main and adjacent diagonals only. Using the Gauss elimination scheme (see Appendix for step-by-step solution by the Gauss scheme) one obtains, after the first time step, $h_2^{0.1}$=6.0 m, $h_3^{0.1}$=5.8 m, and $h_4^{0.1}$=5.0 m. This procedure is repeated using the calculated values $h_i^{0.1}$ to obtain values $h_i^{0.2}$. The solution for head as a function of time for grid (4) is presented in Fig. 6-3 for both implicit and explicit cases.

Analysis has shown that stability is not a problem with the backward difference method and theoretically the selection of space increments Δy and time increments Δt can be made independently of each other. The selection of the values for space and time increments depends upon the user's requirement for accuracy, detail of analysis, and availability of data.

Crank-Nicholson Approximation

In the explicit and implicit methods, the time derivatives are centered at $t+\frac{1}{2}\Delta t$ while the space derivatives are centered at t for the explicit case and $t+\Delta t$ for the implicit case. The Crank-Nicholson approximation centers both the time and space derivatives at $t+\frac{1}{2}\Delta t$ and should provide for a better estimate for the distribution of heads as compared to either of the previous methods, other factors equal. Note that the truncation error in the implicit case is of the order of $[(\Delta x)^2+\Delta t]$ while in the Crank-Nicholson case, the truncation error is smaller being of the order of $[(\Delta x)^2+\Delta t^2]$. Thus, for the same "accuracy" larger time increments may be used for the Crank-Nicholson method (Remsen et al., 1971).

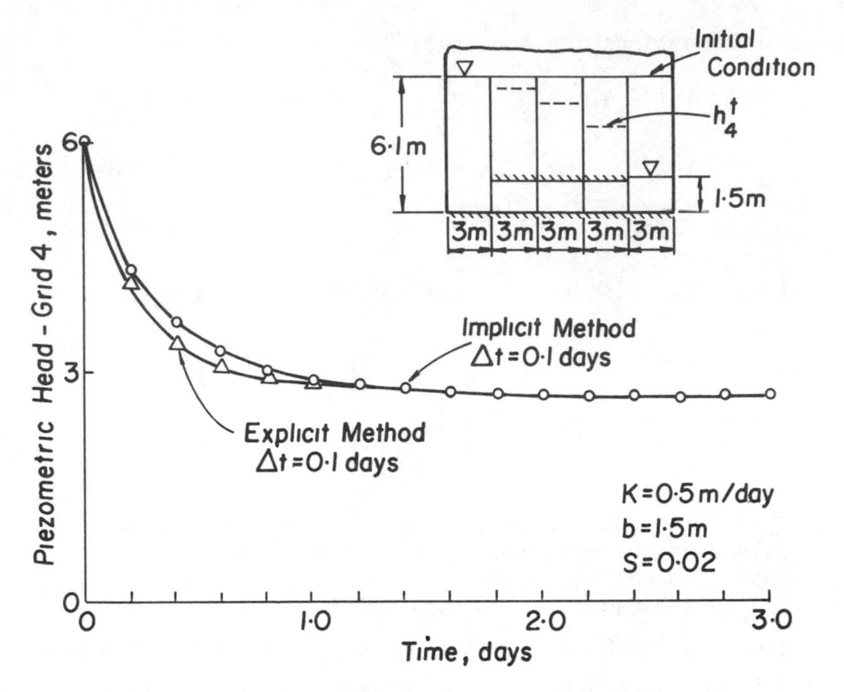

Figure 6-3. Comparison between explicit and implicit method.

The Crank-Nicholson equation for one-dimensional flow in a homogeneous, isotropic, confined aquifer is

$$\frac{1}{2}\left(\frac{h_{i+1}^{t+\Delta t}-2h_i^{t+\Delta t}+h_{i-1}^{t+\Delta t}}{(\Delta y)^2}\right)+\frac{1}{2}\left(\frac{h_{i+1}^{t}-2h_i^{t}+h_{i-1}^{t}}{(\Delta y)^2}\right)$$

$$=\frac{S}{T\Delta t}\,(h_i^{t+\Delta t}-h_i^t)\quad . \tag{6-13}$$

In this form, it can be seen that the Crank-Nicholson approximation is the result of successive application of the implicit method (the first bracketed term in Eq. 6-13) and the explicit method (the second bracketed term in Eq. 6-13). Equation 6-13 may be rearranged to yield

$$h_{i+1}^{t+\Delta t}-[2+2\,\frac{S(\Delta y)^2}{T\Delta t}\,]h_i^{t+\Delta t}+h_{i-1}^{t+\Delta t}$$

$$=-\,\frac{S(\Delta y)^2}{T\Delta t}\,h_i^t-[h_i^t-(2+\frac{S(\Delta y)^2}{T\Delta t}\,)h_i^t+h_{i-1}^t]\quad ,(6\text{-}14)$$

which, except for the two bracketed terms, is identical to the implicit Eq. 6-12. Since the two bracketed quantities are known at the beginning of the time step, the solution procedure is nearly the same as for the implicit case. However, for the same degree of accuracy, larger time increments may be used for the Crank-Nicholson method and, thus, fewer time steps are necessary to compute the value of head for a given elapsed time.

The Crank-Nicholson equation, although stable for all values of Δx and Δt has the potential for severe oscillations of the computed heads. The solution oscillates about the "true" value, and for some (but not all) cases, the oscillations become insignificant after sufficiently long periods of time (Rushton, 1973). The oscillation cannot be predicted in advance but the severity depends upon the boundary and initial conditions, at least. Small values of Δt tend to reduce the severity of oscillation.

6.2 TWO-DIMENSIONAL FLOW - CONFINED AND UNCONFINED AQUIFER

In this section the developments of the previous section are extended to cases in which the flow is two-dimensional and in which the transmissivity can change as a result of changes in the saturated flow depth.

Alternating Direction Implicit Procedure - Two-Dimensional Case

The alternating direction implicit (ADI) solution technique is a perturbation of the Crank-Nicholson method for the two-dimensional case in which the heads are computed in two separate steps. A grid system is established for the aquifer as shown in Fig. 6-4.

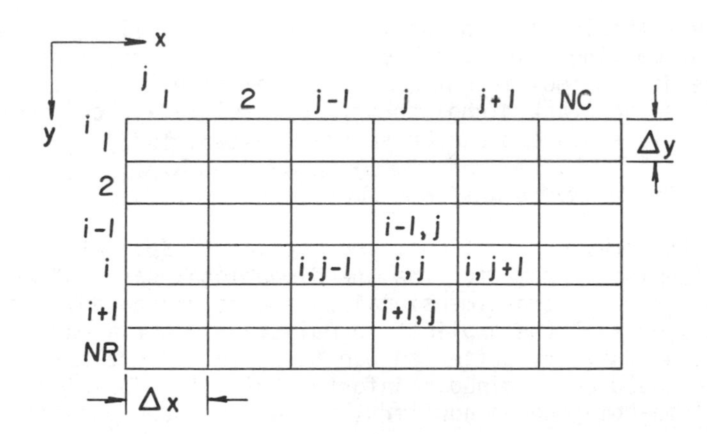

Figure 6-4. Grid overlay system for confined aquifer.

The ADI equations for the homogeneous, isotropic, confined aquifer are:

$$T\left[\frac{h_{i+1,j}^{t+\frac{1}{2}\Delta t}-2h_{i,j}^{t+\frac{1}{2}\Delta t}+h_{i-1,j}^{t+\frac{1}{2}\Delta t}}{(\Delta y)^2}\right]+T\left[\frac{h_{i,j+1}^{t}-2h_{i,j}^{t}+h_{i,j-1}^{t}}{(\Delta x)^2}\right]$$

$$=\frac{S}{(\Delta t/2)}\left(h_{i,j}^{t+\frac{1}{2}\Delta t}-h_{i,j}^{t}\right) \quad , \qquad (6\text{-}15)$$

for the y-direction and

$$T\left[\frac{h_{i+1,j}^{t+\frac{1}{2}\Delta t}-2h_{i,j}^{t+\frac{1}{2}\Delta t}+h_{i-1.j}^{t+\frac{1}{2}\Delta t}}{(\Delta y)^2}\right]+T\left[\frac{h_{i,j+1}^{t+\Delta t}-2h_{i,j}^{t+\Delta t}+h_{i,j-1}^{t+\Delta t}}{(\Delta x)^2}\right]$$

$$=\frac{S}{(\Delta t/2)}\left(h_{i,j}^{t+\Delta t}-h_{i,j}^{t+\frac{1}{2}\Delta t}\right) \quad . \qquad (6\text{-}16)$$

for the x-direction. Equation 6-15 is solved first in the y-direction column by column for $h^{t+\frac{1}{2}\Delta t}$ using known values of heads in the adjacent grids in the x-direction at the beginning of the time step. Consequently, the heads in each column are obtained implicitly within a column and explicitly between columns for the first $\frac{1}{2}\Delta t$ in the same manner as for the one-dimensional case. One obtains a tridiagonal coefficient matrix for each column. For the second $\frac{1}{2}\Delta t$, Eq. 6-16 is solved in the x-direction implicitly, row by row, for $h^{t+\Delta t}$ using known values of heads $h^{t+\frac{1}{2}\Delta t}$ in the adjacent grid in the y-direction. This process of alternating solutions in the y and x directions is repeated until the head at the desired time is obtained.

The ADI solution process is extremely fast since the algorithm for solving tridiagonal matrices is very efficient and the size of the coefficient matrix is relatively small. However, like the Crank-Nicholson scheme, oscillation of the calculated values of head can be severe (Ruston, 1973).

Backward Difference Equation - Two-Dimensional Case

In the previous sections, the implicit, explicit, and Crank-Nicholson scheme for the one-dimensional case and ADI procedure for the two-dimensional case were presented. It was demonstrated that the explicit formulation is not always stable and that a stability criterion had to be met before a stable solution could be obtained. Unfortunately, for two-dimensional flow in non-homogeneous aquifers, it has not been possible to define a general stability requirement. Therefore, it is desirable to use a method where stability is not a problem. Also, since the Crank-Nicholson and ADI procedures require a great deal of user experience in order to obtain a solution in which oscillation is not a problem, we focus attention on the backward difference equation solved by the Gauss scheme for the remainder of this chapter. This technique is fairly simple, completely stable, and is without oscillation problems. The procedure is to overlay the aquifer with a rectangular grid network as shown in Fig. 6-5 and to develop the backward difference equation for each grid. The resulting set of simultaneous equations is then solved for selceted boundary conditions.

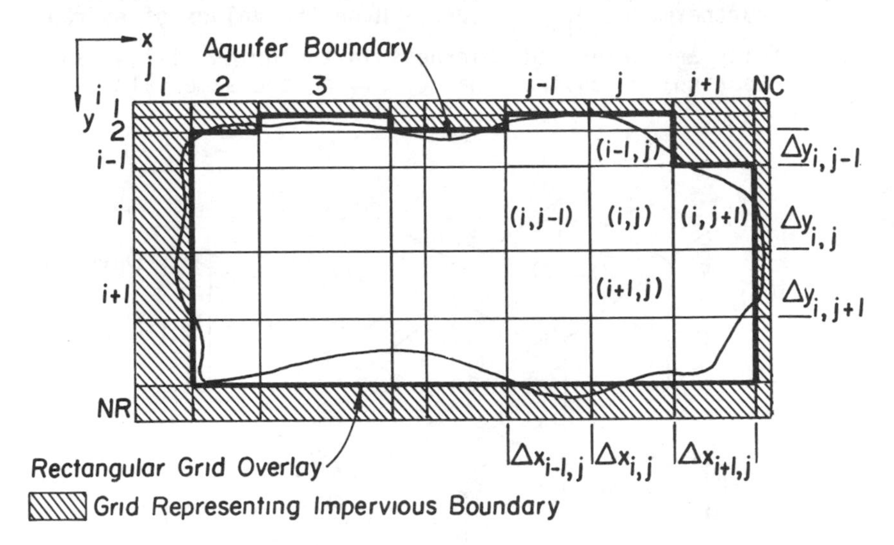

Figure 6-5. Plan view grid overlay and numbering system.

For the general two-dimensional case, the variable parameters of hydraulic conductivity, storage coefficient, and external water sources are treated as average constant values within a grid of constant dimensions. At any time t, each grid may have different values for external water sources. For the unconfined aquifer, the nonlinear equation is effectively linearized by treating the aquifer thickness as a constant for the time increment Δt. The symbol b is used for the saturated thickness to avoid confusion with the head h; b being equal to h only when the floor of the aquifer is horizontal and used as the datum for measuring h. Note that the Dupuit-Forchheimer approximation is used in the following analysis.

Grid (i,j) and its four adjacent grids are shown in Fig. 6-6. Flow from grid (i-1,j) to grid (i,j) is computed by Darcy's equation using as the flow area $\Delta x_{i,j}$ times the present value of the saturated thickness $b^t_{i-\frac{1}{2},j}$ which is located at the boundary between grid (i-1,j) and grid (i,j). Since the hydraulic conductivity may be different in each grid, a weighted average value between the two grids is used.

The development of the equations for flow between adjacent grids is illustrated by developing the equation appropriate for flow between grids (i-1,j) and (i,j). Referring to Fig. 6-7, the flow from grid (i-1,j)to grid (i,j) is $Q_{i-\frac{1}{2},j}$. However, the gradient of piezometric head (slope of the water table) at $(i-\frac{1}{2},j)$ must be computed using values of head at (i-1,j) and (i,j). Furthermore, $Q_{i-\frac{1}{2},j}$ depends upon the values of hydraulic conductivity and saturated thickness in the grids (i-1,j) and (i,j). Continuity requires that $Q_{i-\frac{1}{2},j}$ be the same value,

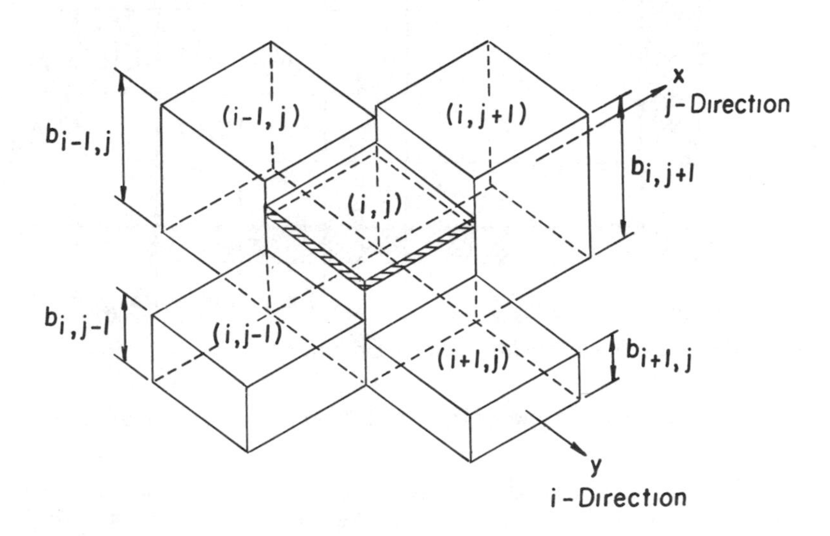

Figure 6-6. Grid system for two-dimensional unconfined aquifer (Eckhardt and Sunada, 1975).

however, whether computed using parameter values in grid (i-1,j) or parameter values in grid (i,j). Thus,

$$Q_{i-\frac{1}{2},j} = K_{i-1,j}\Delta x_{i-1,j} b^t_{i-\frac{1}{2},j} \frac{(h_{i-1,j}-h_{i-\frac{1}{2},j})^{t+\Delta t}}{\frac{\Delta y_{i-1,j}}{2}} \tag{6-17}$$

and

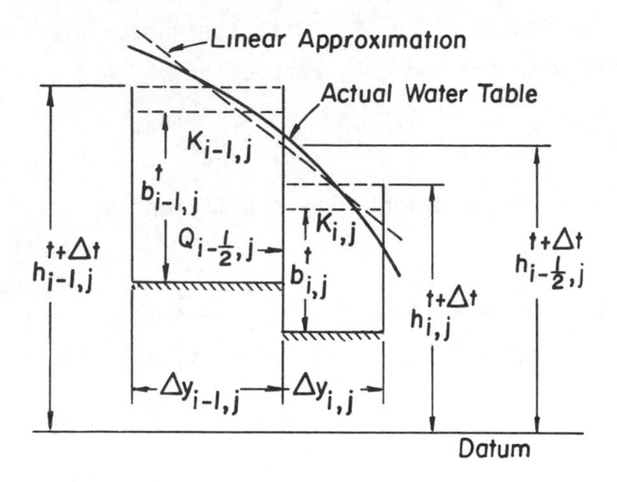

Figure 6-7. Flow between two adjacent grids.

$$Q_{i-\frac{1}{2},j} = K_{i,j}\Delta x_{i,j} b^t_{i-\frac{1}{2},j} \frac{(h_{i-\frac{1}{2},j}-h_{i,j})^{t+\Delta t}}{\frac{\Delta y_{i,j}}{2}} \quad . \tag{6-18}$$

Eliminating the unknown $h^{t+\Delta t}_{i-\frac{1}{2},j}$ in Eqs. 6-17 and 6-18 by solving them simultaneously yields

$$Q_{i-\frac{1}{2},j} = \left\{ \frac{\Delta y_{i,j}}{2K_{i,j}\Delta x_{i,j} b^t_{i-\frac{1}{2},j}} + \frac{\Delta y_{i-1,j}}{2K_{i-1,j}\Delta x_{i-1,j} b^t_{i-\frac{1}{2},j}} \right\}^{-1} (h_{i-1,j}-h_{i,j})^{t+\Delta t} \quad . \tag{6-19}$$

The coefficient multiplying the difference in head in Eq. 6-19 is given the symbol $C^t_{i,j}$ and Eq. 6-19 becomes

$$Q_{i-\frac{1}{2},j} = C^t_{i,j}(h_{i-1,j}-h_{i,j})^{t+\Delta t} \quad . \tag{6-20}$$

In a similar manner

$$Q_{i+\frac{1}{2},j} = D^t_{i,j}(h_{i,j}-h_{i+1,j})^{t+\Delta t} \quad , \tag{6-21}$$

where $D_{i,j}^t$ is identical to $C_{i,j}^t$ with the subscripts i-1 and i-½ replaced with i+1 and i+½, respectively. Also,

$$Q_{i,j-\frac{1}{2}} = A_{i,j}^t (h_{i,j-1} - h_{i,j})^{t+\Delta t} \qquad (6\text{-}22)$$

for flow in the j-direction. The coefficient $A_{i,j}^t$ is

$$A_{i,j}^t = \left\{ \frac{\Delta x_{i,j}}{2K_{i,j}\Delta y_{i,j} b_{i,j-\frac{1}{2}}^t} + \frac{\Delta x_{i,j-1}}{2K_{i,j-1}\Delta y_{i,j-1} b_{i,j-\frac{1}{2}}^t} \right\}^{-1} . \qquad (6\text{-}23)$$

Finally,

$$Q_{i,j+\frac{1}{2}} = B_{i,j}^t (h_{i,j} - h_{i,j+1})^{t+\Delta t} \quad , \qquad (6\text{-}24)$$

where the coefficient $B_{i,j}^t$ is obtained from $A_{i,j}^t$ by replacing the subscripts j-1 and j-½ by j+1 and j+½, respectively.

The rate of change of the volume of water stored in grid (i,j) is

$$\frac{\Delta V_{i,j}}{\Delta t} = S_{i,j}\Delta x_{i,j}\Delta y_{i,j} \left(\frac{h_{i,j}^{t+\Delta t} - h_{i,j}^t}{\Delta t} \right) + Q_{i,j}^t$$

$$= E_{i,j}^t (h_{i,j}^{t+\Delta t} - h_{i,j}^t) + Q_{i,j}^t \quad , \qquad (6\text{-}25)$$

where $Q_{i,j}^t$ is an external source or sink that is used to represent recharge or pumping from the grid. Mass balance for grid (i,j) requires that the sum of the inflows (Eqs. 6-20 and 6-22) minus the sum of the outflows (Eqs. 6-21 and 6-24) be equal to the rate of change of storage (Eq. 6-25). Thus,

$$A_{i,j}^t h_{i,j-1}^{t+\Delta t} + B_{i,j}^t h_{i,j+1}^{t+\Delta t} + C_{i,j}^t h_{i-1,j}^{t+\Delta t} + D_{i,j}^t h_{i+1,j}^{t+\Delta t}$$

$$- (A_{i,j} + B_{i,j} + C_{i,j} + D_{i,j} + E_{i,j})^t h_{i,j}^{t+\Delta t} = Q_{i,j}^t - E_{i,j}^t h_{i,j}^t \qquad (6\text{-}26)$$

is obtained after all known quantities are placed on the right hand side. In matrix notation, Eq. 6-26 is

$$[COEF]^t[h]^{t+\Delta t} = [Q-Eh]^t \quad . \qquad (6\text{-}27)$$

For the confined aquifer, the quantity $b^t_{i,j-\frac{1}{2}}$ in the coefficient $A^t_{i,j}$ (and similar terms in other coefficients) is the aquifer thickness which may vary in space but is constant in time. In the unconfined case, $b^t_{i,j-\frac{1}{2}}$ represents the saturated thickness at the beginning of a time step and is held constant for the time interval Δt. Thus, all of the coefficients are known at the beginning of the time step.

There is a total of (NCxNR) grids for the system shown in Fig. 6-5. The exterior grids are used to introduce the boundary conditions on each boundary segment by putting K=0 for an impervious segment, h=constant for constant head boundary segments, and prescribing the discharge through segments that represent boundaries on which the inflow and outflow is known. Any of the interior grids may be assigned boundary values of constant head or zero hydraulic conductivity.

Since all of the exterior grids must be assigned known boundary values, there are (NR-2)x(NC-2) equations with (NR-2)x(NC-2) unknowns. The coefficient matrix is a square matrix containing $[(NR-2)(NC-2)]^2$ terms most of which are zero. This coefficient matrix is pentadiagonal and can be solved by any one of several available schemes but the Gauss elimination scheme is recommended. It should be noted that, for the one-dimensional case, one obtains a tridiagonal matrix with the terms $D^t_{i,j}$ and $C^t_{i,j}$ being equal to zero since there is no flow in the i-direction. In addition, the coefficient matrix is symmetrical with $B^t_{i,j}=A^t_{i,j+1}$ and $D^t_{i,j}=C^t_{i+1,j}$ and, consequently, only the upper (or lower) terms need to be calculated along with the main diagonal terms. A numerical model based upon the above formulation is presented in Appendix C, together with a user's guide. The use of the two-dimensional model given in Appendix C is illustrated by the following example.

EXAMPLE 6-3

A well is pumped from a confined aquifer surrounded by a constant head boundary as shown in Fig. 6-8. For small time the numerical solution can be conveniently checked with the Theis analytic solution. The values of the aquifer properties are given in the figure.

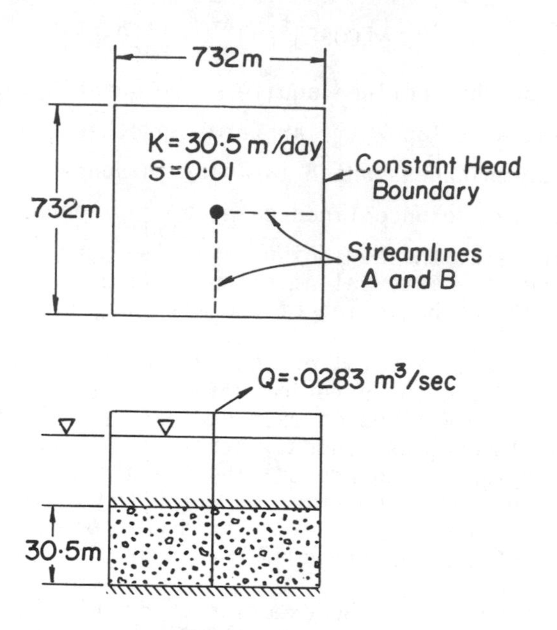

Figure 6-8. Definition sketch for Example 6-3.

To take advantage of symmetry, a quadrant of the flow is analyzed rather than the entire flow field. There is no flow across the streamlines A and B and, therefore, the streamlines A and B can be represented by impermeable boundaries. A variable grid spacing (Fig. 6-9) is selected so that the piezometric head in the vicinity of the well can be adequately approximated by linear distributions of h between adjacent grids. The dimensions of the grids are given in Table 6-1. A time increment of 0.1 days was selected. The computer model presented in the appendix was used to solve this problem.

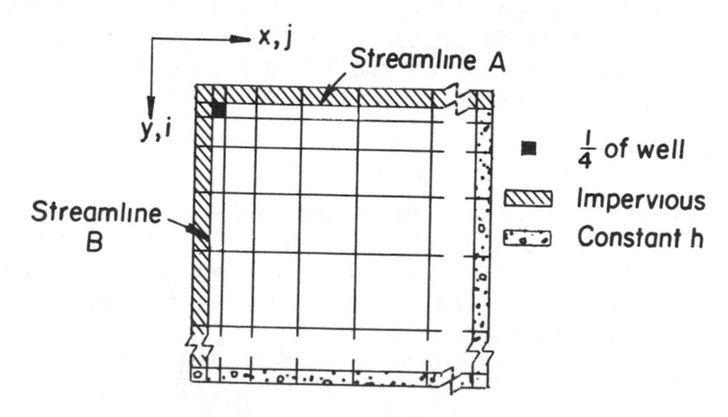

Figure 6-9. Schematic of grid overlay for Example 6-3.

Table 6-1. Grid Dimensions for Example 6-3

Row/Column	1	2	3	4	5	6	7
Δy,m	0.03	0.3	3.0	3.0	6.1	6.1	12.2
Δx,m	0.03	0.3	3.0	3.0	6.1	6.1	12.2

Row/Column	8	9	10	11	12
Δy,m	30.5	61.0	121.0	121.0	0.3
Δx,m	30.5	61.0	121.0	121.0	0.3

Figure 6-10 shows the radial distribution of drawdown at t=0.5 days and a comparison with the Theis analytic solution. It should be noted that, although excellent results were obtained with the variable grid spacing, large differences in grid dimensions between adjacent grids should be avoided to insure that central averages and the assumption of a linear distribution of head between adjacent grids remains appropriate.

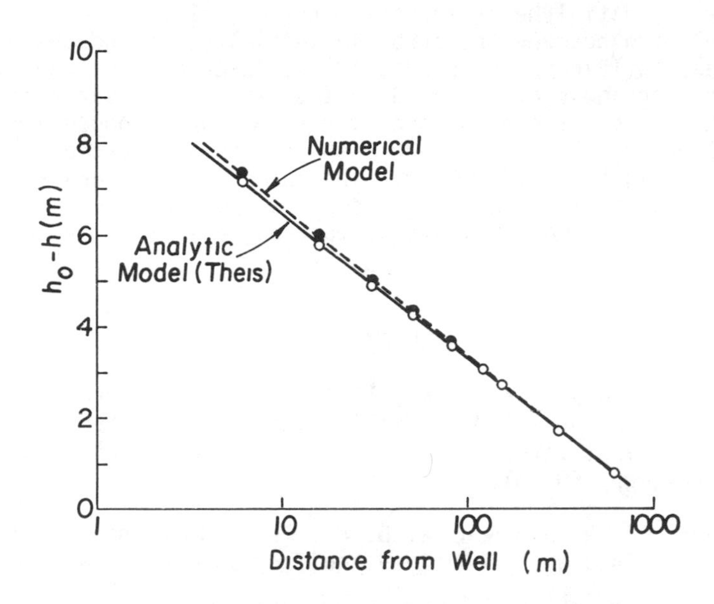

Figure 6-10. Distribution of drawdown along streamline A.

6.3 SENSITIVITY TO SOLUTION PARAMETERS

The analysis of the areal and time distribution of heads in an aquifer by any method requires that values of hydraulic conductivity, storage coefficient (specific yield), aquifer thickness, boundary conditions, hydrologic inputs, and initial heads be known. In many instances, these values are not known and estimates must be made in order to estimate aquifer response to imposed stresses. Consequently, it is desirable to have some indication of the effects of each variable on the solution displayed as a distribution of heads. Although the discussion is restricted to the distribution of heads, one may also want to evaluate the distribution of flows at a particular time or location. It should be pointed out that since mass continuity is always satisfied, errors exist only in the time and space distributions of flow rates. The average error in the computation of flow rates for the entire aquifer is zero (i.e, total volume of water is accounted for).

In a strict sense, the actual effects of errors in input data on the results depends on the particular problem studied and a sensitivity analysis should be conducted for each case studied. However, experience has shown that, in general, some parameters influence the solution more than others.

It has been found that the most important variable is the value of initial heads and that there is almost a one to one correlation between the error in initial heads and the error in results (Bibby and Sunada, 1971). Each of the other variables (storage coefficient S, net inflow Q, and saturated thickness, b) have about the same effect and produce errors about one order of magnitude less than those induced by erroneous values of initial head. The selection of appropriate grid size and time increments depends upon the user's experience and judgment and, often, must be determined by trial and error.

REFERENCES

Bibby, R. and Sunada, D. K., 1971. Statistical Error Analysis of a Numerical Model on Confined Ground-Water Flow. Proc. of First Intern. Symp. Stochastic Hydraulics, Pittsburgh, PA, pp. 591-612.

Eckhardt, J. R. and Sunada, D. K., 1975. Numerical Model of a Confined and Unconfined Aquifer. Colorado State University, 64 p.

Remson, I., Hornberger, G. M., and Molz, R. J., 1971. Numerical Methods in Subsurface Hydrology. Wiley-Interscience, New York, NY. 389 p.

Rushton, K. R., 1973. Discrete Time Steps in Digital Computer Analysis of Aquifers Containing PUmped Wells, Journal of Hydrology, 18:1-19.

Segerlind, L. J., 1976. Applied Finite Element Analysis. Wiley and Sons, New York, NY. 422 p.

von Rosenberg, D. V., 1969. Methods For the Numerical Solution of Partial Differential Equations. No. 16, Elsevier, New York, NY. 128 p.

PROBLEMS AND STUDY QUESTIONS

1. Set up and solve the Crank-Nicholson equation for the problem illustrated in Example 6-1. Select a time increment Δt such that oscillations are not a problem.

2. Show by way of an example that any one-dimensional problem solved by the implicit scheme may be set up to have a tridiagonal coefficient matrix.

3. Set up and solve the explicit equations to compute the distribution of heads and the drawdown as a function of time at r=251 m for the conditions noted in Fig. 5-1.

4. Set up the implicit equations appropriate for the conditions of Example 5-15 and solve for the return flows.

5. Write a computer program for solving the explicit equations for two-dimensional flow toward a pumping well located between an impervious and a constant head boundary as shown in Fig. 4-9.

APPENDIX A

Systems of Units and Conversion Tables

The International System of Units (SI) is being established by international agreement to provide uniform units for measurement throughout the world. The SI, although an update of the metric system, is not the same as the metric system in all respects. The following tables are presented to aid the reader in converting from the English system to the SI.

Symbol of SI Units

Quantity	Name of Unit	SI Symbol
length	metre	m
mass	kilogram	kg
time	second	s
temperature	Kelvin	K
force	Newton	N ($kg \cdot m/s^2$)
pressure	Pascal	Pa (N/m^2)
work	joule	J ($N \cdot m$)
power	watt	W ($N \cdot m/s$)
plane angle	radians	rad

Prefix For SI

Prefix	Symbol	Multiplication Factor
giga	G	10^9
mega	M	10^6
kilo	k	10^3
hecto	h	10^2
deka	da	10^1
deci	d	10^{-1}
centi	c	10^{-2}
milli	m	10^{-3}
micro	μ	10^{-6}

Conversion Table For Common Quantities

Quantity	U.S.English	SI
Length	1 in	25.4 mm
	1 ft	0.3048 m
	1 yard	0.9144 m
Area	1 in^2	645 mm^2
	1 ft^2	0.0929 m^2
	1 acre	4047 m^2 (0.4047 hectare)
Volume	1 in^3	16387 mm^3
	1 ft^3	0.0283 m^3
	1 yd^3	0.765 m^3
	1 acre-ft	1233 m^3
	100 gal	0.3785 m^3
Force	1 pound	4.448 N
Weight/Volume	1 lb/in^3	271.4 kN/m^3
	1 lb/ft^3	157.1 N/m^3
	(62.4 lb/ft^3)	(9.81 N/m^3)
Pressure	1 psi	6.895 kN/m^2 (kPa)
	1 psf	47.88 N/m^2 (Pa)
Mass	1 lb (mass)	0.4536 kg
Temperature	32°F	273 K
Power	1 hp	745.7 W

APPENDIX B

Gauss Elimination Scheme for Example 6-2 in Text

The procedure is to upper triangulate the coefficient matrix

$$\begin{bmatrix} -4.4 & 1 & 0 \\ 1 & -4.4 & 1 \\ 0 & 1 & -4.4 \end{bmatrix}$$

to obtain a matrix equation of the form

$$\begin{bmatrix} 1 & a_{12} & a_{13} \\ 0 & 1 & a_{23} \\ 0 & 0 & 1 \end{bmatrix} \begin{bmatrix} h_2^{0.1} \\ h_3^{0.1} \\ h_4^{0.1} \end{bmatrix} = \begin{bmatrix} c_2 \\ c_3 \\ c_4 \end{bmatrix} ,$$

so that the system of equations

$$\begin{aligned} h_2^{0.1} + a_{12}h_3^{0.1} + a_{13}h_4^{0.1} &= c_2 \\ h_3^{0.1} + a_{23}h_4^{0.1} &= c_3 \\ h_4^{0.1} &= c_4 \end{aligned}$$

can be solved by backward substitution.

Recall that the matrix equation in Chapter VI was

$$\begin{bmatrix} -4.4 & 1 & 0 \\ 1 & -4.4 & 1 \\ 0 & 1 & -4.4 \end{bmatrix} \begin{bmatrix} h_2^{0.1} \\ h_3^{0.1} \\ h_4^{0.1} \end{bmatrix} = \begin{bmatrix} -20.7 \\ -14.6 \\ -16.1 \end{bmatrix} .$$

For this case multiply row 1 by -(1/4.4) and subtract row 1 from row 2 to obtain

$$\begin{bmatrix} 1 & -0.23 & 0 \\ 0 & -4.17 & 1 \\ 0 & 1 & -4.4 \end{bmatrix} \begin{bmatrix} h_2^{0.1} \\ h_3^{0.1} \\ h_4^{0.1} \end{bmatrix} = \begin{bmatrix} 4.7 \\ -19.3 \\ -16.1 \end{bmatrix} .$$

The next step is to multiply row 2 by -(1/4.17) and subtract row 2 from row 3 to obtain

$$\begin{bmatrix} 1 & -0.23 & 0 \\ 0 & 1 & -0.24 \\ 0 & 0 & -4.16 \end{bmatrix} \begin{bmatrix} h_2^{0.1} \\ h_3^{0.1} \\ h_4^{0.1} \end{bmatrix} = \begin{bmatrix} 4.7 \\ 4.6 \\ -20.7 \end{bmatrix}$$

and multiply row 3 by -(1/4.17) to obtain

$$\begin{bmatrix} 1 & -0.23 & 0 \\ 0 & 1 & -0.24 \\ 0 & 0 & 1 \end{bmatrix} \begin{bmatrix} h_2^{0.1} \\ h_3^{0.1} \\ h_4^{0.1} \end{bmatrix} = \begin{bmatrix} 4.7 \\ 4.6 \\ 5.0 \end{bmatrix} \quad .$$

Therefore, for grid (4) the head is

$$h_4^{0.1} = 5.0 \text{ m} \quad .$$

Similarly, the head in grid (3) is

$$0h_2^{0.1} + h_3^{0.1} - 0.24h_4^{0.1} = 4.6 \text{ m} \quad .$$

Or, upon substitution of the value of $h_4^{0.1}$, one obtains

$$h_3^{0.1} = 4.6 + 0.24(5.0) = 5.8 \text{ m} \quad .$$

Finally, for grid (2) the value of $h_2^{0.1}$ may be obtained from

$$h_2^{0.1} - 0.23h_3^{0.1} + 0h_4^{0.1} = 4.7 \text{ m} \quad ,$$

or

$$h_2 = 4.7 + 0.23(5.8) = 6.0 \text{ m} \quad .$$

APPENDIX C

Finite Difference Ground-Water Model

This appendix presents a two-dimensional ground-water model which may be used for analysis of the areal distribution of heads in ground-water aquifers. This model will treat either unconfined or confined ground-water flow problems and will handle the situation when an unconfined aquifer becomes confined or a confined aquifer becomes unconfined. The model is based upon the fully implicit central difference scheme using the Gauss elimination procedure for solving the matrix which was described in the text.

A flow diagram of the model is presented in Fig. C-1. A complete listing of the program is given in the last part of this appendix. The model uses the English system of units.

Procedure for Analysis

The area to be studied is overlain with a grid system. The selection of the space dimensions (DX,DY) for the grids is dependent upon the availability of geologic and hydrologic data, and the desired accuracy and detail of analysis. The accuracy of the solution is enhanced as values for DX and DY are decreased and the smaller values should be used providing the availability of geologic and hydrologic data justifies the additional computation time. The total number of grids selected is also dependent on the storage capacity of the computer being used.

The grid system selected should be oriented to allow for easy boundary approximation, provide for easy adaptation of hydrologic and geologic data, and provide for the desired model accuracy. Space dimensions should be selected such that the geologic and hydrologic conditions may be reasonably assumed uniform over each grid. In areas where detailed values of water level or piezometric head are desired, smaller values of DX and DY may be used. Grids located outside the boundary of the study area must be defined as impermeable grids. Hydraulically connected lakes and rivers should be specified as constant head grids.

Boundary conditions due to geologic and hydrologic influences include (1) impermeable or no flow boundaries, (2) constant head or hydraulic boundaries, and (3) underflow or gradient boundaries. Program GRWATER uses an initial water level coding to identify the type of boundary. H(I,J) is the initial water level or potentiometric head in grid (I,J) and the coding used is:

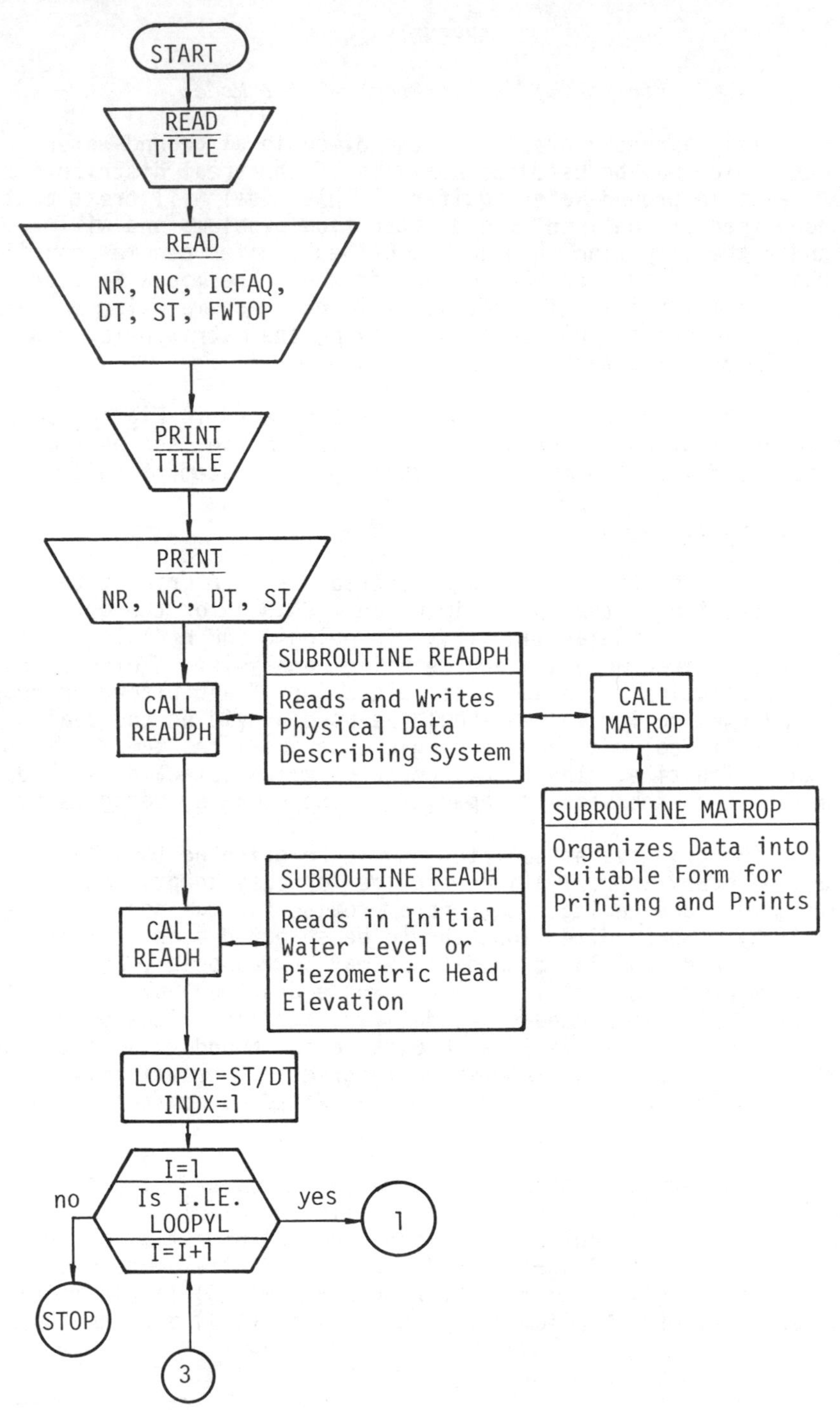

Figure C-1. Flow Chart for Program GRWATER.

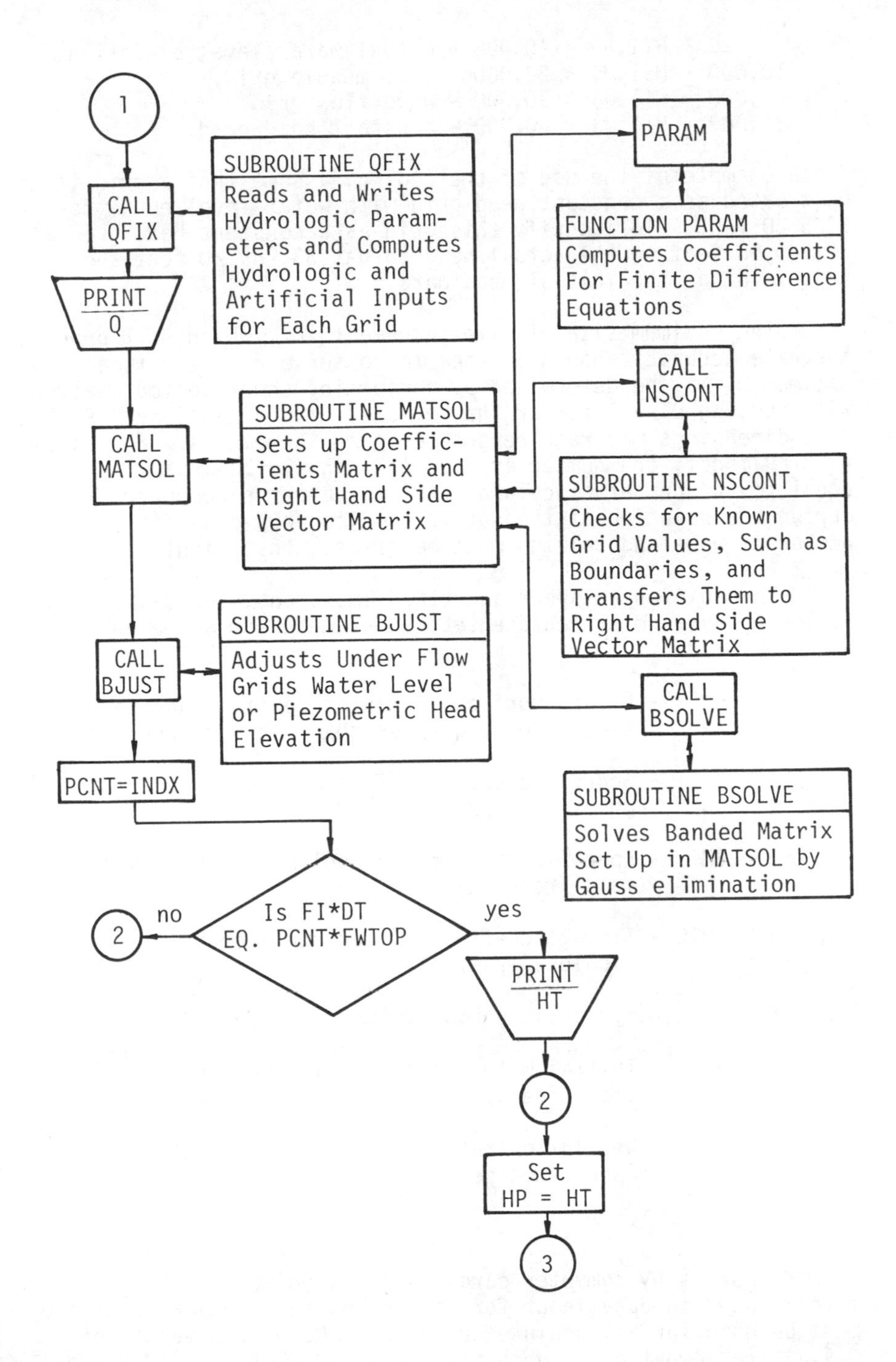

Figure C-1. (Continued).

$0 \leq H(I,J) < 10{,}000$ - actual water level elevation.
$10{,}000 \leq H(I,J) < 20{,}000$ - impermeable grid.
$20{,}000 \leq H(I,J) < 30{,}000$ - underflow grid.
$30{,}000 \leq H(I,J) < 40{,}000$ - constant head grid.

As an example of the use of the coding, assume that a grid is to be treated as a constant head grid with water level equal to 3285.20 feet. To identify this grid as a constant head grid add 30,000.00 feet to the actual head to get 33,285.20 feet and input this value as the initial head data.

The maximum size of time increment, DT, which will provide adequate accuracy should be used to conserve computer time. The optimum DT may be determined by performing short period analyses with varying DT values for the selected grid dimensions. Smaller grid dimensions may require shorter time increments. For this GRWATER model, the number of rows (NR) should be less than or equal to the number of columns (NC) to conserve computer time during the solution of the set of simultaneous equations. The number of rows and columns must be greater than eight.

In addition to space and time dimensions, (DX,DY,DT), the following average or representative parameters must be defined for each grid:

1. G - Ground surface or top of confined aquifer elevation (feet), greater than but not equal to zero.

2. Z - Bedrock elevation (feet), greater than but not equal to zero.

3. PHI - Specific yield for unconfined case (dimensionless fraction).

4. PHIC - Storage coefficient for confined case (dimensionless fraction).

5. FK - Hydraulic conductivity (feet/day).

6. H - Initial water level or piezometric head elevation (feet).

7. Q - Net inflow rate due to pumping, recharge, etc. (acre-feet per day).

Data Input

Input is by computer cards as outlined below. Note that zero is used in data input for coding and that decimal override must be used for all noninteger values. Note that values of Q(I,J) are computed by subroutine QFIX and that the DIMENSION card in the main program must be changed to fit the particular problem (see program).

Card 1: 8A10 Format.
TITLE - Title of the particular run.

Card 2: 3I5,3F10.1 Format.
NR = number of rows in the grid system (NR should be less than NC) > 8.
NC = number of columns in grid system > 8.
ICFAQ = 1 for confined aquifer analysis, otherwise zero.
DT = time increment (days).
ST = total time of analysis (days) (integer multiple of DT).
FWTOP = desired interval of times for printed output (integer multiple of DT).

Card 3: 8F10.1 Format.
DLX = value of uniform DX (zero otherwise) - x-dimension (feet).
DLY = value of uniform DY (zero otherwise) - y-dimension (feet).
FFK = value of uniform FK (zero otherwise) - permeability (feet/day).
ZZ = value of uniform Z (zero otherwise) - bedrock elevation (feet).
GG = value of uniform G (zero otherwise) - ground surface or top of confined aquifer elevation (feet).
PPHI = value of uniform PHI (zero otherwise) - specific yield or storage coefficient (decimal).
PPHIC = value of uniform PHIC (zero otherwise) - confined aquifer (blank for unconfined aquifer analysis) storage coefficient (decimal).

Card 3A: 8F10.1 Format - if DLX is zero, otherwise omit.
DX J=1,NC

Card 3B: 8F10.1 Format - if DLY is zero, otherwise omit.
DY I=1,NR

Card 3C: 8F10.1 Format - if FFK is zero, otherwise omit.
FK I=1,NR*NC by columns (example, the permeability of grid (4,3) for NR=11,NC≥11 is the second entry on the fourth card).

Card 3D: 8F10.1 Format - if ZZ is zero, otherwise omit.
Z I=1,NR*NC by columns (see card 3C).

Card 3E: 8F10.1 Format - if GG is zero, otherwise omit.
G I=1,NR*NC by columns (see card 3C).

Card 3F: 8F10.1 Format - if PPHI is zero, otherwise omit.
PHI I=1,NR*NC by columns (see card 3C).

Card 3G: 8F10.1 Format - if PPHIC is zero, otherwise omit.
PHIC I=1,NR*NC by columns (see card 3C).

Card 4: 5F10.1 Format.
HW = horizontal H - constant value of initial water level or piezometric head elevation (feet) of all grids regardless of boundary conditions (zero otherwise).
*LBC = left boundary code (single value along left boundary).
*RBC = right boundary code (single value along right boundary).
*TBC = top boundary code (single value along top boundary).
*BBC = bottom boundary code (single value along bottom boundary).

*Note: Corner grids are not critical. Use 10,000.00 for impermeable boundary, 20,000.00 for underflow boundary, and 30,000.00 for constant head boundary, to denote outer boundary conditions. Inner boundary conditions must be specified by using card 4A. Specify 0.0 if coding is to be done in card 4A for each grid.

Card 4A: 8F10.1 Format - if HW is zero, otherwise omit.
H I=1, NR*NC by columns (coded H values for boundary conditions).

Description of Subprograms

Subroutine READPH

This subroutine reads and writes the physical data describing the study area. The following variables are read and printed: DX, DY, FK, Z, G, PHI, and PHIC. Only one data card is required if all variables are uniform for each grid, otherwise each parameter that is variable must be read in matrix form. Variables DX and DY require only NC and NR values respectively.

Called From: Main Program
Subprograms Used: MATROP
Important Variables: DX DY FK Z G PHI PHIC .

Subroutine READH

This subroutine reads the initial coded water level or piezometric head elevations. H is decoded and set equal to HT and HP. One data card is required if the initial water level is horizontal, otherwise the entire H-matrix must be read.

Called From: Main Program
Subprograms Used: None
Important Variables: H HT HP .

Subroutine MATSOL

This subroutine sets up the coefficient matrix, CMATRX, and the right hand side vector matrix, CR. CMATRX is a reduced matrix containing only the band of known values in the left side of the difference equations and is written vertically rather than diagonally. Its dimensions are (NR-2)*(NC-2) by 2*NR-3. The coefficients are computed using Function PARAM and checked for adjacent boundary values of H in subroutine NSCONT. MATSOL treats known grid values of H. BSOLVE is used to solve the matrix equation set up.

Called From: Main Program
Subprograms Used: PARAM NSCONT BSOLVE
Important Variables: CMATRX CR .

Function PARAM

This subprogram computes the coefficients in the left side of the finite difference equation. For confined aquifer analysis, saturated thickness is compared to aquifer thickness and the smallest of the two is used to calculate the coefficient.

Called From: MATSOL BYFLOW
Subprograms Used: None
Important Variables: PARAM

Subroutine NSCONT

This subroutine transfers the coefficients, in CMATRX, multiplied by their respective H-value, to the right hand side vector matrix in case of adjacent head or known boundary conditions. It also sets coefficients equal to zero in case of adjacent impermeable grids.

Called From: MATSOL
Subprograms Used: None
Important Variables: None .

Subroutine BSOLVE

This subroutine solves the matrix equation set up in MATSOL by Gauss elimination. BSOLVE is designed specifically for a diagonal matrix that results from analysis of groundwater systems.

Called From: MATSOL
Subprograms Used: None
Important Variables: None .

Subroutine BJUST

This subroutine adjusts the underflow boundary water level elevations. Gradients are calculated three grids in from the exterior boundary grids and the gradients are projected back to the exterior boundary grids to obtain new water level elevations. This calculation is performed at even time steps. At odd time steps the water level elevations are held constant and the exterior boundary grids are treated as constant head grids.

Called From: Main Program
Subprograms Used: None
Important Variables: H HT .

Subroutine MATROP

This subroutine organizes data or results into a suitable form for printing and then prints.

Called From: READPH, Main Program
Subprograms Used: None
Important Variables: NR=NOROW NC=NOCOL .

Subroutine QFIX

This subroutine sets the value of net inflow rates to each grid for each time period due to pumping, recharge, irrigation, etc. units are ac-ft/day. *This subroutine should be modified for user's requirements.*

Called From: Main Program
Subprograms Used: None
Important Variables: Q .

Computer Program Listing

On the following pages a listing of the above described computer program is provided.

PROGRAM GRWATER

```
      PROGRAM GRWATER (INPUT,OUTPUT,TAPE5=INPUT,TAPE6=OUTPUT,TAPE7=101)
C     THIS MODEL IS BASED UPON WTSHED DEVELOPED AT COLO. STATE UNIV.
C     NR AND NC MUST BE GREATER THAN OR EQUAL TO 8 FOR THIS MODEL
C     CONTROL VARIABLES
C       NR=NUMBER OF ROWS
C       NC=NUMBER OF COLUMNS
C       NR SHOULD ALWAYS BE LESS THAN OR EQUAL TO NC
C       ICFAQ=1 FOR CONFINED AQUIFER ANALYSIS, OTHERWISE ZERO
C       DT=TIME INCREMENT (DAYS)
C       ST=TOTAL TIME OF ANALYSIS (DAYS)
C       FWTOP=DESIRED TIME OF OUTPUT (MULTIPLE OF DT)
C     DIMENSION OF CMATRX IS  ((NR-2)*(NC-2),2*NR-3) AND FOR CR
C     ((NC-2)*(NR-2)) AND OF ALL OTHER VARIABLES (NR,NC)
C
C       NA=NUMBER OF ROWS IN REDUCED BAND MATRIX
C       NB=NUMBER OF COLUMNS IN REDUCED BAND MATRIX
C
      DIMENSION FK(12,12), PHI(12,12), PHIC(12,12), G(12,12),
     1Z(12,12), H(12,12),HT(12,12), HP(12,12), HF(12,12), DX(12,12),
     2DY(12,12), Q(12,12), A(12,12), B(12,12), CMATRX(100,21), CR(100)
C
      DIMENSION TITLE(8)
C
      COMMON/BLK1/DT,ST,ICFAQ,FWTOP
C
      READ (5,200) TITLE
      READ (5,220) NR,NC,ICFAQ,DT,ST,FWTOP
      NA=(NR-2)*(NC-2)
      NB=2*NR-3
      WRITE (6,210) TITLE
  110 IF (ICFAQ.LE.0) GO TO 120
      WRITE (6,240)
      GO TO 130
  120 WRITE (6,260)
  130 WRITE (6,190) NR,NC,DT,ST
C
      CALL READPH (NR,NC,FK,PHI,Z,G,DX,DY,PHIC)
      CALL READH (NR,NC,H,HP,HT,HF)
      LOOPUL=ST/DT
      INDX=1
C
      DO 170 K=1,LOOPUL
         FI=K
      T2=FI*DT
      CALL QFIX (NR,NC,Q)
      WRITE(6,530)
      CALL MATROP(NR,NC,Q)
      CALL MATSOL (NR,NC,NA,NB,FK,PHI,H,HT,Z,DX,DY,C,CMATRX,CR,
     1A,B,G,PHIC,HP,HF)
      CALL BJUST (NR,NC,H,HT,HP,DX,DY,K)
         PCNT=INDX
         IF ((FI*DT).NE.(PCNT*FWTOP)) GO TO 150
         INDX=INDX+1
         WRITE (6,180) T2
      CALL MATROP (NR,NC,HT)
  150 DO 160 I=1,NR
      DO 160 J=1,NC
      HP(I,J)=HT(I,J)
  160 CONTINUE

  170 CONTINUE
      STOP
  180 FORMAT (1H1,44X, 22HHEAD MAP AT TIME LEVEL,F10.2,/1H ,52X, 18H(FEE
     1T ABOVE DATUM))
  190 FORMAT (15H-ROW DIMENSION=I4,21H     COLUMN DIMENSION=I4,24H    TIM
     1E INCREMENTED BY G9.2,5H DAYS,27H     TOTAL TIME OF ANALYSIS G9.2,5
     2H DAYS)
  200 FORMAT (8A10)
  210 FORMAT (1H1,/////,27X,8A10)
  220 FORMAT (3I10,3F10.3)
  240 FORMAT (1H-,47X, 25HCONFINED AQUIFER ANALYSIS)
  260 FORMAT (1H-,46X, 27HUNCONFINED AQUIFER ANALYSIS)
  530 FORMAT (1H1,40X,46HQ - MAP IN ACRE-FEET/DAY (PUMPING IS NEGATIVE))
C
      END
```

SUBROUTINE QFIX

```
      SUBROUTINE QFIX (NR,NC,Q)
C     Q(I,J) MUST BE CALCULATED OR READ IN FOR EACH TIME STEP
C     UNITS OF Q(I,J) MUST BE IN ACRE-FEET PER DAY
C     THE SUBROUTINE MAY BE MODIFIED FOR USERS REQUIREMENTS
      DIMENSION Q(NR,NC)
      DO 100 J=1,NC
      DO 100 I=1,NR
      Q(I,J)=0.0
  100 CONTINUE
C     FOR THIS PROBLEM Q(2,2) WILL BE SET EQUAL TO 0.5 AC-FEET/DAY
C     THIS IS EQUAL TO A WELL DISCHARGE OF 1 CUBIC FEET PER SECOND
      Q(2,2)=0.5
      RETURN
      END
```

SUBROUTINE READPH

```
      SUBROUTINE READPH (NR,NC,FK,PHI,Z,G,DX,DY,PHIC)
C
C
C     THIS SUBROUTINE READS AND WRITES THE PHYSICAL DATA DESCRIBING
C       THE SYSTEM.
C       FK=PERMEABILITY (FEET/DAY)
C       PHI=EFFECTIVE POROSITY
C       Z=BEDROCK ELEVATION (FEET)
C       G=GROUND SURFACE ELEVATION, OR TOP OF CONFINED AQUIFER (FEET)
C       DX=X-DIMENSION OF GRID (FEET)
C       DY=Y-DIMENSION OF GRID (FEET)
C       PHIC=CONFINED AQUIFER STORAGE COEFFICIENT
C       DLX=UNIFORM DX
C       DLY=UNIFORM DY
C       FFK=UNIFORM FK
C       ZZ=UNIFORM Z
C       GG=UNIFORM G
C       PPHI=UNIFORM PHI
C       PPHIC=UNIFORM PHIC
C
      DIMENSION FK(NR,NC), PHI(NR,NC), Z(NR,NC), G(NR,NC), DX(NR,NC), DY
     1(NR,NC),  PHIC(NR,NC)
C
      COMMON/BLK1/DT,ST,ICFAQ,FWTOP
C
      DO 110 J=1,NR
      DO 110 K=1,NC
        FK(J,K)=0.0
        Z(J,K)=0.0
        G(J,K)=0.0
        PHI(J,K)=0.0
        PHIC(J,K)=0.0
  110 CONTINUE
C
      READ (5,440) DLX,DLY,FFK,ZZ,GG,PPHI,PPHIC
C
      IF (DLX.LE.0.0) GO TO 130
      DO 120 J=1,NC
      DO 120 I=1,NR
  120 DX(I,J)=DLX
      GO TO 150
  130 READ (5,440) (DX(1,J),J=1,NC)
      DO 140 I=2,NR
      DO 140 J=1,NC
  140 DX(I,J)=DX(1,J)
  150 CONTINUE
C
      IF (DLY.LE.0.0) GO TO 170
      DO 160 J=1,NC
      DO 160 I=1,NR
  160 DY(I,J)=DLY
      GO TO 190
  170 READ (5,440) (DY(I,1),I=1,NR)
      DO 180 J=2,NC
      DO 180 I=1,NR
  180 DY(I,J)=DY(I,1)
  190 CONTINUE
C
      IF (FFK.LE.0.0) GO TO 210
```

```
      DO 200 J=1,NR
      DO 200 K=1,NC
  200 FK(J,K)=FFK
      GO TO 220
  210 READ (5,440) FK
C
  220 IF (ZZ.LE.0.0) GO TO 250
  230 DO 240 J=1,NR
      DO 240 K=1,NC
  240 Z(J,K)=ZZ
      GO TO 260
  250 READ (5,440) Z
C
  260 IF (GG.LE.0.0) GO TO 290
  270 DO 280 J=1,NR
      DO 280 K=1,NC
  280 G(J,K)=GG
      GO TO 300
  290 READ (5,440) G
C
  300 IF (PPHI.LE.0.0) GO TO 320
      DO 310 J=1,NR
      DO 310 K=1,NC
  310 PHI(J,K)=PPHI
      GO TO 370
  320 READ (5,440) PHI
C
  370 IF (ICFAQ.LE.0) GO TO 400
      IF (PPHIC.LE.0.0) GO TO 390
      DO 380 I=1,NR
      DO 380 J=1,NC
  380 PHIC(1,J)=PPHIC
      GO TO 400
  390 READ (5,440) PHIC
  400 CONTINUE
C
      WRITE (6,450)
      CALL MATROP (NR,NC,DX)
      WRITE (6,460)
      CALL MATROP (NR,NC,DY)
      IF (ICFAQ.GT.0) GO TO 410
      WRITE (6,470)
      GO TO 420
  410 WRITE (6,510)
  420 CALL MATROP (NR,NC,G)
      WRITE (6,480)
      CALL MATROP (NR,NC,Z)
      WRITE (6,490)
      CALL MATROP (NR,NC,PHI)
      IF (ICFAQ.LE.0) GO TO 430
      WRITE (6,520)
      CALL MATROP (NR,NC,PHIC)
  430 WRITE (6,500)
      CALL MATROP (NR,NC,FK)
      RETURN
C
  440 FORMAT (8F10.1)
  450 FORMAT (1H1,40X, 50HDELTA-X MAP,SPACING ACROSS IN J-DIRECTION  (FE
     1ET) ,/)
  460 FORMAT (1H1,41X, 47HDELT-Y MAP, SPACING DOWN IN I-DIRECTION  (FEET
     1),/)
  470 FORMAT (1H1,44X, 41HSURFACE ELEVATION MAP  (FEET ABOVE DATUM),/)
  480 FORMAT (1H1,44X, 41HBEDROCK ELEVATION MAP  (FEET ABOVE DATUM),/)
  490 FORMAT (1H1,55X, 18HSPECIFIC YIELD MAP,/)
  500 FORMAT (1H1,46X, 28HPERMEABILITY MAP  (FEET/DAY),/)
  510 FORMAT (1H1,30X, 57HTOP OF CONFINED AQUIFER ELEVATION MAP  (FEET A
     1BOVE DATUM),/)
  520 FORMAT (1H1,39X, 40HCONFINED AQUIFER STORAGE COEFFICIENT MAP,/)
C
      END
```

SUBROUTINE READH

```
      SUBROUTINE READH (NR,NC,H,HP,HT,HF)
C
C
C     THIS SUBROUTINE READS IN AN INITIAL WATER TABLE ELEVATION OR
C       HEAD FOR COMPARING WATER LEVEL CHANGES.
C       H=INITAL WATER TABLE ELEVATION OR HEAD (FEET)
C       HT=PRESENT WATER TABLE ELEVATION OR HEAD (FEET)
C       HP=WATER TABLE ELEVATION OR HEAD AT PREVIOUS TIME LEVEL (FEET)
C       HW=HORIZONTAL WATER LEVEL
C       LBC=LEFT BOUNDARY CODE
C       RBC=RIGHT BOUNDARY CODE
C       TBC=TOP BOUNDARY CODE
C       BBC=BOTTOM BOUNDARY CODE
C     IDENTIFICATION OF BOUNDARY VALUES OF H.
C       H(I,J) LESS THAN 10,000 - WATER TABLE ELEVATION (NO BOUNDARY)
C       H(I,J) GREATER THAN 10000 BUT LESS THAN 20000 - IMPERMEABLE
C       H(I,J) GREATER THAN 20000 BUT LESS THAN 30000 - UNDERFLOW
C       H(I,J) GREATER THAN 30000 BUT LESS THAN 40000 - CONSTANT HEAD
C
      DIMENSION H(NR,NC), HT(NR,NC), HP(NR,NC), HF(NR,NC)
      REAL LBC
C
      READ (5,220) HW,LBC,RBC,TBC,BBC
C
      IF (HW.LE.0.0) GO TO 140
      DO 110 I=1,NR
      DO 110 J=1,NC
  110 H(I,J)=HW
      DO 120 I=1,NR
         H(I,1)=LBC+HW
         H(I,NC)=RBC+HW
  120 CONTINUE
      DO 130 J=1,NC
         H(1,J)=TBC+HW
         H(NR,J)=BBC+HW
  130 CONTINUE
      GO TO 150
C
  140 READ (5,220) H
  150 DO 210 J=1,NC
      DO 210 I=1,NR
         KK=H(I,J)/10000.+1
         GO TO (160,170,180,190), KK
  160    HT(I,J)=H(I,J)
         GO TO 200
  170    HT(I,J)=H(I,J)-10000.
         GO TO 200
  180    HT(I,J)=H(I,J)-20000.
         GO TO 200
  190    HT(I,J)=H(I,J)-30000.
  200    HP(I,J)=HT(I,J)
         HF(I,J)=HT(I,J)
  210 CONTINUE
      WRITE (6,590)
      CALL MATROP (NR,NC,H)
      RETURN
C
  220 FORMAT (8F10.1)
  590 FORMAT (1H1,37X, 46HINITIAL HEAD ELEVATION MAP  (FEET ABOVE DATUM)
     1,/)
C
      END
```

SUBROUTINE MATSOL

```
      SUBROUTINE MATSOL (NOROW,NOCOL,IP,IR,FK,PHI,H,HT,Z,DELX,DELY,Q,CMA
     1TRX,CR,A,B,G,PHIC,HP,HF)
C
C
C     THIS SUBROUTINE SETS UP THE COEFFICIENT MATRIX AND RIGHT HAND
C       SIDE VECTOR MATRIX.
C       CMATRX=COEFFICIENT MATRIX
C       CR=RIGHT HAND SIDE VECTOR MATRIX
C
      DIMENSION FK(NOROW,NOCOL), PHI(NOROW,NOCOL), H(NOROW,NOCOL), HT(NO
     1ROW,NOCOL), Z(NOROW,NOCOL), DELX(NOROW,NOCOL), DELY(NOROW,NOCOL),
     2Q(NOROW,NOCOL), CMATRX(IP,IR), CR(IP), A(NOROW,NOCOL), B(NOROW,NOC
```

```
     3CL), G(NCROW,NCCOL), PHIC(NOROW,NOCCL), HF(NOROW,NOCOL), HF(NOROW,
     4NCCCL)
C
      COMMON/BLK1/DT,ST,ICFAQ,FWTOP
C
      DELT=DT
      DO 110 J=1,IP
      DO 110 I=1,IP
        CMATRX(I,J)=0.0
  110 CONTINUE
      DO 120 I=1,NOROW
      DO 120 J=1,NOCCL
        A(I,J)=0.0
        B(I,J)=0.0
  120 CONTINUE
      NT=0
      NC1=NOCCL-1
      NR1=NORCW-1
      IB=NCROW-2
      IM=IB+1
      IC=IM+1
      ID=2*IB+1
      DO 160 J=2,NC1
      DO 160 I=2,NR1
         NT=NT+1
         CR(NT)=0.0
         IF (H(I,J).GE.10000.0) GO TO 150
         JA=I
         JD=I
C
C                         LEFT(A)
C
         CMATRX(NT,1)=PARAM(FK(JA,J-1),FK(I,J),HT(JA,J-1),HT(I,J),Z(JA,J
     1   -1),Z(I,J),DELX(JA,J-1),DELX(I,J),DELY(JA,J-1),DELY(I,J),G(JA,J
     2   -1),G(I,J))
C
C                         TOP(B)
C
         CMATRX(NT,IB)=PARAM(FK(I-1,J),FK(I,J),HT(I-1,J),HT(I,J),Z(I-1,J
     1   ),Z(I,J),DELY(I-1,J),DELY(I,J),DELX(I-1,J),DELX(I,J),G(I-1,J),G
     2   (I,J))
C
C                         BOTTOM(C)
C
         CMATRX(NT,IC)=PARAM(FK(I+1,J),FK(I,J),HT(I+1,J),HT(I,J),Z(I+1,J
     1   ),Z(I,J),DELY(I+1,J),DELY(I,J),DELX(I+1,J),DELX(I,J),G(I+1,J),G
     2   (I,J))
         A(I,J)=CMATRX(NT,IC)
C
C                         RIGHT(D)
C
         CMATRX(NT,ID)=PARAM(FK(JD,J+1),FK(I,J),HT(JD,J+1),HT(I,J),Z(JD,
     1   J+1),Z(I,J),DELX(JD,J+1),DELX(I,J),DELY(JD,J+1),DELY(I,J),G(JD,
     2   J+1),G(I,J))
         B(I,J)=CMATRX(NT,ID)
C
         CALL NSCONT (H(JA,J-1),HT(JA,J-1),HT(I,J),Z(JA,J-1),Z(I,J),CMAT
     1   RX(NT,1),CMATRX(NT,IM),CR(NT))
         CALL NSCONT (H(I-1,J),HT(I-1,J),HT(I,J),Z(I-1,J),Z(I,J),CMATRX(
     1   NT,IB),CMATRX(NT,IM),CR(NT))
         CALL NSCONT (H(I+1,J),HT(I+1,J),HT(I,J),Z(I+1,J),Z(I,J),CMATRX(
     1   NT,IC),CMATRX(NT,IM),CR(NT))
         CALL NSCONT (H(JD,J+1),HT(JD,J+1),HT(I,J),Z(JD,J+1),Z(I,J),CMAT
     1   RX(NT,ID),CMATRX(NT,IM),CR(NT))
C
C                         (E)
C
         IF (ICFAQ.LE.0) GO TO 130
         IF (HT(I,J).LE.G(I,J)) GO TO 130
         STCOEF=PHIC(I,J)
         GO TO 140
  130    STCOEF=PHI(I,J)
  140    CONTINUE
         CMATRX(NT,IM)=CMATRX(NT,IM)-(CMATRX(NT,1)+CMATRX(NT,IB)+CMATRX(
     1   NT,IC)+CMATRX(NT,ID)+(STCOEF*DELX(I,J)*DELY(I,J))/DELT)
         CR(NT)=CR(NT)-(HT(I,J)*STCOEF*DELX(I,J)*DELY(I,J))/DELT-G(I,J)*
     1   43560.0
C
         GO TO 160
  150    CMATRX(NT,IM)=1.0
         CR(NT)=HT(I,J)
```

```
  160 CONTINUE
      REWIND 7
      WRITE (7) CMATRX,CR
      CALL BSCLVE (CMATRX,IP,IR,CR)
      NT=0
      DO 170 J=2,NC1
      DO 170 I=2,NR1
         NT=NT+1
         HT(I,J)=CR(NT)
         HF(I,J)=CR(NT)
  170 CONTINUE
      IF (ICFAQ.LE.0) GO TO 230
      REWIND 7
      READ (7) CMATRX,CR
      ICAC=0
      NT=0
      DO 210 J=2,NC1
      DO 210 I=2,NR1
         NT=NT+1
         IF (HT(I,J).LE.G(I,J)) GO TO 180
         IF (HP(I,J)-G(I,J)) 190,190,210
  180    IF (HP(I,J)-G(I,J)) 210,210,200
  190 ERRCR=(G(I,J)*(PHI(I,J)-PHIC(I,J)))*DELX(I,J)*DELY(I,J)/DELT
         CR(NT)=CR(NT)+ERRCR
         CMATRX(NT,NR1)=CMATRX(NT,NR1)+(PHI(I,J)-PHIC(I,J))*DELX(I,J)*DE
     1   LY(I,J)/DELT
         WRITE (6,240) I,J
         ICAC=1
         GC TC 210
  200 ERROR=(G(I,J)*(PHI(I,J)-PHIC(I,J)))*DELX(I,J)*DELY(I,J)/DELT
         CR(NT)=CR(NT)-ERRCR
         CMATRX(NT,NR1)=CMATRX(NT,NR1)+(PHIC(I,J)-PHI(I,J))*DELX(I,J)*DE
     1   LY(I,J)/DELT
         WRITE (6,250) I,J
         ICAC=1
  210 CONTINUE
      IF (ICAC.EQ.0) GO TC 230
      CALL BSCLVE (CMATRX,IP,IR,CR)
      NT=0
      DO 220 J=2,NC1
      DO 220 I=2,NR1
         NT=NT+1
         HT(I,J)=CR(NT)
         HF(I,J)=CR(NT)
  220 CONTINUE
  230 RETURN
C
  240 FORMAT (1H ,43X,  4HGRID,2I5,5X, 22HUNCONFINED TO CONFINED)
  250 FORMAT (1H ,43X,  4HGRID,2I5,5X, 22HCONFINED TO UNCONFINED)
C
      END
```

FUNCTION PARAM

```
      FUNCTION PARAM(AK1,AK2,AHT1,AHT2,AZ1,AZ2,AX1,AX2,AY1,AY2,AG1,AG2)
C
C
C     THIS SUBPROGRAM COMPUTES THE COEFFICIENTS USED IN MATSOL AND
C       PYFLOW.  IT IS APPLICABLE TO CASES OF VARIABLE DX, DY, FK,
C       AND SATURATED THICKNESS.
C
      COMMON/BLK1/DT,ST,ICFAQ,FWTOP
C
      IF (ICFAQ.LE.0) GO TC 110
      A=AMIN1(AHT1,AG1)
      B=AMIN1(AHT2,AG2)
      SATHCK=AMAX1(A,B)-AMAX1(AZ1,AZ2)
      GO TC 120
  110 SATHCK=AMAX1(AHT1,AHT2)-AMAX1(AZ1,AZ2)
  120 PARAM=(2.*AK1*AK2*AY1*AY2*SATHCK)/((AX1*AK2*AY2)+(AX2*AK1*AY1))
C
      RETURN
C
      END
```

SUBROUTINE NSCONT

```
      SUBROUTINE NSCONT (HA,HTA,HTM,ZA,ZM,CRXA,CRXM,CRL)
C
C
C     THIS SUBROUTINE TRANSFERS THE COEFFICIENTS, MULTIPLIED BY THEIR
C       RESPECTIVE H-VALUE, TO THE RIGHT HAND SIDE VECTOR MATRIX IN
C       CASE OF ADJACENT CONSTANT HEAD OR KNOWN BOUNDARY CONDITIONS.
C       IT ALSO SETS COEFFICIENTS EQUAL TO ZERO IN CASE OF ADJACENT
C       IMPERMEABLE BOUNDARIES.
C
      IF (HA.LT.20000.0) GO TO 110
      CRL=CRL-CRXA*HTA
      CRXM=CRXM-CRXA
      CRXA=0.0
      GO TO 120
  110 IF (HA.GE.10000.0) GO TO 130
  120 IF ((HTM-ZM).LE.1.0.AND.HTM.GT.HTA) GO TO 130
      IF ((HTA-ZA).GT.1.0.OR.HTA.LE.HTM) GO TO 140
  130 CRXA=0.0
C
  140 RETURN
C
      END
```

SUBROUTINE BJUST

```
      SUBROUTINE BJUST (NR,NC,H,HT,HP,DX,DY,I)
C
C
C     THIS SUBROUTINE ADJUSTS THE UNDERFLOW BOUNDARY WATER TABLE OR
C       HEAD ELEVATIONS.  BOUNDARY ELEVATIONS ARE HELD CONSTANT FOR
C       ODD TIME STEPS.  GRADIENTS ARE COMPUTED THREE GRIDS IN AND
C       PROJECTED BACK TO OBTAIN NEW WATER LEVEL ELEVATIONS OR
C       HEAD ELEVATIONS AT BOUNDARIES.
C
      DIMENSION H(NR,NC), HT(NR,NC), HP(NR,NC), DX(NR,NC), DY(NR,NC)
C
      IF ((I/2*2).NE.I) RETURN
C
      NR1=NR-1
      NC1=NC-1
C
      DO 120 I2=2,NC1
         IF (H(1,I2).GE.30000.0) GO TO 110
         IF (H(1,I2).LT.20000.0) GO TO 110
         DITP=(HT(2,I2)-HT(3,I2))*(DY(1,I2)+DY(2,I2))/(DY(2,I2)+DY(3,I2)
     1   )
         HT(1,I2)=HT(2,I2)+DITP
  110    IF (H(NR,I2).GE.30000.0) GO TO 120
         IF (H(NR,I2).LT.20000.0) GO TO 120
         DIBT=(HT(NR-2,I2)-HT(NR-1,I2))*(DY(NR,I2)+DY(NR-1,I2))/(DY(NR-2
     1   ,I2)+DY(NR-1,I2))
         HT(NR,I2)=HT(NR1,I2)-DIBT
  120 CONTINUE
C
      DO 140 I1=2,NR1
         IF (H(I1,1).GE.30000.0) GO TO 130
         IF (H(I1,1).LT.20000.0) GO TO 130
         DILT=(HT(I1,2)-HT(I1,3))*(DX(I1,1)+DX(I1,2))/(DX(I1,2)+DX(I1,3)
     1   )
         HT(I1,1)=HT(I1,2)+DILT
  130    IF (H(I1,NC).GE.30000.0) GO TO 140
         IF (H(I1,NC).LT.20000.0) GO TO 140
         DIRT=(HT(I1,NC-2)-HT(I1,NC-1))*(DX(I1,NC)+DX(I1,NC-1))/(DX(I1,N
     1   C-2)+DX(I1,NC-1))
         HT(I1,NC)=HT(I1,NC1)-DIRT
  140 CONTINUE
C
      RETURN
C
      END
```

SUBROUTINE BSOLVE

```
      SUBROUTINE BSOLVE (D,N,M,V)
C
C
C     THIS SUBROUTINE SOLVES THE MATRIX, SET UP IN MATSOL, BY GAUSS
C       ELIMINATION.
C
      DIMENSION D(N,M), V(N)
C
      LR=(M-1)/2
      DO 120 L=1,LR
         IM=LR-L+1
      DO 120 I=1,IM
         DO 110 J=2,M
  110    D(L,J-1)=D(L,J)
         KN=N-L
         KM=M-I
         D(L,M)=0.0
  120 D(KN+1,KM+1)=0.0
      LR=LR+1
      IM=N-1
      DO 190 I=1,IM
         NPIV=I
         LS=I+1
         DO 130 L=LS,LR
            IF (ABS(D(L,1)).GT.ABS(D(NPIV,1))) NPIV=L
  130    CONTINUE
         IF (NPIV.LE.I) GO TO 150
         DO 140 J=1,M
            TEMP=D(I,J)
            D(I,J)=D(NPIV,J)
  140.   D(NPIV,J)=TEMP
         TEMP=V(I)
         V(I)=V(NPIV)
         V(NPIV)=TEMP
  150    V(I)=V(I)/D(I,1)
         DO 160 J=2,M
  160    D(I,J)=D(I,J)/D(I,1)
         DO 180 L=LS,LR
            TEMP=D(L,1)
            V(L)=V(L)-TEMP*V(I)
            DO 170 J=2,M
  170       D(L,J-1)=D(L,J)-TEMP*D(I,J)
  180    D(L,M)=0.0
         IF (LR.LT.N) LR=LR+1
  190 CONTINUE
      V(N)=V(N)/D(N,1)
      JM=2
      DO 210 I=1,IM
         L=N-I
         DO 200 J=2,JM
            KM=L+J
  200    V(L)=V(L)-D(L,J)*V(KM-1)
         IF (JM.LT.M) JM=JM+1
  210 CONTINUE
C
      RETURN
C
      END
```

SUBROUTINE MATROP

```
      SUBROUTINE MATROP (NOROW,NOCOL,B)
C
C
C     THIS SUBROUTINE ORGANIZES DATA OR RESULTS INTO A SUITABLE FORM
C       FOR PRINTING AND PRINTS.
C
      DIMENSION B(NOROW,NOCOL)
C
      NOCOLM=NOCOL
      ICONT=1
      NO1=NOCOLM
      IF (NOCOLM.GT.12) NO1=12
  110 NO2=NOCOLM-12
      WRITE (6,140) (JJ,JJ=ICONT,NO1)
      DO 120 I=1,NOROW
  120 WRITE (6,150) I,(B(I,J),J=ICONT,NO1)
      IF (NO2.LE.0) RETURN
      NOCOLM=NOCOLM-12
      ICONT=ICONT+12
      IF (NOCOLM.LE.12) GO TO 130
      NO1=ICONT+11
      GO TO 110
  130 NO1=ICONT-1+NOCOLM
      GO TO 110
C
  140 FORMAT (1H ,//,3X,12(7X,  1HX,I2)/)
  150 FORMAT (1H ,  1HY,I2,12F10.3)
C
      END
```

APPENDIX D

Additional Solutions

This appendix contains several solutions, in addition to those in the text, that are of interest in ground-water hydrology. All of the solutions satisfy

$$\frac{\partial^2 s}{\partial x^2} + \frac{\partial^2 s}{\partial y^2} = \frac{1}{\alpha}\frac{\partial s}{\partial t}$$

expressed in the (x,y) coordinates or in the radial coordinate r as appropriate.

Problem: Recharge From a Rectangular Source - Slug Injection That Produces an Instantaneous Mound of Height H.

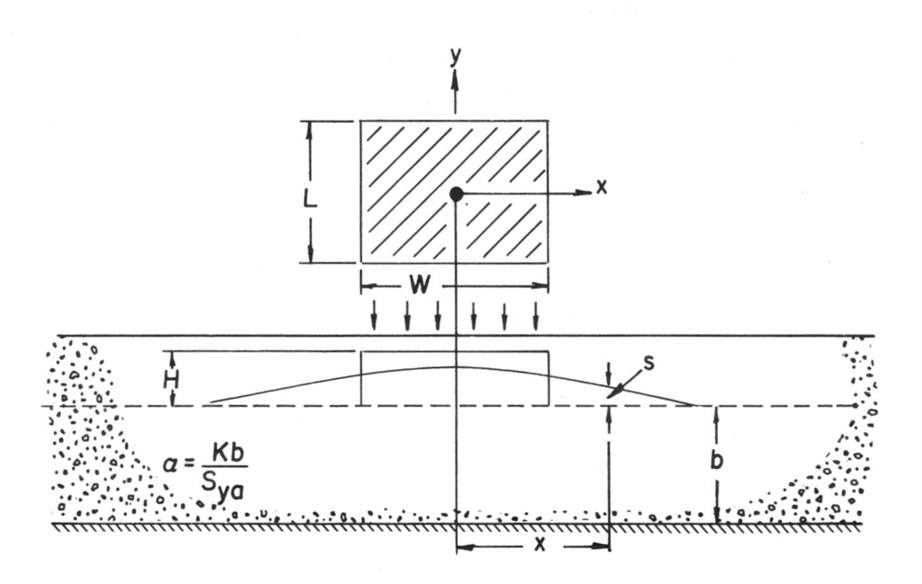

Solution:

$$s = \frac{H}{4}\,(\text{erf } u_2 - \text{erf } u_1)(\text{erf } v_2 - \text{erf } v_1)$$

where

$$u_1 = \frac{x - W/2}{\sqrt{4\alpha t}} \qquad v_1 = \frac{y - L/2}{\sqrt{4\alpha t}}$$

$$u_2 = \frac{x + W/2}{\sqrt{4\alpha t}} \qquad v_2 = \frac{y + L/2}{\sqrt{4\alpha t}}$$

Reference: Glover, R. E., 1960. Mathematical Derivations As Pertain To Ground-Water Recharge. Agricultural Research Service, USDA, Fort Collins, Colorado.

Problem: Recharge From a Rectangular Source - Continuous Constant Recharge Rate That Causes the Water Table to Rise at Constant Rate R = dH/dt in the Absence of Spreading.

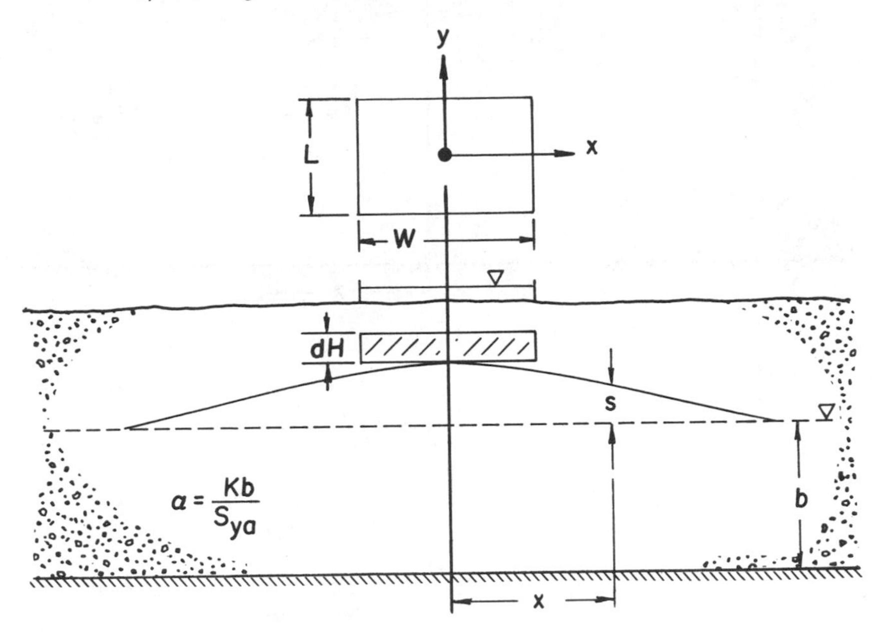

Solution:

$$s = \frac{Rt}{4} \int_0^1 \left\{ \frac{2}{\sqrt{\pi}} \int_{u_1}^{u_2} \exp(-u^2)du \right\} \left\{ \frac{2}{\sqrt{\pi}} \int_{u_3}^{u_4} \exp(-u^2)du \right\} d\tau$$

$$u_1 = \frac{x - W/2}{\sqrt{4\alpha t(1-\tau)}} \qquad u_2 = \frac{x + W/2}{\sqrt{4\alpha t(1-\tau)}}$$

$$u_3 = \frac{y - L/2}{\sqrt{4\alpha t(1-\tau)}} \qquad u_4 = \frac{y + L/2}{\sqrt{4\alpha t(1-\tau)}}$$

Reference: Glover, R. E., 1960. Mathematical Derivations As Pertain To Ground-Water Recharge. Agricultural Research Service, USDA, Fort Collins, Colorado.

Problem: Recharge From a Long Strip of Width W - Slug Injection That Produces an Instantaneous Mound of Height H.

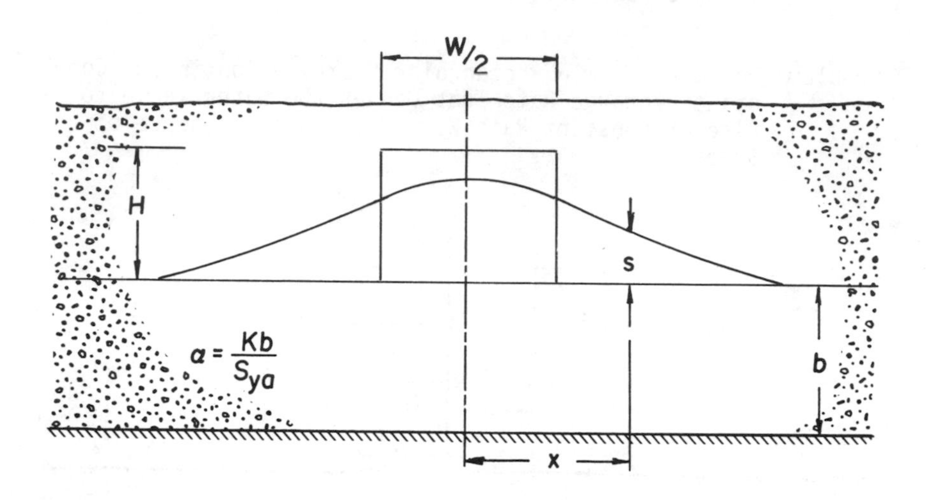

Solution:

$$s = H(\text{erf } u_2 - \text{erf } u_1)$$

$$u_1 = \frac{x - W/2}{\sqrt{4\alpha t}} \qquad u_2 = \frac{x + W/2}{\sqrt{4\alpha t}}$$

Reference: Glover, R. E., 1960. Mathematical Derivations As Pertain To Ground-Water Recharge. Agricultural Research Service, USDA, Fort Collins, Colorado.

Problem: Recharge From a Long Strip of Width W - Continuous Constant Recharge Rate That Causes the Water Table to Rise at Constant Rate $R = dH/dt$ in the Absence of Spreading.

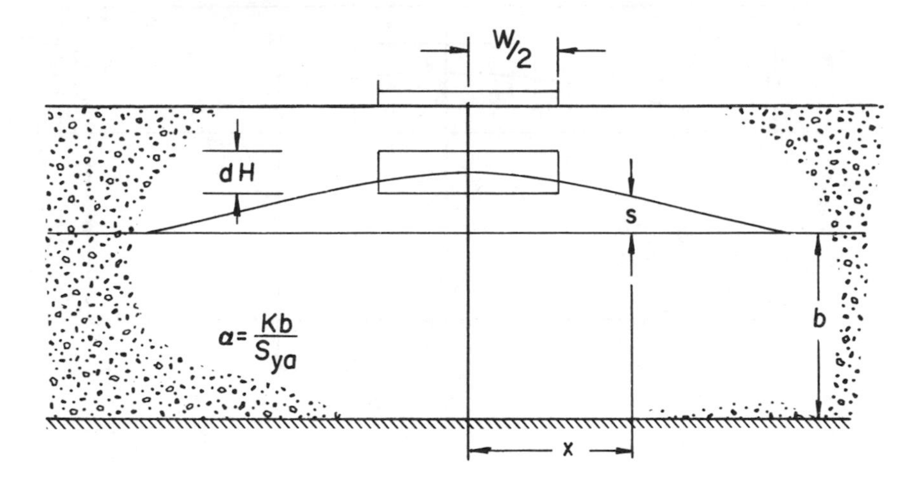

Solution:

$$s = \frac{R}{2} \int_0^t (\text{erf } u_2 - \text{erf } u_1) d\tau$$

$$u_1 = \frac{x - W/2}{\sqrt{4\alpha\ (t-\tau)}} \qquad u_2 = \frac{x + W/2}{\sqrt{4\alpha\ (t-\tau)}}$$

Reference: Glover, R. E., 1960. Mathematical Derivations As Pertain To Ground-Water Recharge. Agricultural Research Service, USDA, Fort Collins, Colorado.

Problem: Recharge From a Circular Source - Slug Injection That Produces an Instantaneous Mound Height H.

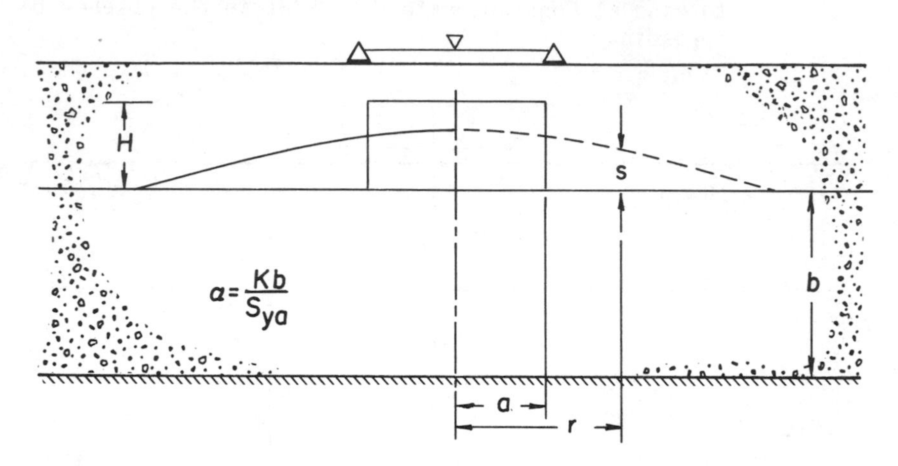

Solution:

$$s = \frac{H}{2\alpha t} \int_0^a r_1 \exp\{-(r^2+r_1{}^2)/4\alpha t\}\, I_o\left(\frac{rr_1}{2\alpha t}\right) dr_1$$

where I_o is the modified Bessel function of the first kind and order zero.

Reference: Glover, R. E., 1960. Mathematical Derivations As Pertain To Ground-Water Recharge. Agricultural Research Service, USDA, Fort Collins, Colorado.

INDEX

Ahmed, 70
Alternating direction implicit, 249
Anisotrophy, 87
Apparent specific yield, 15, 28
  example, 31-33, 200, 202, 205
Aquiclude, 8
Aquifer, 7
  anisotropic, 87, 90
  artesian, 9
  coastal, 160
  confined, 9, 90, 99, 240, 243
  homogeneous, 82
  isotropic, 87
  non-homogeneous, 82
  leaky, 77, 99, 100, 142
  unconfined, 9, 11, 146, 250
Aquifer testing, considerations, 195
  examples, 200, 202, 203, 204, 205
  hydrogeologic boundaries, 213
  Jacob method, 204
  recovery, 206
  slug, 206
  Theis method, 198
Aquifuge, 8
Aquitard, 8
Available water, 54

Bank storage, 214, 218
  examples, 217, 220
Balmer, 210
Barometric efficiency, 40
Barotropic fluids, 74
Bear, 36, 66, 97, 147, 156, 158, 162, 163, 164
Bessel function, 145
  table of, 145
Bibby, 258
Boltzman variable, 178
Boulton, 187, 188, 190
Bossinesq, 98
Bossinesq equation, 97
  linear, 98
  nonlinear, 98
Brooks, 98

Capillary fringe, 1, 26, 27
  pressure, 20, 26
  pressure-saturation curve, 22
    measurements, 24
    example, 24
Carbonate rock, 7
Carr, 217
Chandler, 158
Characteristic dimension, 68, 70
Churchill, 110
Collector wells, 122
Collins, 67
Compressibility of aquifer, 35
  bulk, 36
  pore volume, 35
  value of, 36
  water, 37
  water, value of, 38
Connate water, 9
Continuity, 35
  of mass, 35, 91, 97
  of volume, 35
Cooper, 38, 194, 216
Conservative tracer, 66
Convective acceleration, 68-69
Counter lift, 2
Corapciglu, 36
Crank-Nicholson, 247
Cut-section, 17

Dagan, 158, 162, 163, 164
Daly, 194
Darcy, 5
Darcy's Law, 65-66, 73
  validity, 70
  velocity, 65, 67, 111
  examples of, 71, 72, 78
Delayed yield, 187
Desaturation, 20
Desorptive curve, 22
DeWiest, 5, 38, 71
Doorenbas, 53
Drains, parallel, 127, 221
  steady state, 127
    example, 130
  unsteady state, 221
    examples, 224, 225, 227
  varying recharge, 226
Drawdown, 94
  relation with discharge, 153
Duke, 28, 98
Dupuit-Forchheimer assumption, 96

Eckhardt, 252
Effective radius of influence, 182

example, 183
Error function, 211
Equipotential, 110
boundary conditions, 113
definition, 111
examples, 113, 128
relation with Laplace Eq., 114
Evanson, 89
Evaporation, 51
pan, 53
pan coefficients, 54
Evapotranspiration, 47, 52
effect on water table, 47
example, 56
potential, 52
Explicit technique, 243
example, 244
Exponential integral, 179
table of, 180

Fair-Hatch equation, 81
Fatt, 36
Ferris, 126, 127, 194, 208, 217, 219
Finite difference model, 265
Finite element, 240
Flow equations, 92
confined, 92, 115
Laplace, 95
unconfined, 95
vertical leakage, 99
Flow net, 136
anisotropic case, 138
example, 140
homogeneous aquifers, 136
non-homogeneous aquifer, 141
Forces,
driving, 67
inertia, 69
potential, 73
resistance, 67
Forchheimer, 69
Fresh water lens, 165
example, 166
Friction factor, 69

Gauss scheme, 247, 263
Geertsma, 36
Ghyben-Herzberg, 159, 160
example, 161
Glover, 160, 161, 210, 222,
Grain-size distribution, 81
effective, 81
geometric mean, 81
Ground water, definition, 1
budget, example, 56-58
extraction, 2
geology, 6
hydrology, 5
model, 265
observation, 11
occurance, example, 10
vessels, 6
Ground water and wells, 1, 116

Hagen, 5
Hall, 216
Hansen, 29, 44
Hantush, 145, 152, 189, 191
Harleman, 81
Harmonic functions, 110
Hydraulic conductivity, 7, 75
example, 76
laboratory determination, 79
stratified aquifers, 82, 83, 84, 85
stratified aquifer example, 82, 86
table of, 82
tensor form, 87
Hydraulic diffusivity, 178
Hydrologic boundaries, 124, 209, 210, 213
Hydrologic budget, 48
equation for, 50
soil water component, 51
Hubbert, 68, 71, 73, 75
Hunt, 152

Igneous rocks, 8
Image, drains, 127
examples, 128, 130
recharge wells, 125, 126
Implicit solution, 246
one-dimensional example, 246
two-dimensional example, 255
two-dimensional, general, 250
Infiltration, 51
Interflow, 11

Jacob, 38, 189
Jensen, 53, 55
Jensen-Haise, 52
Johnson, 18, 31, 82

Kanats, 5
Klaer, 123, 124

Knowles, 194, 208

Laplace, 110
Laplace equation, 95, 110
  one-dimensional, 115
  radial coordinates, 117
Leakage factor, 100
Leaky aquifer, 77, 99, 100, 142
  example, 145
  function, 189
  function, table of, 191
  steady radial flow, 143
  unsteady radial flow, 188
Line source, 178

Maasland, 224
Marcus, 89
Mariotte, 5
Mass balance, 35, 91, 97
McWhorter, 23, 98, 158, 163, 224
Moench, 216
Meinzer, 6, 48
Mercado, 158, 162
Metamorphic rocks, 8
Mikels, 122, 124
Moisture characteristics, 22
Morel-Seytoux, 194
Morris, 18, 31, 82
Muskat, 134, 162

Neuman, 187
Non-homogeneous fluids, 74, 156
  aquifers, 82
Numerical methods, 240

Part remaining, 223, 224
Partially penetrating well, 134
Papadopulos, 194, 207
Peck, 43, 44
Penman, 53
Permanent wilting, 55
Permeability, intrinsic, 69, 81
  tensor, 88
Permeameters, constant head, 79
  falling head, 80
Perrault, 5
Phreatic surface, 9
Phreatophytes, 47, 48
Piezometer, 11, 12
Piezometric head, 73, 74
  surface, 9
Plane source, 220
  example, 220
Plant water use,
  coefficient, 54
  values of, 55
Poiseuille, 5
Poland, 39
Polubarinova-Kochina, 139, 140
Pontin, 187
Porosity, 7, 15, 28
  effective, 28
  example, 19, 20
  factors affecting, 18
  table of, 18
Pressure, atmospheric, 40
  barometric, 40, 43
  barometric effect on water
    table, 43
  capillary, 20, 26
  head, 26
Pruitt, 53
Principal direction, 88
Psuedo-steady state, 184
  example, 185

Radial flow, example, 120, 123,
    130, 154, 179
  near hydrogeologic bound-
    aries, 124
  partially penetrating well,
    134
  steady state well field, 121
  steady state well flow, 116
  unsteady fully penetrating
    well, 177
Recovery test, 206
  example, 207
Refraction of velocity vector,
    83
Regolith, 6
Remson, 247
Representative volume element,
    15-18, 65
Reynolds number, 69
Robinson, 48
Rorabough, 216
Rose, 70,
Rumer, 67
Rushton, 249, 250

Safe yield, 49
Sahni, 162
Schmorak, 158, 162
Seepage face, 148
Segerlind, 240
Sensitivity analysis, 257

Shames, 119
Slug test, 206
  example, 207, 208
Specific capacity, 135
  example, 135
Specific retention, 20, 23, 28
  example, 39
Specific storage, 15, 35
  definition, 38
  example, 39
Specific yield, 28
  example, 30
  table of, 31
Stability of solutions, 243, 245
Statistical methods, 17
Storage coefficient, 38
  example, 39
Stratagraphic column, 6
Stream functions, 111
Streamlines, 110
  boundary conditions, 113
  definition, 111
  example, 113
  relative with Laplace Eq., 114
  velocity vector, 112
Streamtube, 136
Streltsova, 187, 188
Stress, intergranular, 34, 41
  total, 34
Sunada, 69, 70, 252, 258
Superposition, 121-136
  example, 122, 128, 133
  images, 125
  in time, 192
  well field, 121
  well in uniform flow, 131

Theis, well solution, 179, 198
  example, 200
Tidal efficiency, 46
Todd, 70, 80
Tolman, 2, 5
Transmissivity, 94
Transpiration, 51
Two fluids, 74, 156
  coastal aquifers, 160
  example, 161
  fresh water lens, 165
  interface, 157

Unconfined flow, 146, 185, 214
  example, 150
  parallel channels, 146
  radial flow, 151
  with accretion, 148
Unit vectors, 74
Upconing, salt water, 162
  example, 164

Vallentine, 119
van Der Kamp, 217
van Schilfgaarde, 98
Variable pumping rates, 190
Varying recharge, 226
Velocity, example, 75, 76
  potential, 73, 75, 110, 114
  seepage, 66
Voids, interconnected, 15
  isolated, 15
  variation with element size, 16
Volume flux, 65
von Rosenberg, 240
Vugs, 6

Walton, 7, 135, 181, 187
Ward, 71
Warner, 98
Water compressibility, 37, 38
Water content, distribution, 30
  measurement of, 51
  volumetric, 21
Water ladder, 3
Water retention curve, 22
Water tables, fluctuations of, 40, 48
  effects of external loads, 45-47
  levels, 9, 45
Wells, collector, 123
  drilled, 2
  dug, 2
  field of, 121
  function, 179, 189
  losses, 116
  observation, 11, 12
  partially penetrating, 134
Wenze , 180
Wood, 3, 4
Wymore, 54, 55

Youngs, 28